Universitext

Universitext is a series of textbooks that presents material from a wide variety of mathematical disciplines at master's level and beyond. The books, often well class-tested by their author, may have an informal, personal, or even experimental approach to their subject matter. Some of the most successful and established books in the series have evolved through several editions, always following the evolution of teaching curricula, into very polished texts.

Thus as research topics trickle down into graduate-level teaching, first textbooks written for new, cutting-edge courses may find their way into *Universitext*.

Filippo Santambrogio

A Course in the Calculus of Variations

Optimization, Regularity, and Modeling

 Springer

Filippo Santambrogio 🆔
Université Claude Bernard Lyon 1
Villeurbanne, France

ISSN 0172-5939 ISSN 2191-6675 (electronic)
Universitext
ISBN 978-3-031-45035-8 ISBN 978-3-031-45036-5 (eBook)
https://doi.org/10.1007/978-3-031-45036-5

Mathematics Subject Classification: 49J45, 35B65, 46N10

This Springer imprint is published by the registered company Springer Nature Switzerland AG
The registered company address is: Gewerbestrasse 11, 6330 Cham, Switzerland

Paper in this product is recyclable.

To Maud and Iris,
two new characters
who appeared in a recent chapter
of the book of my life,
with the clear intent of making
the rest of it much funnier

Preface

This book arose from the classes on the calculus of variations which I gave several times in Orsay (first in a PDE master's program, and then jointly with an optimization program), in Pavia, and in Lyon. During these classes I used to write a webpage with a precise schedule, and for each lecture I would point the students toward various references, which could be existing books, papers, or, occasionally, short notes that I wrote for them. The main goal of this book is to collect all this material together, present it in the form of an exploratory journey through the calculus of variations, and of course take the time to develop related topics, both for completeness and for advertisement.

Many important books already exist on the subject (let me just mention those which I used the most with my students, such as [95, 101] or [67]) and, naturally, the intersection with them will be non-negligible. Yet, besides collecting together various parts of the theory which, so far, have not always appeared together in the same volume, what I also want to do is to present the material from a point of view which is not often found in other texts. This is a particular perspective coming from optimization. On the one hand, a very important role is played in this book by convex analysis, convex optimization, and convex duality; on the other hand, many examples are taken from applications coming from what in France is called "mathématiques de la décision" (an expression which appears, by the way, in the name of the first Mathematics Department in which I held a position, the CEREMADE in Paris-Dauphine, as well as in the scientific society—actually, a part of the French Society for Applied and Industrial Mathematics—which I led in recent years, the group MODE). Indeed, although I would undoubtedly classify this book as a piece of mathematical analysis, many parts of it are inspired by applications and modeling, and among the models which come from decision sciences, let me mention the Ramsey model in economic growth (Sect. 1.5), equilibrium issues in traffic congestion (Sects. 1.3 and 1.6), the optimal shape of a transportation network (Sect. 6.3), the optimal location of facilities (Sect. 7.3), ...

In order to compare this book with other existing texts or other existing approaches, let me start with a brief list of what it does *not* discuss.

First of all, the calculus of variations is not presented as a tool to do geometry. Of course, some geometric optimization problems are detailed, such as the isoperimetric problem (it is one of the most classical examples of a variational problem and it deserves to be addressed at least briefly in every treatise on the subject, in my opinion). Geodesic curves also play an important role in the book, but it is important to note that this term is only used here to denote curves which minimize the length, and not curves which solve a certain differential equation; moreover, except in a few examples (for instance, in the case of the Euclidean sphere), geodesics on manifolds do not play a particular role, and many results are directly presented in the case of metric spaces. In the Euclidean case, the geodesic curves for the standard Euclidean distance are trivial, motivating us to concentrate instead on curves which minimize a weighted length, which is the same as using a Riemannian distance, but only for metrics which are a multiple of the identity. This, however, already covers many cases which are interesting for applications (in traffic congestion, for instance). Coming back to the general connections with geometry, this book features many connections with geometric measure theory, but not with the branch of analysis called geometric analysis (minimal surfaces, geometric flows...).

As we have already pointed out concerning geodesics, no critical point which is not a global minimizer is considered in this book. In optimization, we are given a set of competitors and a function to minimize, and if we want to exploit optimality conditions, we are not bound to the choice of a specific notion of neighborhood or perturbations. Hence, we understand what a global minimizer is, but the notion of a local minimizer is not so clear (and, indeed, many classical texts distinguish between strong and weak minima, something which is absent from this book, and quite absent from modern research), not to mention critical points which are not even local minimizers! As a choice, in the book, the differentiability of functions defined on infinite-dimensional spaces will never be defined or discussed, so that critical points cannot be considered, but only minimizers. This also reflects a personal choice in the relation between the calculus of variations and the theory of those partial differential equations which arise as Euler–Lagrange equations of some minimization problems: in general, we minimize a functional, and then use a PDE to describe the properties of the minimizers, instead of choosing a PDE and using a suitable functional to produce one (or many) solution(s). We do not develop a functional setting to define a notion of differential for the functionals we minimize because, often, the function space in which the variational problem is posed is not canonical (we may sometimes decide to minimize over $W^{1,p}$ a functional which only takes finite values on a smaller space). Hence, it is a better choice in my opinion to obtain optimality conditions by ad hoc perturbations for each variational problem. In particular, we will not try to characterize the optimality in terms of generalized notions of gradients or differentials, as is done in non-smooth analysis. In this respect, this approach differs from that taken by another fundamental book, [66].

Finally, I made the choice (for reasons of simplicity) to exclude the vector calculus of variations from the topics which are dealt with in the book. I have the impression that, for an introductory book, there are already sufficiently many interesting and challenging examples and questions to consider without having to move to the case of vector-valued functions of several variables. This means that in the variational problems which are described here, either the dimension of the domain or that of the codomain will be one: we consider scalar functions of arbitrary many variables, or curves valued in arbitrary spaces (including abstract metric spaces), but we exclude the optimization of vector fields (except in the divergence-constrained optimization problems of Sects. 4.3, 4.4, and 4.6, where the vector field plays the role of the dual of the gradient of a scalar function, and anyway the derivatives of such a vector field do not enter the functional). In particular, all the notions related to quasi-convexity are sketched in a discussion section, but are not fully developed, and the applications coming from material sciences and continuum mechanics are reduced to simple scalar examples. This is, for instance, an important difference compared to the book by F. Rindler [172]. Also, the regularity theory using elliptic equations is presented avoiding the theory of systems which, at least, makes it much easier to read (and is a difference compared to [95] or [97]).

Some of my colleagues who commented on the preliminary versions of this book have observed that an existing book, [20], is quite close in spirit to this one (which is not surprising, since one of its authors was my PhD advisor). Besides appearing more than 15 years later, and hence including novel material, there is a difference in the size of the two books, as this one is meant to be more concise. Of course, this has a price, and it cannot be fully self-contained. Among the prerequisites that I assume the reader will have, the most important ones are functional analysis and measure theory, even if most of the required results from these theories are recalled throughout the book. For pedagogical reasons (despite being suggested to do so by some referees), I decided not to include a chapter or appendix summarizing the required preliminaries. Instead, I recall the notions I need at the very moment when I need them, using some special boxes called *Memo* or *Important notion*. As I already pointed out in my book on optimal transport (the box system was essentially the same), the difference between these two kinds of boxes is just that *I* would consider the results covered by *Important notion* boxes as a memo, but I saw that *students* in general do not agree. A third type of box called *Good to know!* provides extra notions that are not usually part of the background of non-specialized graduate students in mathematics. For *Important notion* and *Good to know!* boxes some references are provided, while for *Memo* boxes any classical book on the topic (in most cases, [48]) will do the job. The density of the boxes decreases as the chapters go by.

The book is composed of seven chapters, which I will briefly describe in a while. When teaching a master's course of 30 hours (10 lessons of 3 hours), I used to cover one chapter per lesson, except for Chaps. 4, 5, and 7 to which two classes were devoted (in Chap. 4 one for duality theory and one for regularity via duality, in Chap. 5 one for the Schauder estimates and one for the De Giorgi–Moser theorem, and in Chap. 7 one for the general theory and a first example and one for other examples). Of course many parts of each chapter have been added to the book for completeness and were impossible to treat in 3 hours. On the other hand, it was also impossible to include in each chapter all possible developments, or to treat everything in an exhaustive way. In many sections, the choice was made to stick to the simplest case, just directing the interested reader to more complete references. For the same reason (the impossibility of covering everything within a reasonable length), some topics are completely absent from the book. Some have not been included because they are connected with the calculus of variations but are in my opinion a different subject (for instance, gradient flows, which are first-order evolution equations with a variational structure: the partial differential equations which are considered in this book are all "static" and arise from optimization as optimality conditions, and not as steepest descent flows). For some others, instead, the reason is different: this book is meant as introductory and exploratory, and I found some subjects too specialized (for instance, it was suggested to include Young measures, but I believe a master's student and many researchers can live without them).

Each chapter concludes with a discussion section, where I chose a topic related to the rest of the chapter and presented it with fewer (or no) details, and a list of exercises with hints. Many of these exercises come from the exams or the problem sheets I gave to my students.

Chapter 1 is concerned with one-dimensional problems, which are the first examples of variational problems and, historically, the first to have been analyzed. Some classical examples are presented, the first being that of geodesic curves. We start from the example of great circles on the sphere and then move to a more abstract analysis of curves valued in metric spaces, which also allows us to obtain a short presentation of another classical problem in the 1D calculus of variations, namely heteroclinic connections for multi-well potentials. The other classical problems which are presented as examples are the brachistochrone curve, which is not fully solved but elements are proposed—in the text or in the exercises— for the reader to find the characterization of the solutions, and the Ramsey model for the optimal growth of an economic activity. This last problem is studied in a rigorous and original setting in the space of functions with bounded variation, and introduces the role of 1D BV functions which will also be used in other problems. Many examples studied here go beyond what is done in the classical reference on the subject, [53]. Of course the chapter includes the main tools to prove or disprove the existence of the minimizers and to study them via the Euler–Lagrange equation; the discussion section is devoted to optimal control problems with a formal discussion of the Hamilton–Jacobi equation solved by the value function and of the Pontryagin maximum principle. The whole chapter is written with the idea that the general procedures employed to study a variational problem will be clearer to the reader later on in the book, and that the 1D case is a way to taste them in a simpler setting; some memo boxes are included in the chapter, but some are postponed to later chapters, when I considered them to be more meaningful in a multi-dimensional setting.

Chapter 2 extends this analysis to the higher-dimensional case, presenting what changes (not so much) for the existence and the characterization of the solutions. Since the most studied example is the minimization of the Dirichlet energy $\int |\nabla u|^2 \, dx$, a section is devoted to the properties of harmonic functions, which are indeed the minimizers of this energy. We prove in particular that every harmonic distribution is actually an analytic function. This section follows classical texts on the topic such as [110]. The discussion section is devoted to the p-power variant, i.e. p-harmonic functions, including the limit cases $p = 1$ and $p = \infty$.

The main tool to prove the existence of a solution of a variational problem is the so-called direct method in the calculus of variations, which is essentially an application of the Weierstrass theorem. This requires the lower semicontinuity of the functional to be minimized, typically with respect to weak convergences since in infinite dimension we usually do not have strong compactness. Hence, Chap. 3 is devoted to several lower semicontinuity results. In particular, we prove, following the presentation in [101] (up to minor modifications), the lower semicontinuity of integrals of the form $\int L(x, u, v) \, \mathrm{d}x$ w.r.t. the strong convergence of u and weak of v (assuming convexity in v and continuity in u: we do not consider the case where we only have lower semicontinuity in u, as is done in [52]). A section is also devoted to the semicontinuity of functionals defined on measures w.r.t. the weak-* convergence. Convexity plays a very important role in these semicontinuity results, and in this section some advanced tools from convex analysis are needed which are only detailed in the next chapter. The discussion section is devoted to a brief summary of the semicontinuity theory in the case of vector-valued maps, introducing the notion of quasi-convexity without complete proofs.

Chapter 4 is devoted to the role of convexity and convex analysis in the calculus of variations. Section 4.1 is a little different than the rest of the chapter. It deals with the tools to prove sufficient optimality conditions and/or uniqueness of the minimizers, first in an abstract setting, then translating it to the case of classical integral functionals. The rest of the chapter is concerned with notions from convex analysis: Fenchel–Legendre transforms, subdifferentials, duality.... An original theory of *regularity via duality*, easy to explain and capable of recovering some classical results, is presented in Sect. 4.4. This theory was introduced to study time-dependent and less regular problems such as the incompressible Euler equation or Mean-Field Games, but here we stick to simpler, static, problems. The discussion section is devoted to a very short presentation of several connected problems which are of significant interest nowadays: optimal transport, Wardrop equilibria, and Mean-Field Games, emphasizing the role that convex duality plays in their analysis. This chapter is highly reminiscent of [81] in spirit, but, of course, since it has been written almost half a century later, includes new material and examples.

Chapter 5 is less variational than the others as it is mainly focused on regularity properties of elliptic PDEs, independently of their variational origin. Yet, it is crucial in the calculus of variations since what often happens is that we face a model where we need to optimize an integral functional of the form $\int L(x, u, \nabla u) \, \mathrm{d}x$ and we do not know which function space to choose. Most likely, due to the presence of the gradient, we would like to use $u \in C^1$, but existence results in C^1 are essentially impossible to obtain in a direct way because of lack of compactness. We then extend the problem to a suitable Sobolev space, where it is possible to prove the existence of a minimizer, and we secretly hope that this minimizer is indeed a C^1 function. In order to do this, we need to prove a regularity result, and this is often done by using the regularity theory for elliptic PDEs. In this chapter we first develop the tools, based on the integral characterization of $C^{0,\alpha}$ functions, in order to study the regularity in the case of smooth coefficients (following a standard approach which can be found in many classical books, such as those by Giaquinta and co-authors, [95–97] or [11]). Then, motivated by the regularity question contained in Hilbert's 19th problem, we move to the famous De Giorgi result about equations where the coefficients in the equation are only assumed to be bounded, following Moser's proof. Since all the results of the chapter are related to the Hölder regularity in the interior of the domain, the discussion section presents a brief summary of the ideas behind the L^p theory, and sketches how to extend the results to the boundary. Note that in many master's programs the calculus of variations is taught together with elliptic equations, and this is another reason for including a chapter on the most classical results on elliptic PDEs.

Chapter 6 is concerned with a different class of variational problems, those where the unknown, instead of being a function, is a set. The most well-known problem of this class is the isoperimetric one, where we prove that the ball minimizes the perimeter among sets with prescribed volume (or maximizes the volume under prescribed perimeter). A full proof of this result is provided, following De Giorgi's strategy and the presentation in [142]. This is then applied to another shape optimization problem, the optimization of the first Dirichlet eigenvalue λ_1 under volume constraints. Another section is then devoted to those optimization problems where the set to be optimized is a 1D connected set and its length comes into play: we start with an informal discussion about some variational problems of this kind, including motivations from applications (for instance, in the planning of a transportation network), but then we move to the mathematically most challenging part. We prove the lower semicontinuity of the length under connectedness constraints, which is known as Gołąb's theorem. This is a very nice result which requires some non-trivial prerequisites on geometric measure theory. These are not detailed, but just recalled, and the need for geometric measure theory characterizes almost the entire chapter, which also needs to introduce in detail BV functions of several variables. The discussion section is devoted to variants, alternative proofs, and quantitative extensions of the isoperimetric inequality, which have generated a huge and celebrated literature in recent years.

The last Chap. 7, deals with a very natural question in optimization: What about limits of optimization problems? Which notion of convergence on the functionals should we use to guarantee the convergence of the minimal values and/or of the minimizers? This notion of convergence is called Γ-convergence and its general theory is presented in detail in Sect. 7.1, followed by some sections discussing classical and less classical examples. One of the most important of these is the Modica–Mortola result about the approximation of the perimeter (the same functional which plays an important role in Chap. 6), but we also present in detail a more original example, the asymptotics of the optimal location problem (where to put N facilities in a town, knowing the population density, when $N \to \infty$?), a problem which can also be interpreted in terms of quantization of a continuous density. The discussion section is devoted to other geometric problems (including the Steiner minimal connection problem, already presented in Chap. 6, and the

Mumford–Shah model from image segmentation) which admit approximations in the same spirit as the Modica–Mortola result.

The intention was to produce a book which could be used for master's-level courses on the calculus of variations, and for the courses which I personally gave this text would do the job. I hope and I think it could be useful to other students and other colleagues, both as a reference for some classical results and as a starting point for the exploration of some modern areas of this fascinating subject.

Lyon, France   Filippo Santambrogio
July 2023

Acknowledgments

It is almost impossible not to meet the calculus of variations when studying mathematics in Pisa, and the first people I should thank are indeed those who gave me the taste for this wonderful topic during my undergraduate and graduate studies.

First of all I feel that I must thank Paolo Tilli, who introduced me to the calculus of variations, to the work by De Giorgi, and to the fascinating theory of regularity, starting from an undergraduate project where he supervised me, and later during our joint work which led to one of my first publications. If he had not played such a significant role during my first years as a mathematician more than 20 years ago, this book would probably not exist.

I would also like to thank my PhD advisor Giuseppe Buttazzo and my closest collaborator Guillaume Carlier, with whom I started to work during my PhD. Both of them had, and still have, a strong influence on my mathematics, in particular in what concerns modeling and economic applications. Giuseppe taught me how to master the theory of semicontinuity and how to be protected against the worst enemy of existence (oscillations :-), and I owe to Guillaume almost everything I know in convex analysis, convex duality, and their applications in the calculus of variations. His comments and suggestions have been very useful during the preparation of this manuscript as well.

The role of these three influential friends and colleagues can really be found in the three words of the subtitle of this book.

After being a student, I later started supervising students myself, and the discussions with my own students have always been extremely useful to me. I would like to thank all of them. Throughout this book I found the occasion to present works, ideas, and computations that I first discussed with my former students Lorenzo Brasco, Nicolas Bonnotte, Jean Louet, Alpár Mészáros, Antonin Monteil, Paul Pegon, Samer Dweik, Hugo Lavenant, Alexey Kroshnin, and Clément Sarrazin, as well as with my current students Annette Dumas and Thibault Caillet, and I would like to thank them all for this. Alpár, Paul, Hugo, Annette, and Thibault also helped me by carefully reading some parts of this manuscript; I'm also grateful that I could rely on my former postdoc Guilherme Mazanti, who read and proofchecked many parts of the book impressively quickly.

I also want to thank my experienced colleagues who accepted to read other parts of the book, and gave precious comments on preliminary versions, suggesting important improvements: Antonin Chambolle, Chloé Jimenez, Antoine Lemenant, and Francesco Maggi.

Many other colleagues were involved in discussions about some topics in the calculus of variations which appear in this book, or about precise points in their presentation, and it's impossible to mention them all, but let me list at least Nicholas Alikakos, Luigi Ambrosio, Yann Brenier, Alessio Figalli, Nicola Gigli, Quentin Mérigot, Petru Mironescu, and Athanasios Yannacopoulos.

Finally, I wish to thank all the people around me, both at work and at home, for their patience during the last months of my work on this manuscript, since the effort I spent on it has, on some occasions, significantly reduced the time available for the rest of my activities.

Notation

The following are standard symbols used throughout the book without always recalling their meaning.

- "domain": this word is used in a very informal way and in general means "the set that we want to consider"; domains are assumed to be non-empty subsets of $\mathbb{R}^d$ contained in the closure of their interior; when we need them to be open, bounded, compact, connected, or smooth, we usually state this explicitly; warning: in convex analysis, the word "domain" also refers to the set where a function defined on the whole space is finite.
- $\mathbb{R}_+$: the set of non-negative real numbers, i.e. $[0, +\infty)$.
- log: the natural Naperian logarithm of a positive number.
- $\lim_n$, $\liminf_n$, $\limsup_n$ (but n could be replaced by $k, h, j \ldots$): limit, inferior limit (liminf), superior limit (limsup) as $n \to \infty$ (or $k, h, j \cdots \to \infty$).
- ∇ and $\nabla \cdot$ denote gradient and divergence, respectively.
- Δ denotes the Laplacian: $\Delta u := \nabla \cdot (\nabla u)$ (and not minus it).
- Δ_p denotes the p-Laplacian: $\Delta_p u := \nabla \cdot (|\nabla u|^{p-2} \nabla u)$.
- $D^2 u$: Hessian of the scalar function u.
- $\mathcal{M}(X)$, $\mathcal{M}_+(X)$, $\mathcal{M}^d(X)$: the spaces finite measures, positive finite measures, and vector measures (valued in $\mathbb{R}^d$) on X.
- $\mathbb{R}^d$, $\mathbb{T}^d$, $\mathbb{S}^d$: the d-dimensional Euclidean space, flat torus, and sphere ($\mathbb{S}^d$ is the unit sphere in $\mathbb{R}^{d+1}$).
- $C(X), C_0(X)$: continuous (resp. continuous vanishing at infinity) functions on X.
- δ_a: the Dirac mass concentrated at the point a.
- $\mathbb{1}_A$: the indicator function of a set A, equal to 1 on A and 0 on A^c.
- I_A: the indicator function in the sense of convex analysis, 0 on A and $+\infty$ on A^c.
- $\mathbb{I}$: the identity matrix.
- id the identity map.
- $\fint_A f(x) \, dx$: the average value of f on the set A, equal to $\frac{1}{|A|} \int_A f(x) \, dx$:
- $|A|$, $\mathcal{L}^d(A)$: the Lebesgue measure of a set $A \subset \mathbb{R}^d$; integration w.r.t. this measure is denoted by dx.
- ω_d: the measure of the unit ball in $\mathbb{R}^d$.

- $\mathcal{H}^k$: the k-dimensional Hausdorff measure.
- Per(A): the perimeter of a set A in $\mathbb{R}^d$ (equal to $\mathcal{H}^{d-1}(\partial A)$ for smooth sets A).
- Per(A; Ω): the perimeter of a set A inside Ω (excluding the part of ∂A in $\partial\Omega$).
- $\mu \ll \nu$: the measure μ is absolutely continuous w.r.t. ν.
- $f \cdot \mu$: the measure with density f w.r.t. μ.
- $T_\#\mu$: the image measure of μ through the map T.
- $\mathrm{M}^{k\times h}$: the set of all real matrices with k rows and h columns.
- M^t: the transpose of the matrix M.
- Cof(M): the cofactor matrix of M, such that $\mathrm{Cof}(M) \cdot M = \det(M)\mathrm{I}$.
- $a^i_j, a^{ij}_{kh} \ldots$: superscripts are components (of vectors, matrices...) and subscripts derivatives. No distinction between vectors and covectors is made.
- $a\otimes b$: the rank-one matrix equal to the tensor product of a and b: $(a\otimes b)^{ij} = a^i b^j$.
- Tr(A): the trace of a matrix A.
- Tr[u]: the trace (on the boundary) of a Sobolev function u.
- $x_1, \ldots, x_d$: coordinates of points in the Euclidean space are written as subscripts.
- $|x|$: the Euclidean norm of a vector $x \in \mathbb{R}^d$.
- v^α: the vector with norm $|v|^\alpha$ and the same orientation as v, i.e. $|v|^{\alpha-1}v$.
- Length(ω), $L_K(\omega)$: the length or weighted length (with coefficient K) of the curve ω.
- $e_t : C \to \Omega$: the evaluation map defined on a set C of curves, i.e. $e_t(\omega) := \omega(t)$.
- $\mathbf{n}$: the outward normal vector to a given domain.
- Lip_L: the set of $L-$Lipschitz functions.
- X': the topological dual of the normed vector space X.
- $d(x, A)$: the distance from a point to a set, i.e. $d(x, A) := \inf\{d(x, y) : y \in A\}$.
- A_r: the r-neighborhood of the set A, i.e. $A_r = \{x : d(x, A) < r\}$.
- $d(A, B)$: the minimal distance between the sets A and B given by $d(A, B) := \inf\{d(x, y) : x \in A, y \in B\}$.
- $d_H(A, B)$: the Hausdorff distance between two sets, defined via $d_H(A, B) := \inf\{r > 0 : A \subset B_r, B \subset A_r\}$.
- p'; the dual exponent of p, characterized by $\frac{1}{p} + \frac{1}{p'} = 1$, i.e. $p' = \frac{p}{p-1}$.

The following, instead, are standard choices of notations.

- The dimension of the ambient space is d; we use $\mathbb{R}^N$ for the target space of a map which is supposed to produce vectors and not points, or for a finite-dimensional vector space which could be replaced by an abstract vector space X. The index n is kept for the subscripts of sequences.
- ω is usually a curve (but sometimes a modulus of continuity; when we need the modulus of continuity of a curve, we try to call the curve γ).
- α is the symbol usually used for arbitrary family of indices (instead of $i, j, \ldots$ which are kept for countable families).
- Ω is usually a domain in $\mathbb{R}^d$, and general metric spaces are usually called X. General normed vector spaces are called X.
- A vector-valued variable can be called v, but a vector field $\mathbf{v}$.

Contents

Chapter 1
One-Dimensional Variational Problems

This chapter is devoted to one of the most classical classes of variational problems, the one-dimensional case. It consists in minimizing an integral cost, typically of the form

$$\int_0^T L(t, \gamma(t), \gamma'(t)) \, dt,$$

among functions $\gamma : [0, T] \to X$ defined on a one-dimensional interval and valued in a suitable space X. The one-dimensional aspect of the competitor makes them objects to be interpreted as trajectories in the space X, and the independent variable t is often interpreted as time (even if sometimes it could represent a horizontal coordinate or another parameterization of a one-dimensional object). The above integral cost can be completed with some penalizations on the values of γ on the boundary, thus obtaining a total cost of the form

$$\int_0^T L(t, \gamma(t), \gamma'(t)) \, dt + g(\gamma(0), \gamma(T)),$$

and the minimization of an energy like this one under possible constraints on the curve γ is known as the *Bolza* problem. As a very natural alternative, the values of γ at $t = 0, T$ could be prescribed instead of penalized (which is typical in some geometric problems such as in the case of geodesic curves) or we could prescribe the value at $t = 0$ and penalize the one at $t = T$ (which is natural in many applications where we do know the state of the system now and try to optimize the future evolution).

This chapter will present a detailed discussion of some classes of examples, starting with the most classical ones. Their analysis may sometimes require refined approaches which will become standard after reading the rest of the book.

© The Author(s), under exclusive license to Springer Nature Switzerland AG 2023
F. Santambrogio, *A Course in the Calculus of Variations*, Universitext,
https://doi.org/10.1007/978-3-031-45036-5_1

1.1   Optimal Trajectories

We present here three very classical examples of one-dimensional variational problems.

1.1.1   Three Classical 1D Examples

The Geodesic Problem   One of the most classical problems in the calculus of variations and geometry is that of finding minimal-length curves connecting two given points in a given space M (which can be, for instance, a manifold, a submanifold of Euclidean space, or even a more abstract metric space).

We recall that the length of a curve $\omega : [0, 1] \to \mathbb{R}^d$ is defined via

$$\text{Length}(\omega) := \sup \left\{ \sum_{k=0}^{n-1} |\omega(t_k) - \omega(t_{k+1})| \ : \ n \geq 1, \ 0 = t_0 < t_1 < \cdots < t_n = 1 \right\}.$$

It is easy to see that the length does not change under a reparameterization of the curve, so we can always assume that the interval of definition of ω is $[0, 1]$.

If $\omega \in C^1$ it is well known that we have

$$\text{Length}(\omega) = \int_0^1 |\omega'|(t) \, dt.$$

The same representation also holds when ω is only Lipschitz or absolutely continuous, which will be proven in Sect. 1.4. Given two points $x_0, x_1 \in M \subset \mathbb{R}^d$, the geodesic problem is then

$$\min \{ \text{Length}(\omega) \ : \ \omega \in AC([0, T]; M), \omega(0) = x_0, \omega(1) = x_1 \},$$

where $AC([0, T]; M)$ stands for the set of absolutely continuous curves defined on $[0, T]$ and valued in M (see Sect. 1.4.1).

The most classical example of a geodesic problem, which is actually the one which justifies its name (*geo* standing for the Earth), is the one where M is the two-dimensional unit sphere in $\mathbb{R}^3$, which we will analyze later in this section.

The Brachistochrone Problem   Another very classical problem in the one-dimensional calculus of variations is the problem of finding the optimal shape of a curve such that a point only subject to the force of gravity goes from one endpoint to the other in minimal time. We can consider it as the problem of finding an optimal playground slide with given height and ground length. When kids are approximated by material points and no friction is considered, the goal is that they arrive at the bottom of the slide in minimal time. The name brachistochrone also comes from Greek and derives from the phrase "quickest time". This is an example of a problem where the one-dimensional variable—which we will call x instead of t—does not represent time but the horizontal coordinate, and the function $x \mapsto z(x)$ does not represent the evolution of the state of a system but its graph represents a shape. We consider then all possible smooth functions z with $z(0) = H$, the number $H > 0$ being the height of the slide on the side where children start sliding, and $z(L) = 0$, the number $L > 0$ being the horizontal extension of the slide on the ground. Given the profile z we need to compute the time required for a material point to go from $(0, H)$ to $(L, 0)$. We assume that the velocity at $(0, H)$ is zero, so that the conservation of energy implies that when the material point is at $(x, z(x))$ its velocity $v(x)$ satisfies

$$\frac{1}{2}m|v(x)|^2 + mgz(x) = mgH,$$

where m is the mass of the point. This provides $|v(x)| = \sqrt{2g(H - z(x))}$ and implies that the time to move from $(x, z(x))$ to $(x + dx, z(x + dx))$ equals

$$\frac{d((x + dx, z(x + dx)), (x, z(x)))}{v(x)} = \frac{\sqrt{(dx)^2 + (z(x + dx) - z(x))^2}}{v(x)}$$

$$= \frac{\sqrt{1 + |z'(x)|^2}}{v(x)}\,dx = \sqrt{\frac{1 + |z'(x)|^2}{2g(H - z(x))}}\,dx.$$

As a consequence, the total time $\mathcal{T}$ to slide down is given by

$$\mathcal{T}(z) = \int_0^L \sqrt{\frac{1 + |z'(x)|^2}{2g(H - z(x))}}\,dx,$$

and the brachistochrone problem is thus

$$\min\left\{\mathcal{T}(z) \,:\, z \in C^1([0, L]), z(0) = H, z(L) = 0\right\}.$$

The Ramsey Model for the Optimal Growth of an Economic Activity In this model we describe the financial situation of an economic activity (say, a small firm owned by an individual) by a unique number representing its capital, and aggregating all relevant information such as money in the bank account, equipment, properties, workers... This number can evolve in time and will be denoted by $k(t)$. The evolution of k depends on how much the firm produces and how much the owner decides to "consume" (put in his pocket or give as dividends to the share-holders). We denote by $f(k)$ the production when the capital level is k, and we usually assume that f is increasing and often concave. We also assume that capital depreciates at a fixed rate $\delta > 0$ and we call $c(t)$ the consumption at time t. We then have $k'(t) = f(k(t)) - \delta k(t) - c(t)$. The goal of the owner is to optimize the consumption (which is required to make an income out of the activity) bearing in mind that reducing the consumption for some time allows the capital to be kept high, letting it grow even further, to finally be consumed later. More precisely, the owner has a utility function $U : \mathbb{R} \to \mathbb{R}$, also increasing and concave, and a discount rate $r > 0$, and his goal is to optimize

$$\int_0^T e^{-rt} U(c(t)) \, dt,$$

where T is fixed time horizon. One can take $T = \infty$ or $T < \infty$ and possibly add a final pay-off $\psi(k(T))$. The maximization problem can be considered as a maximization over c, and $k(T)$ can be deduced from c, or everything can be expressed in terms of k. Positivity constraints are also reasonably added on c (no money can be inserted in the activity besides the initial capital and its production) and on k (if the capital becomes negative there is default). The problem thus becomes a classical calculus of variations problem of the form

$$\max \left\{ \int_0^T e^{-rt} U(f(k) - \delta k - k') \, dt + \psi(k(T)) \ : \ \begin{matrix} k(0) = k_0, k \geq 0, \\ c := f(k) - \delta k - k' \geq 0 \end{matrix} \right\}.$$

1.1.2 Geodesics on the Sphere

We consider here the most classical problem concerning minimal-length curves: we want to find geodesic curves in the set $M = \{x \in \mathbb{R}^3 \ : \ |x| = 1\}$, the unit sphere of $\mathbb{R}^3$. Given two vectors $e_0, e_1 \in M$ such that $e_0 \cdot e_1 = 0$ we define the curve

$$\gamma(t) = e_0 \cos(t) + e_1 \sin(t).$$

This curve is an arc of a great circle on the sphere. It is a C^∞ curve, with constant unit speed since $\gamma'(t) = -e_0 \sin(t) + e_1 \cos(t)$ and

$$|\gamma'(t)|^2 = |e_0|^2 \sin^2(t) + |e_1|^2 \cos^2(t) - 2e_0 \cdot e_1 \sin(t) \cos(t) = \sin^2(t) + \cos^2(t) = 1.$$

Moreover it satisfies $\gamma''(t) = -\gamma(t)$.

We want to prove that, if T is small enough, then the curve γ has minimal length among those which share the same initial and final point and are valued in M. We first use the following lemma.

Lemma 1.1 *Assume that $\gamma : [0, T] \to M$ is a curve parameterized by unit speed such that the following optimality holds:*[1]

$$\int_0^T |\gamma'(t)|^2 \, dt \le \int_0^T |\tilde{\gamma}'(t)|^2 \, dt$$

for every other smooth curve $\tilde{\gamma} : [0, T] \to M$ with $\tilde{\gamma}(0) = \gamma(0)$ and $\tilde{\gamma}(T) = \gamma(T)$. Then we also have $\mathrm{Length}(\gamma) \le \mathrm{Length}(\tilde{\gamma})$ *for these same curves.*

Proof As we will see in Proposition 1.12, given a curve $\tilde{\gamma} : [0, T] \to M$ and a number $\varepsilon > 0$, it is possible to reparameterize the curve into a new curve $\hat{\gamma}$ with the same length, same initial and final points, and $|\hat{\gamma}'| \le (\mathrm{Length}(\tilde{\gamma}) + \varepsilon)/T$. We then have

$$\int_0^T |\hat{\gamma}'(t)|^2 \, dt \le T \left(\frac{\mathrm{Length}(\tilde{\gamma}) + \varepsilon}{T} \right)^2 = \frac{(\mathrm{Length}(\tilde{\gamma}) + \varepsilon)^2}{T}.$$

Then, using $|\gamma'(t)| = 1$, we obtain

$$T = \int_0^T |\gamma'(t)|^2 \, dt \le \int_0^T |\hat{\gamma}'(t)|^2 \, dt = \frac{(\mathrm{Length}(\tilde{\gamma}) + \varepsilon)^2}{T},$$

which implies $\mathrm{Length}(\tilde{\gamma}) + \varepsilon \ge T = \int_0^T |\gamma'(t)| \, dt = \mathrm{Length}(\gamma)$. Since $\varepsilon > 0$ is arbitrary, this shows $\mathrm{Length}(\tilde{\gamma}) \ge \mathrm{Length}(\gamma)$ ☐

With this lemma in mind we can prove the desired result.

Proposition 1.2 *The arcs of a great circle defined above are geodesic if T is small enough.*

Proof We just need to take γ of the form $\gamma(t) = e_0 \cos(t) + e_1 \sin(t)$ and $\tilde{\gamma} : [0, T] \to M$ of the form $\tilde{\gamma} = \gamma + \eta$ with $\eta(0) = \eta(T) = 0$, and prove $\int_0^T |\gamma'|^2 \le \int_0^T |\gamma' + \eta'|^2$. Imposing $\tilde{\gamma} \in M$ means $|\gamma + \eta|^2 = 1$, i.e. $|\eta|^2 + 2\gamma \cdot \eta = 0$.

We have

$$\int_0^T |\gamma' + \eta'|^2 \, dt = \int_0^T |\gamma'|^2 \, dt + \int_0^T |\eta'|^2 \, dt + \int_0^T 2\gamma' \cdot \eta' \, dt.$$

We can integrate the last term by parts, using $\eta(0) = \eta(T) = 0$, and then use $\gamma'' = -\gamma$ and $|\eta|^2 + 2\gamma \cdot \eta = 0$, so that we obtain

$$\int_0^T 2\gamma' \cdot \eta' \, dt = -\int_0^T 2\gamma'' \cdot \eta \, dt = \int_0^T 2\gamma \cdot \eta \, dt = -\int_0^T |\eta|^2 \, dt.$$

---

[1] Note that the assumption that γ has unit speed is useless and we could deduce that it is parameterized by constant speed from the optimality $\int_0^T |\gamma'(t)|^2 \, dt \le \int_0^T |\tilde{\gamma}'(t)|^2 \, dt$.

We then have

$$\int_0^T |\tilde{\gamma}'|^2 \, dt = \int_0^T |\gamma'|^2 \, dt + \int_0^T |\eta'|^2 \, dt - \int_0^T |\eta|^2 \, dt.$$

The result follows by the next lemma, which provides $\int_0^T |\eta'|^2 \geq \int_0^T |\eta|^2$ for T small enough.   □

Lemma 1.3 *There exists a universal constant C such that, if $\eta : [0, T] \to \mathbb{R}^d$ is a C^1 function with $\eta(0) = \eta(T) = 0$, then we have $\int_0^T |\eta|^2 \, dt \leq CT^2 \int_0^T |\eta'|^2 \, dt$.*

Proof We will prove the result with $C = 1$. We have $\eta(t) = \int_0^t \eta'(s) \, ds$, hence

$$|\eta(t)| \leq \int_0^t |\eta'(s)| \, ds \leq \int_0^T |\eta'(s)| \, ds \leq \sqrt{T} \sqrt{\int_0^T |\eta'(s)|^2 \, ds}.$$

This implies

$$\int_0^T |\eta(t)|^2 \, dt \leq T \cdot T \cdot \int_0^T |\eta'(s)|^2 \, ds$$

and proves the claim with $C = 1$.   □

The above inequality is known as the *Poincaré inequality* and the optimal constant is not $C = 1$. Indeed

- If instead of writing $\int_0^t |\eta'(s)| \, ds \leq \int_0^T |\eta'(s)| \, ds$ we keep the dependence in t we obtain $|\eta(t)| \leq \sqrt{t} \sqrt{\int_0^t |\eta'(s)|^2 \, ds} \leq \sqrt{t} \sqrt{\int_0^T |\eta'(s)|^2 \, ds}$ and, integrating in t, we gain a factor 2.
- If we also exploit $\eta(T) = 0$ we can use the above argument in order to estimate $\eta(t)$ for $t \leq T/2$ but integrate backwards starting from $t = T$ for $t \geq T/2$. We then obtain $|\eta(t)| \leq \sqrt{\min\{t, T - t\}} \sqrt{\int_0^T |\eta'(s)|^2 \, ds}$, which provides another factor 2 when integrating, since $\int_0^T \min\{t, T - t\} = T^2/4$.
- The optimal constant is in fact $(1/\pi)^2$ as explained in Box 1.1. This shows the optimality of all arcs of a great circle as soon as $T \leq \pi$, i.e. when the arc is at most half of the circumference, which is absolutely natural (since arcs longer than half of the circle are not optimal, as the other half has smaller length).

Box 1.1 Memo—*The Poincaré Inequality in 1D*

2π-Periodic functions which are L^2_{loc} on $\mathbb{R}$ can be written via Fourier series: given such a function η there are coefficients $a_n \in \mathbb{C}$ such that

$$\eta(t) = \sum_{n \in \mathbb{Z}} a_n e^{int},$$

where the convergence of the series is (at least) in the L^2 sense. The well-known *Parseval identity* provides $\int_0^{2\pi} |\eta|^2 = \sum |a_n|^2$. In terms of the Fourier coefficients we have $\eta' \in H^1_{loc}$ if and only if $\sum_{n \in \mathbb{Z}} n^2 |a_n|^2 < +\infty$, and we have $\int_0^{2\pi} |\eta'|^2 = \sum n^2 |a_n|^2$.

The Poincaré inequality for zero-mean functions (Poincaré–Wirtinger): if η is 2π-periodic and $\int_0^{2\pi} \eta = 0$ then $a_0 = 0$ and we clearly have $\int_0^{2\pi} |\eta|^2 \leq \int_0^{2\pi} |\eta'|^2$ as a consequence of $n^2 \geq 1$ for $n \neq 0$. The only equality case in this inequality is given by functions which are linear combinations of $e^{\pm it}$ or, equivalently, of $\sin(t)$ and $\cos(t)$. Note that a similar inequality exists, but with a different constant, for zero-mean functions without the periodicity conditions (in particular, without requiring the function to be in H^1_{loc} when extended by periodicity), as one can see from the example $\eta(t) = \sin(t/2)$ on $[-\pi, \pi]$ for which we have $\int_0^{2\pi} |\eta|^2 = 4 \int_0^{2\pi} |\eta'|^2$.

The Poincaré inequality for H^1_0 functions: if $\eta \in H^1([0, \pi])$ is such that $\eta(0) = \eta(\pi) = 0$ we can define $\tilde{\eta}(t) = -\eta(t - \pi)$ for $t \in [\pi, 2\pi]$ and then extend by 2π-periodicity. We can see that $\tilde{\eta}$ is in H^1_{loc}, has zero mean, and satisfies $\int_0^{2\pi} |\tilde{\eta}|^2 = \int_0^{\pi} |\eta|^2$ together with $\int_0^{2\pi} |\tilde{\eta}'|^2 = \int_0^{\pi} |\eta'|^2$. We thus obtain $\int_0^{\pi} |\eta|^2 \leq \int_0^{\pi} |\eta'|^2$. This inequality is also sharp, and only attained as an equality for $\eta(t) = C \sin(t)$.

Alternative proofs of this last statement can be found in Exercises 1.19 and 1.20.

All these inequalities can be extended to other intervals $[0, T]$ instead of $[0, 2\pi]$ or $[0, \pi]$ by considering $\tilde{u}(t) = u(2\pi t/T)$ or $\tilde{u}(t) = u(\pi t/T)$ and the Poincaré constant is proportional to T^2 (see also Box 5.2).

1.2 Examples of Existence and Non-existence

One of the first questions that we should address in optimization, when the set of competitors is not finite, is the existence of a minimizer. This question is already non-trivial when optimizing over a continuum of competitors, as we do when minimizing a function defined on $\mathbb{R}$ or $\mathbb{R}^N$, but becomes even trickier when the optimization is performed in infinite dimension, in a class of functions, as is typical in the calculus of variations.

1.2.1   Existence

We will see in this section how to prove the existence for a very standard example of variational problem, the technique being useful to understand how to attack more general problems. We will consider a functional involving first derivatives of the unknown u, but the existence of the minimizers will not be proven in the space C^1. We will instead set the problem in a Sobolev space, since Sobolev spaces have better compactness properties and their norm is more adapted to study variational problems.

Box 1.2 Memo—*Sobolev Spaces in 1D*

Given an open interval $I \subset \mathbb{R}$ and an exponent $p \in [1, +\infty]$ we define the Sobolev space $W^{1,p}(I)$ as

$$W^{1,p}(I) := \left\{ u \in L^p(I) \ : \ \begin{array}{l} \exists g \in L^p(I) \text{ s.t.} \\ \int u\varphi' \, dt = -\int g\varphi \, dt \text{ for all } \varphi \in C_c^\infty(I) \end{array} \right\}.$$

The function g, if it exists, is unique, and will be denoted by u' since it plays the role of the derivatives of u in the integration by parts.

The space $W^{1,p}$ is endowed with the norm $||u||_{W^{1,p}} := ||u||_{L^p} + ||u'||_{L^p}$. With this norm $W^{1,p}$ is a Banach space, separable if $p < +\infty$, and reflexive if $p \in (1, \infty)$.

All functions $u \in W^{1,p}(I)$ admit a continuous representative, which moreover satisfies $u(t_0) - u(t_1) = \int_{t_1}^{t_0} u'(t) \, dt$. This representative is differentiable a.e. and the pointwise derivative coincides with the function u' a.e. Moreover, for $p > 1$ the same representative is also Hölder continuous, of exponent $\alpha = 1 - \frac{1}{p} > 0$. The injection from $W^{1,p}$ into $C^0(I)$ is compact if I is bounded.

If $p = 2$ the space $W^{1,p}$ can be given a Hilbert structure, choosing as a norm $\sqrt{||u||_{L^2}^2 + ||u'||_{L^2}^2}$, and is denoted by H^1.

Higher-order Sobolev spaces $W^{k,p}$ can also be defined for $k \in \mathbb{N}$ by induction as follows: $W^{k+1,p}(I) := \{u \in W^{k,p}(I) \ : \ u' \in W^{k,p}(I)\}$ and the norm in $W^{k+1,p}$ defined as $||u||_{W^{k,p}} + ||u'||_{W^{k,p}}$. In the case $p = 2$ the Hilbert spaces $W^{k,2}$ are also denoted by H^k.

We consider an interval $I = [a, b] \subset \mathbb{R}$ and a continuous function $F : I \times \mathbb{R} \to \mathbb{R}$ that we assume bounded from below.

Theorem 1.4 *Let us consider the problem*

$$\min \left\{ J(u) := \int_a^b \left(F(t, u(t)) + |u'(t)|^2 \right) dt \; : \; \begin{array}{l} u \in H^1(I), \\ u(a) = A, \; u(b) = B \end{array} \right\}. \quad (1.1)$$

This minimization problem admits a solution.

Proof Take a minimizing sequence u_n such that $J(u_n) \to \inf J$. The functional J is composed of two terms (the one with F and the one with $|u'|^2$) and their sum is bounded from above. Since they are both bounded from below, we can deduce that they are also both bounded from above. In particular, we obtain an upper bound for $\|u_n'\|_{L^2}$. Since the boundary values of u_n are fixed, applying the Poincaré inequality of Lemma 1.3 to the functions $t \mapsto u_n(t) - \frac{B-A}{b-a}(t - a) - A$, we obtain a bound on $\|u_n\|_{H^1}$ (i.e. the L^2 norms of u_n and not only of u_n' are bounded).

Hence, $(u_n)_n$ is a bounded sequence in H^1 and we can extract a subsequence which weakly converges in H^1 to a function u (see Box 1.4). In dimension one, the weak convergence in H^1 implies the uniform convergence, and in particular the pointwise convergence on the boundary. We then deduce from $u_n(a) = A$ and $u_n(b) = B$ that we have $u(a) = A$ and $u(b) = B$, i.e. that u is an admissible competitor for our variational problem. We just need to show $J(u) \leq \liminf_n J(u_n) = \inf J$ in order to deduce $J(u) = \inf J$ and the optimality of u.

This strategy is very general and typical in the calculus of variations. It is called the *direct method* (see Box 1.3) and requires a topology (or a notion of convergence) to be found on the set of admissible competitors such that

- there is compactness (any minimizing sequence admits a convergent subsequence, or at least a properly built minimizing sequence does so),
- the functional that we are minimizing is lower semicontinuous, i.e. $J(u) \leq \liminf_n J(u_n)$ whenever u_n converges to u.

$\square$

Box 1.3 Memo—*Weierstrass Criterion for the Existence of Minimizers, Semicontinuity*

The most common way to prove that a function admits a minimizer is called the "direct method in the calculus of variations". It simply consists in the classical Weierstrass Theorem, possibly replacing continuity with semicontinuity.

(continued)

Box 1.3 (continued)

Definition On a metric space X, a function $f : X \to \mathbb{R} \cup \{+\infty\}$ is said to be lower semicontinuous (l.s.c. in short) if for every sequence $x_n \to x$ we have $f(x) \leq \liminf_n f(x_n)$. A function $f : X \to \mathbb{R} \cup \{-\infty\}$ is said to be upper-semicontinuous (u.s.c. in short) if for every sequence $x_n \to x$ we have $f(x) \geq \limsup_n f(x_n)$.

Definition A metric space X is said to be compact if from any sequence x_n we can extract a convergent subsequence $x_{n_k} \to x \in X$.

Theorem (Weierstrass) *If $f : X \to \mathbb{R} \cup \{+\infty\}$ is lower semicontinuous and X is compact, then there exists an $\bar{x} \in X$ such that $f(\bar{x}) = \min\{f(x) : x \in X\}$.*

Proof Define $\ell := \inf\{f(x) : x \in X\} \in \mathbb{R} \cup \{-\infty\}$ ($\ell = +\infty$ only if f is identically $+\infty$, but in this case any point in X minimizes f). By definition there exists a minimizing sequence x_n, i.e. points in X such that $f(x_n) \to \ell$. By compactness we can assume $x_n \to \bar{x}$. By lower semicontinuity, we have $f(\bar{x}) \leq \liminf_n f(x_n) = \ell$. On the other hand, we have $f(\bar{x}) \geq \ell$ since ℓ is the infimum. This proves $\ell = f(\bar{x}) \in \mathbb{R}$ and this value is the minimum of f, realized at $\bar{x}$.

In our case, the uniform convergence $u_n \to u$ implies $\int_a^b F(t, u_n(t)) \, dt \to \int_a^b F(t, u(t)) \, dt$ and proves the continuity of the first integral term.

We now observe that the map $H^1 \ni u \mapsto u' \in L^2$ is continuous and hence the weak convergence of u_n to u in H^1 implies $u_n' \rightharpoonup u'$ in L^2. An important property of the weak convergence in any Banach space is the fact that the norm itself is lower semicontinuous, so that $\|u'\|_{L^2} \leq \liminf_n \|u_n'\|_{L^2}$ and hence

$$\int_a^b |u'(t)|^2 \, dt \leq \liminf_n \int_a^b |u_n'(t)|^2 \, dt.$$

This concludes the proof. $\square$

Box 1.4 Memo—*Hilbert Spaces*

A Hilbert space is a Banach space whose norm is induced by a scalar product: $\|x\| = \sqrt{x \cdot x}$.

Theorem (Riesz) *If $\mathcal{H}$ is a Hilbert space, for every $\xi \in \mathcal{H}$ there is a unique vector $h \in \mathcal{H}$ such that $\langle \xi, x \rangle = h \cdot x$ for every $x \in \mathcal{H}$, and the dual space $\mathcal{H}'$ is isomorphic to $\mathcal{H}$.*

In a Hilbert space $\mathcal{H}$ we say that x_n weakly converges to x and we write $x_n \rightharpoonup x$ if $h \cdot x_n \to h \cdot x$ for every $h \in \mathcal{H}$. Every weakly convergent sequence

(continued)

> **Box 1.4** (continued)
> is bounded, and if $x_n \rightharpoonup x$, using $h = x$ we find $||x||^2 = x \cdot x = \lim_n x \cdot x_n \leq \liminf_n ||x|| \, ||x_n||$, i.e. $||x|| \leq \liminf_n ||x_n||$.
>
> In a Hilbert space $\mathcal{H}$, every bounded sequence x_n admits a weakly convergent subsequence (see Box 2.2).

Variants The same argument can be applied if the constraints on $u(a)$ and/or $u(b)$ are removed and replaced by penalizations:

$$\min\left\{ J(u) := \int_a^b \left(F(t, u(t)) + |u'(t)|^2 \right) dt + g(u(a)) + h(u(b)) \; : \; u \in H^1(I) \right\}.$$

If g and h are continuous, the uniform convergence of u_n to u implies the continuity of the extra terms. Yet, we need to guarantee that the sequence is still bounded in H^1. This can be obtained in many cases, for instance:

- If $\lim_{s \to \pm\infty} g(s) = +\infty$ and h is bounded from below: in this case we obtain a bound on $|u_n(a)|$. Together with the bound on $||u'_n||_{L^2}$ and $u_n(t) = u_n(a) + \int_0^t u'_n(s)\,ds$, we also obtain a bound on $||u_n||_{L^\infty}$. Analogously, if we have $\lim_{s \to \pm\infty} h(s) = +\infty$ and g bounded from below, we would obtain a bound on $|u_n(b)|$ and then on $||u_n||_{L^\infty}$.
- If $F(t, s) \geq c(|s|^2 - 1)$ for some constant $c > 0$ and g, h are bounded from below. In this case the bound on $J(u_n)$ also implies a bound on $||u_n||_{L^2}$.
- Actually, much less is needed on F. Assume that there exists a subset $A \subset [a, b]$ of positive measure and a function $\omega : \mathbb{R} \to \mathbb{R}$ with $\lim_{s \to \pm\infty} \omega(s) = +\infty$ such that $F(t, s) \geq \omega(s)$ for every $t \in A$. In this case we are also able to obtain a bound on $||u_n||_{L^\infty}$ from the bound on J. Indeed, the bound on $||u'_n||_{L^2}$ implies

$$|u_n(t) - u_n(s)| \leq (b - a)^{1/2} ||u'_n||_{L^2} \leq C$$

for every n. In case we had, on a subsequence, $||u_n||_{L^\infty} \to \infty$, we would also have the existence of a point $t_n \in [a, b]$ such that $|u_n(t_n)| = ||u_n||_{L^\infty} \to \infty$ and hence we would have $|u_n(t)| \geq ||u_n||_{L^\infty} - C \to \infty$ for every $t \in A$, and hence $\int_a^b F(t, u_n(t))\,dt \geq \int_A \omega(||u_n||_{L^\infty} - C) - C \to \infty$.

We also note that the same argument would work if the continuity of F, g, h was replaced by their lower semicontinuity

Non-quadratic Dependence in u' The term $\int |u'(t)|^2$ could be easily replaced by a term $\int |u'(t)|^p$, $p > 1$, and the results would be exactly the same using the space $W^{1,p}$ instead of $H^1 = W^{1,2}$, but this would require the use of weak convergence in Banach spaces which are not Hilbert spaces (see Boxes 2.2 and 2.3). Luckily, for $p \in (1, \infty)$ the spaces L^p and $W^{1,p}$ are reflexive, which allows us to act almost as if we were in a Hilbert setting. The case would be different for $p = 1$ since the space $W^{1,1}$ is not reflexive and it is in general not possible to extract weakly convergent subsequences from bounded sequences. It is possible to use a larger space, i.e. the BV space, but in this case the limit would not necessarily belong to $W^{1,1}$. This is the reason why the geodesic problem, which indeed involves a term $\int |u'(t)|\, dt$, is treated in a different way. We will re-discuss this very particular problem in Sect. 1.4 and consider the minimization of $\int (|u'(t)| + F(t, u(t)))\, dt$ in Sect. 1.6

On the other hand, it would be possible to handle terms of the form $\int H(u'(t))\, dt$ where H is convex but not necessarily a power. In order to obtain compactness we need a lower bound of the form $H(v) \geq c(|v|^p - 1)$ for $p > 1$ and $c > 0$. In order to handle the lower semicontinuity, observe that $L^p \ni v \mapsto \int H(v)$ is l.s.c. for the weak convergence for the following reason (see Sect. 3.1 for details): it is l.s.c. for the strong convergence in L^p and it is convex, which makes it weakly l.s.c.

Minimization Among Smooth Functions In many natural questions, coming from geometry or from applied modeling, we do not look for a minimizer among H^1 or $W^{1,p}$ functions, but the problem is set in the space C^1. In this case a natural strategy is the following. First, extend the minimization problem to a larger Sobolev space, chosen so that the functional guarantees weak compactness of a minimizing sequence (in practice, if the integrand has growth of order p in terms of u' we usually choose $W^{1,p}$) and lower semicontinuity. Then, prove that this extended problem admits a minimizer, i.e. a $W^{1,p}$ function which is better than any other $W^{1,p}$ function. Finally, write the optimality conditions of the problem—as we will see in Sect. 1.3—and hope that these conditions (which are a differential equation on the solution) help in proving that the optimizer is actually a smooth function. We will discuss the Euler–Lagrange equation in Sect. 1.3 but a simple example is the following: when solving (1.1), if we assume that F is C^1 then the Euler–Lagrange equation reads $2u''(t) = F'(t, u(t))$ (where F' is the derivative w.r.t. the second variable). Since the right-hand side is continuous (due to $F \in C^1$ and $u \in H^1 \subset C^0$) then we even obtain $u \in C^2$.

1.2.2 Non-existence

We consider in this section a very classical example of a variational problem in 1D which has no solution. With this aim, let us consider

$$\inf\left\{J(u) := \int_0^1 \left(\left||u'(t)|^2 - 1\right| + |u(t)|^2\right) dt \; : \; u \in H^1([0, 1]), u(0) = u(1) = 0\right\}.$$

It is clear that for any admissible u we have $J(u) > 0$: indeed, J is composed of two non-negative terms, and they cannot both vanish for the same function u. In order to have $\int_0^1 |u(t)|^2 \, dt = 0$ one would need $u = 0$ constantly, but in this case we would have $u' = 0$ and $J(u) = 1$. On the other hand, we will prove $\inf J = 0$, which proves that the minimum cannot be attained.

To do so, we consider the following sequence of Lipschitz functions u_n: we first define $U : [0, 1] \to \mathbb{R}$ as $U(t) = \frac{1}{2} - |t - \frac{1}{2}|$. It satisfies $U(0) = U(1) = 0$, $|U| \leq \frac{1}{2}$, and $|U'| = 1$ a.e. (U is Lipschitz continuous but not C^1). We then extend U as a 1-periodic function on $\mathbb{R}$, that we call $\tilde{U}$, and then set $u_n(t) = \frac{1}{n} \tilde{U}(nt)$. The function u_n is $\frac{1}{n}$-periodic, satisfies $u_n(0) = u_n(1) = 0$ again, and $|u_n'| = 1$ a.e. We also have $|u_n| \leq \frac{1}{2n}$. If we compute $J(u_n)$ we easily see that we have $J(u_n) \leq \frac{1}{4n^2} \to 0$, which shows $\inf J = 0$.

The above example is very useful to understand the relation between compactness and semicontinuity. Indeed, the sequence u_n which we built is such that u_n converges uniformly to 0 and u_n' is bounded in L^∞. This means that we also have $u_n' \stackrel{*}{\rightharpoonup} 0$ in L^∞ (indeed, a sequence which is bounded in L^∞ admits a weakly-* convergent subsequence and the limit in the sense of distributions—see Box 2.5— of u_n' can only be the derivative of the limit of u_n, i.e. 0). This means that, if we use weak convergence in Sobolev spaces (weak convergence in L^∞ implies weak convergence in L^2, for instance), then we have compactness. Yet, the limit is the function $u = 0$ but we have $J(u) = 1$ while $\lim_n J(u_n) = 0$, which means that semicontinuity fails. This is due to the lack of convexity of the double-well function $W(v) = ||v|^2 - 1|$ (see Chap. 3): indeed, the 0 derivative of the limit function u is approximated through weak convergence as a limit of a rapidly oscillating sequence of functions u_n' taking values ± 1, and the values of W at ± 1 are better than the value at 0 (which would have not been the case if W was convex). On the other hand, it would have been possible to choose a stronger notion of convergence, for instance strong H^1 convergence. In this case, if we have $u_n \to u$ in H^1, we deduce $u_n' \to u'$ in L^2 and $J(u_n) \to J(u)$. We would even obtain continuity (and not just semicontinuity) of J. Yet, what would be lacking in this case is the compactness of minimizing sequences (and the above sequence u_n, which is a minimizing sequence since $J(u_n) \to 0 = \inf J$ proves that it is not possible to extract strongly convergent subsequences). This is hence a clear example of the difficult task of choosing a suitable convergence for applying the direct method of calculus of variations : not too strong, otherwise there is no convergence, not too weak, otherwise lower semicontinuity could fail.

We finish this section by observing that the problem of non-existence of the minimizer of J does not depend on the choice of the functional space. Indeed, even if we considered

$$\inf\left\{ J(u) := \int_0^1 \left(\left| |u'(t)|^2 - 1 \right| + |u(t)|^2 \right) dt \ : \ u \in C^1([0, 1]), u(0) = u(1) = 0 \right\}$$

we would have $\inf J = 0$ and $J(u) > 0$ for every competitor, i.e. no existence. To show this, it is enough to modify the above example in order to produce a sequence of C^1 functions. In this case it will not be possible to make the term $\int_0^1 \left| |u'(t)|^2 - 1 \right| dt$ exactly vanish, since this requires $u' = \pm 1$, but for a C^1 function this means either $u' = 1$ everywhere or $u' = -1$ everywhere, and neither choice is compatible with the boundary data. On the other hand, we can fix $\delta > 0$, use again the Lipschitz function U introduced above, and define a function $U_\delta : [0, 1] \to \mathbb{R}$ such that

$$U_\delta = U \text{ on } [\delta, \frac{1}{2} - \delta] \cup [\frac{1}{2} + \delta, 1 - \delta], \quad U_\delta(0) = U_\delta(1) = 0, \quad |U_\delta| \le \frac{1}{2}, \quad |U_\delta'| \le 2.$$

We then extend U_δ to a 1-periodic function $\tilde{U}_\delta$ defined on $\mathbb{R}$ and set $u_{n,\delta}(t) := \frac{1}{n} \tilde{U}_\delta(nt)$. We observe that we have $|u_{n,\delta}'| \le 2$ and $|u_{n,\delta}'| = 1$ on a set

$$A_{n,\delta} = \bigcup_{k=0}^{n-1} \left([\frac{k}{n} + \frac{\delta}{n}, \frac{2k+1}{2n} - \frac{\delta}{n}] \cup [\frac{2k+1}{2n} + \frac{\delta}{n}, \frac{k+1}{n} - \frac{\delta}{n}] \right)$$

whose measure is $1 - 4\delta$. We then have

$$J(u_{n,\delta}) = \int_{[0,1] \setminus A_{n,\delta}} \left| |u_{n,\delta}'(t)|^2 - 1 \right| dt + \int_0^1 |u_{n,\delta}(t)|^2 \, dt \le 12\delta + \frac{1}{4n^2}.$$

This shows $\inf J \le 12\delta$ and, $\delta > 0$ being arbitrary, $\inf J = 0$.

1.3   Optimality Conditions

We consider here the necessary optimality conditions for a typical variational problem. The result will be presented in 1D but we will see in Sect. 2.2 that the procedure is exactly the same in higher dimensions. We consider a function $L : [a, b] \times \mathbb{R}^d \times \mathbb{R}^d \to \mathbb{R}$. The three variables of L will be called t (time), x (position), and v (velocity). We assume that L is C^1 in (x, v) for a.e. $t \in [a, b]$.

1.3.1   The Euler–Lagrange Equation

We start from the minimization problem

$$\min \left\{ J(u) := \int_a^b L(t, u(t), u'(t)) \, dt \; : \; u \in X, \, u(a) = A, \, u(b) = B \right\},$$

where X is a functional space which could be, in most cases, a Sobolev space. We assume anyway $X \subset W^{1,1}([a, b]; \mathbb{R}^d)$ which, at the same time, guarantees both the existence of a suitably defined derivative, and the continuity of all functions in X. We also assume that for every $u \in X$ the negative part of $L(\cdot, u, u')$ is integrable, so that J is a well-defined functional from X to $\mathbb{R} \cup \{+\infty\}$. We assume that u is a solution of such a minimization problem and that $u + C_c^{\infty}((a, b)) \subset X$. This means that for every $\varphi \in C_c^{\infty}((a, b))$ we have $J(u) \le J(u + \varepsilon\varphi)$ for small ε.

We now fix the minimizer u and a perturbation φ and consider the one-variable function

$$j(\varepsilon) := J(u_\varepsilon), \quad \text{where} \quad u_\varepsilon := u + \varepsilon\varphi,$$

which is defined in a neighborhood of $\varepsilon = 0$, and minimal at $\varepsilon = 0$. We now want to compute $j'(0)$.

In order to so, we will assume that for every $u \in X$ with $J(u) < +\infty$ there exists a $\delta > 0$ such that we have

$$t \mapsto \sup \left\{ |\nabla_x L(t, x, v)| + |\nabla_v L(t, x, v)| \; : \; \begin{array}{l} x \in B(u(t), \delta), \\ v \in B(u'(t), \delta) \end{array} \right\} \in L^1([a, b]).$$

$$(1.2)$$

We will discuss later some sufficient conditions on L which guarantee that this is satisfied. If this is the case, then we can differentiate w.r.t. ε the function $\varepsilon \mapsto L(t, u_\varepsilon, u'_\varepsilon)$, and obtain

$$\frac{d}{d\varepsilon} L(t, u_\varepsilon, u'_\varepsilon) = \nabla_x L(t, u_\varepsilon, u'_\varepsilon) \cdot \varphi + \nabla_v L(t, u_\varepsilon, u'_\varepsilon) \cdot \varphi'.$$

Since we assume $\varphi \in C_c^{\infty}((a, b))$, both φ and φ' are bounded, so that for ε small enough we have $\varepsilon|\varphi(t)|, \varepsilon|\varphi'(t)| \le \delta$ and we can apply the assumption in (1.2) in order to obtain domination in L^1 of the pointwise derivatives. This shows that, for small ε, we have

$$j'(\varepsilon) = \int_a^b \left(\nabla_x L(t, u_\varepsilon, u'_\varepsilon) \cdot \varphi + \nabla_v L(t, u_\varepsilon, u'_\varepsilon) \cdot \varphi' \right) dt.$$

In particular, we have

$$j'(0) = \int_a^b \left(\nabla_x L(t, u, u') \cdot \varphi + \nabla_v L(t, u, u') \cdot \varphi' \right) dt.$$

Imposing $j'(0) = 0$, which comes from the optimality of u, means precisely that we have, in the sense of distributions, the following differential equation, known as the *Euler–Lagrange* equation:

$$\frac{\mathrm{d}}{\mathrm{d}t}\left(\nabla_v L(t, u, u')\right) = \nabla_x L(t, u, u').$$

This is a second-order differential equation (on the right-hand side we have the derivative in t of a term already involving u') and, as such, requires the choice of two boundary conditions. These conditions are $u(a) = A$ and $u(b) = B$, which means that we are not facing a Cauchy problem (where we would prescribe $u(a)$ and $u'(a)$).

Energy Conservation When speaking of variational principles in mathematical physics, the word *energy* has multiple meanings due to an ambiguity in the physical language, which is increased by the mathematicians' use of the notion. Sometimes we say that the motion should minimize the total energy (and we think in this case of an integral in time of a cost involving kinetic energy and potential energy; a more adapted name which can be found in the literature is *action*), while on other occasions we can say that the energy is preserved along the evolution (and in this case we think of a quantity computed at every time t).

Mathematically, this can be clarified in the following way: assume the integrand in a variational problem is independent of time and of the form $L(t, x, v) = \frac{1}{2}|v|^2 + V(x)$. The Euler–Lagrange equation of the corresponding minimization problem would be $\gamma'' = \nabla_x V(\gamma)$. If we take the scalar product with γ' we obtain

$$\frac{\mathrm{d}}{\mathrm{d}t}\left(\frac{1}{2}|\gamma'(t)|^2\right) = \gamma'(t) \cdot \gamma''(t) = \gamma'(t) \cdot \nabla_x V(\gamma(t)) = \frac{\mathrm{d}}{\mathrm{d}t}(V(\gamma(t))).$$

This shows that the difference $\frac{1}{2}|\gamma'|^2 - V(\gamma)$ is constant in time. We see that the minimization of the integral of $\frac{1}{2}|\gamma'|^2 + V(\gamma)$ (i.e. the sum of the kinetic energy and of V) implies that the difference $\frac{1}{2}|\gamma'|^2 - V(\gamma)$ is constant. Which quantity should be called energy is then a matter of convention (or taste).

This very same result could also be obtained by looking at the optimality of the curve γ in terms of the so-called "internal variations", i.e. taking $\gamma_\varepsilon(t) = \gamma(t + \varepsilon\eta(t))$ (see Exercise 1.13).

In the very particular case $V = 0$, this result reads as "minimizers of the integral of the square of the speed have constant speed", and is a well-known fact for geodesics (see also Sect. 1.1.2).

Remark 1.5 The energy conservation principle described above is a particular case of the so-called *Beltrami formula*, which is valid whenever the integrand L does not depend explicitly on time. In this case, from the Euler–Lagrange equation $\frac{\mathrm{d}}{\mathrm{d}t}(\nabla_v L(\gamma(t), \gamma'(t))) = (\nabla_x L(\gamma(t), \gamma'(t)))$ we can deduce

$$L(\gamma(t), \gamma'(t))) - \gamma'(t) \cdot \nabla_v L(\gamma(t), \gamma'(t)) = const.$$

This can be proven by differentiating in time, as we have

$$\frac{d}{dt}\left(L(\gamma, \gamma')\right) - \gamma' \cdot \nabla_v L(\gamma, \gamma')$$

$$= \nabla_x L(\gamma, \gamma') \cdot \gamma' + \nabla_v L(\gamma, \gamma') \cdot \gamma'' - \gamma'' \cdot \nabla_v L(\gamma, \gamma') - \gamma' \cdot \frac{d}{dt}(\nabla_v L(\gamma, \gamma')) = 0.$$

Example 1.6 The Euler–Lagrange equation was a key point in the development of the calculus of variations and one of the most important examples revealing how useful it can be is that of the brachistochrone curve. In this case, we saw in Sect. 1.1.1 that we have to use $L(z, v) = \sqrt{(1 + |v|^2)/(H - z)}$ and the Euler–Lagrange equation, after simplification, becomes

$$\frac{z''}{1 + |z'|^2}(H - z) = \frac{1}{2}.$$

It is interesting that one of the first to study the brachistochrone problem was Galileo Galilei, who understood that a straight line was not optimal, but he believed that the solution was an arc of a circle (Third Day, Theorem 22, Proposition 36 in [93]). Curiously enough, if one inserts $z(x) = H - \sqrt{1 - |x|^2}$ in the left-hand side of the above equation, the result is a constant, but not $1/2$ (the result would be 1). The true solution, after the problem was well-formalized by Johann Bernoulli ([26]) and correctly solved, is an arc of a cycloid. In this book we do not want to provide a complete discussion of the solution of this problem, which is just an example for us. Instead, we refer, for instance, to [53]. However, the reader will find two useful exercises on the brachistochrone problem, Exercises 1.1 and 1.2, at the end of the chapter.

1.3.2 Examples of Growth Conditions on L

We discuss now under which conditions on L and X we can guarantee (1.2). We will not give an exhaustive classification of all possible cases, which is probably impossible to do. We will only consider three examples

1. The case where L has growth of order p in terms of v and is of the form

$$L(t, x, v) = c|v|^p + F(t, x),$$

which is the example we considered for existence in the case $p = 2$. Note that in this case a natural choice for the space X is $X = W^{1,p}([a, b])$ since any minimizing sequence for J will be bounded in X and the arguments of Sect. 1.2 prove that a minimizer exists.

2. The case where L has growth of order p in terms of v but it has a multiplicative form

$$L(t, x, v) = a(t, x)|v|^p$$

for $a : [0, T] \times \mathbb{R}^d \to \mathbb{R}$ bounded from below and above by positive constants. In this case we will also choose $X = W^{1,p}([a, b])$ since minimizing sequences will also be bounded, but we observe that Sect. 1.2 does not yet provide a proof of existence since the semicontinuity of the functional $u \mapsto \int L(t, u(t), u'(t)) \, dt$ still has to be proven (see Chap. 3).

3. A less standard example such as

$$L(t, x, v) = e^{h(v)} + F(t, x),$$

where $h : \mathbb{R}^d \to \mathbb{R}$ is a C^1, Lipschitz continuous and convex function such that $\lim_{|v| \to \infty} h(v) = +\infty$.

We start from the first case. Let us assume that F is C^1 w.r.t. the variable x and that $\nabla_x F$ is continuous in (t, x). In this case we have

$$\nabla_x L(t, x, v) = \nabla_x F(t, x), \quad \nabla_v L(t, x, v) = pc|v|^{p-2}v.$$

For every $u \in X$ we have $u \in C^0$ and $u' \in L^p$. In particular, assuming $|u| \leq M$, we have

$$\sup \left\{ |\nabla_x L(t, x, v)| \ : \ x \in B(u(t)), \delta), v \in B(u'(t)), \delta) \right\}$$

$$\leq \sup \left\{ |\nabla_x F(t, x)| \ : t \in [a, b], \ |x| \leq M + \delta \right\} < +\infty.$$

Concerning $\nabla_v L$, we have

$$\sup \left\{ |\nabla_v L(t, x, v)| \ : \ x \in B(u(t)), \delta), v \in B(u'(t)), \delta) \right\} \leq C(|u'(t)|^{p-1} + \delta^{p-1}),$$

and this is integrable in t as soon as $u' \in L^{p-1}$, which is satisfied since we even have $u' \in L^p$.

In the second case we assume that a is C^1 in x and that $\nabla_x a$ is continuous in (t, x). We have

$$\nabla_x L(t, x, v) = \nabla_x a(t, x)|v|^p, \quad \nabla_v L(t, x, v) = pa(t, x)|v|^{p-2}v.$$

Thus, using again $u \in C^0$ and assuming $|u| \leq M$, we have

$$\sup \left\{ |\nabla_x L(t, x, v)| \ : \ x \in B(u(t)), \delta), v \in B(u'(t)), \delta) \right\}$$

$$\leq \sup \left\{ |\nabla_x a(t, x)| \ : t \in [a, b], \ |x| \leq M + \delta \right\} \left(|u'(t)| + \delta \right)^p \leq C(|u'(t)|^p + \delta^p).$$

This is integrable in t as soon as $u' \in L^p$, which is true for every $u \in X$.

The third case is trickier because of the non-standard growth of L in v. In this case a natural choice for the space X would be a Sobolev–Orlicz space which imposes integrability of $e^{|u'|}$ but it is not necessary to do so. Indeed, one can take for arbitrary p, $X = W^{1,p}([a, b])$. Thanks to the growth of L, any minimizing sequence will be bounded in such a space, and the semicontinuity results of Chap. 3 will also allow us to prove the existence of a solution. The difference with respect to the previous cases is that simply using $u \in X$ will not be enough to guarantee condition (1.2), but one will really need to use $J(u) < +\infty$. The condition on $\nabla_x L$ will be treated in the same standard way since we still have $u \in C^0$. As for the condition on $\nabla_v L$, we have

$$\nabla_v L(t, x, v) = \nabla h(v) e^{h(v)}.$$

We then observe that we have $|\nabla_v L(t, x, v)| \le (\text{Lip } h) e^{h(v)}$ and then

$$\sup \left\{ |\nabla_v L(t, x, v)| \; : \; \begin{matrix} x \in B(u(t)), \delta), \\ v \in B(u'(t)), \delta) \end{matrix} \right\} \le (\text{Lip } h) e^{h(u') + \text{Lip } h\delta} = C e^{h(u')},$$

so that its integrability is equivalent to that of $e^{h(u')}$, which is true when $J(u) < +\infty$.

1.3.3 Transversality Conditions

We consider now the case where the Dirichlet boundary conditions on $u(a)$ and $u(b)$ are replaced by penalizations on the values of u at $t = a, b$. Let us consider for instance the problem

$$\min \{ J(u) + \psi_0(u(a)) + \psi_1(u(b)) \; u \in X \},$$

where, again, we set $J(u) := \int_a^b L(t, u(t), u'(t)) \, dt$. We assume that u is a minimizer and we set $u_\varepsilon := u + \varepsilon \varphi$, but we do not assume $\varphi \in C_c^\infty((a, b))$, since we are no longer obliged to preserve the values at the boundary points. Let us set $J_\psi(u) := J(u) + \psi_0(u(a)) + \psi_1(u(b))$ and $j_\psi(\varepsilon) := J_\psi(u_\varepsilon)$. The optimality of u provides $j_\psi(0) \le j_\psi(\varepsilon)$ and we want to differentiate j_ψ in terms of ε. The computation is exactly the same as before (and requires the very same assumptions) concerning the term $j(\varepsilon) = J(u_\varepsilon)$ and is very easy for the boundary terms. We then obtain

$$j'_\psi(0) = \int_a^b \left(\nabla_x L(t, u, u') \cdot \varphi + \nabla_v L(t, u, u') \cdot \varphi' \right) dt$$
$$+ \nabla \psi_0(u(a)) \cdot \varphi(a) + \nabla \psi_1(u(b)) \cdot \varphi(b).$$

This derivative should vanish for arbitrary φ, and it is possible to first consider $\varphi \in C_c^\infty((a, b))$. In this case we obtain exactly as before the Euler–Lagrange equation

$$\frac{d}{dt} \left(\nabla_v L(t, u, u') \right) = \nabla_x L(t, u, u').$$

We now assume that u and L are such that $t \mapsto \nabla_v L(t, u(t), u'(t))$ is L^1. This implies that $t \mapsto \nabla_v L(t, u(t), u'(t))$ is $W^{1,1}$ and in particular continuous. Moreover, the term $\int_a^b \nabla_v L(t, u, u') \cdot \varphi' \, dt$ can be integrated by parts, thus obtaining

$$j'_\psi(0) = \int_a^b \nabla_x L(t, u, u') \cdot \varphi \, dt - \int_a^b \left(\nabla_v L(t, u, u') \right)' \cdot \varphi \, dt$$

$$+ \left(\nabla \psi_0(u(a)) - \nabla_v L(a, u(a), u'(a)) \right) \cdot \varphi(a)$$

$$+ \left(\nabla \psi_1(u(b)) + \nabla_v L(b, u(b), u'(b)) \right) \cdot \varphi(b).$$

Since the first two integrals coincide thanks to the Euler–Lagrange equation, we are finally only left with the boundary terms. Their sum should vanish for arbitrary φ, which provides

$$\nabla \psi_0(u(a)) - \nabla_v L(a, u(a), u'(a)) = 0, \quad \nabla \psi_1(u(b)) + \nabla_v L(b, u(b), u'(b)) = 0.$$

These two boundary conditions, called *transversality conditions*, replace in this case the Dirichlet boundary conditions on $u(a)$ and $u(b)$, which are no longer available, and allow us to complete the equation. Of course, it is possible to combine the problem with fixed endpoints and the Bolza problem with penalization on the boundary, fixing one endpoint and penalizing the other. The four possible cases, with their Euler–Lagrange systems, are the following.

- For the problem $\min \left\{ \int_a^b L(t, u(t), u'(t)) \, dt \ : \ u(a) = A, u(b) = B \right\}$, the Euler–Lagrange system is, as we already saw,

$$\begin{cases} \frac{d}{dt} \left(\nabla_v L(t, u, u') \right) = \nabla_x L(t, u, u') & \text{in } (a, b), \\ u(a) = A, \\ u(b) = B. \end{cases}$$

- For the problem $\min \left\{ \int_a^b L(t, u(t), u'(t)) \, dt + \psi_0(u(a)) \ : \ u(b) = B \right\}$, the Euler–Lagrange system is

$$\begin{cases} \frac{d}{dt} \left(\nabla_v L(t, u, u') \right) = \nabla_x L(t, u, u') & \text{in } (a, b), \\ \nabla_v L(a, u(a), u'(a)) = \nabla \psi_0(u(a)), \\ u(b) = B. \end{cases}$$

- For the problem $\min\left\{\int_a^b L(t, u(t), u'(t))\,\mathrm{d}t + \psi_1(u(b)) \ : \ u(a) = A\right\}$ (the most classical choice in applications, where the initial point is known and the final point is free), the Euler–Lagrange system is

$$\begin{cases} \frac{\mathrm{d}}{\mathrm{d}t}\left(\nabla_v L(t, u, u')\right) = \nabla_x L(t, u, u') & \text{in } (a, b), \\ u(a) = A, \\ \nabla_v L(b, u(b), u'(b)) = -\nabla\psi_1(u(b)). \end{cases}$$

- For the problem $\min\left\{\int_a^b L(t, u(t), u'(t))\,\mathrm{d}t + \psi_0(u(a)) + \psi_1(u(B))\right\}$, the Euler–Lagrange system is

$$\begin{cases} \frac{\mathrm{d}}{\mathrm{d}t}\left(\nabla_v L(t, u, u')\right) = \nabla_x L(t, u, u') & \text{in } (a, b), \\ \nabla_v L(a, u(a), u'(a)) = \nabla\psi_0(u(a)), \\ \nabla_v L(b, u(b), u'(b)) = -\nabla\psi_1(u(b)). \end{cases}$$

1.4   Curves and Geodesics in Metric Spaces

In this section we generalize the study of the problem of geodesics to the setting of metric spaces. This will show that most of the results do not depend on the Euclidean structure of $\mathbb{R}^d$ and will allow a more abstract and intrinsic point of view to be adopted for some problems in $\mathbb{R}^d$, such as the search for geodesics for a weighted distance.

1.4.1   Absolutely Continuous Curves and Metric Derivative

A curve ω is a continuous function defined on a interval, say $[0, 1]$, and valued in a metric space (X, d). As it is a map between metric spaces, one can meaningfully ask whether it is Lipschitz or not, but its speed $\omega'(t)$ has no meaning, unless X is a vector space.

Surprisingly, it is possible to give a meaning to the modulus of the velocity, $|\omega'|(t)$.

Definition 1.7 If $\omega : [0, 1] \to X$ is a curve valued in the metric space (X, d) we define the *metric derivative* of ω at time t, denoted by $|\omega'|(t)$, as

$$|\omega'|(t) := \lim_{h \to 0} \frac{d(\omega(t + h), \omega(t))}{|h|},$$

provided this limit exists.

Definition 1.8 A curve $\omega : [0, 1] \to X$ is said to be *absolutely continuous* whenever there exists a $g \in L^1([0, 1])$ such that $d(\omega(t_0), \omega(t_1)) \leq \int_{t_0}^{t_1} g(s)\,ds$ for every $t_0 < t_1$. The set of absolutely continuous curves defined on $[0, 1]$ and valued in X is denoted by $AC([0, 1]; X)$.

We will prove that any absolutely continuous curve ω admits a metric derivative at a.e. t, and also that we have $\text{Length}(\omega) = \int_0^1 |\omega'|(t)\,dt$.

A first useful lemma is the following. It involves the Banach space ℓ^∞, which is defined as the space of bounded sequences $y = (y[0], y[1], \dots)$ endowed with the norm $||y||_\infty := \sup_i |y[i]|$, and thus with the distance

$$d_\infty(y, z) := ||y - z||_\infty = \sup_i |y[i] - z[i]|.$$

Lemma 1.9 *Every compact metric space (X, d) can be isometrically embedded in ℓ^∞.*

Proof Any compact metric space is separable, so let us take a countable dense subset of X and let us order its points, obtaining a dense sequence $x_0, x_1, \dots$. We then define a map $T : X \to \ell^\infty$ as follows

$$T(x) := (d(x, x_0), d(x, x_1), \dots),$$

which means that the map T is defined via $T(x)[i] = d(x, x_i)$. This sequence is bounded (which we need, for otherwise we would not have $T(x) \in \ell^\infty$) since X is compact and hence bounded. For every pair $x, x' \in X$ and every i we have, using the triangle inequality on X,

$$|T(x)[i] - T(x')[i]| = |d(x, x_i) - d(x', x_i)| \leq d(x, x')$$

and, passing to the sup in i, we get $d_\infty(T(x), T(x')) \leq d(x, x')$. Moreover, for any $\varepsilon > 0$ we can find i such that $d(x_i, x') < \varepsilon$, so that

$$|T(x)[i] - T(x')[i]| = |d(x, x_i) - d(x', x_i)| \geq d(x, x') - 2\varepsilon,$$

which implies $d_\infty(T(x), T(x')) \geq d(x, x') - 2\varepsilon$. Since $\varepsilon > 0$ is arbitrary, we finally obtain $d_\infty(T(x), T(x')) = d(x, x')$, which shows that T is an isometry from X to its image. $\quad\square$

Proposition 1.10 *Given a metric space (X, d) and an absolutely continuous curve $\omega : [0, T] \to X$ the metric derivative $|\omega'|(t)$ exists for a.e. t and we have $\text{Length}(\omega) = \int_0^1 |\omega'|(t)\,dt$.*

Proof It is not restrictive to assume that X is compact, since the image $\omega([0, T])$ is indeed compact as a continuous image of a compact interval. Thanks to Lemma 1.9 we can then assume $X \subset \ell^\infty$ and we consider the curves $\omega_i : [0, T] \to \mathbb{R}$ which are the coordinates of ω. Using $|\omega_i(t_0) - \omega_i(t_1)| \leq d(\omega(t_0), \omega(t_1)) \leq \int_{t_0}^{t_1} g(t)\,dt$ we see

that each coordinate is indeed a real-valued absolutely continuous function, and is
hence differentiable a.e. (since it belongs to $W^{1,1}$). The set of points t where all the
ω_i are differentiable is of full measure (since its complement is a countable union
of negligible sets) and on this set we define $m(t) := \sup_i |\omega_i'(t)|$. At every Lebesgue
point of g (see, for instance, the Box 5.1) we have $|\omega_i'(t)| \leq g(t)$ and hence $m \leq g$
a.e. In particular, m is an L^1 function. We want to prove $|\omega'|(t) = m(t)$ a.e. and,
more precisely, at all Lebesgue points of m. Let us choose one such point t. First we
observe that we have

$$\frac{d(\omega(t+h), \omega(t))}{|h|} \geq \frac{|\omega_i(t+h) - \omega_i(t)|}{|h|} \to |\omega_i'(t)|.$$

This shows $\liminf_h d(\omega(t+h), \omega(t))/|h| \geq |\omega_i'(t)|$ and, taking the sup over i, we
obtain $\liminf_h d(\omega(t+h), \omega(t))/|h| \geq m(t)$.

For the opposite inequality we first note that we have, for every $t_0 < t_1$, the
inequality $|\omega_i(t_0) - \omega_i(t_1)| \leq \int_{t_0}^{t_1} |\omega_i'(s)| \, ds \leq \int_{t_0}^{t_1} m(s) \, ds$, and, taking the sup over
i, we obtain

$$d(\omega(t_0), \omega(t_1)) \leq \int_{t_0}^{t_1} m(s) \, ds. \qquad (1.3)$$

Using this, we obtain

$$\frac{d(\omega(t+h), \omega(t))}{|h|} \leq \fint_t^{t+h} m(s) \, ds \to m(t),$$

where we used the fact that t is a Lebesgue point for m. This shows
$\limsup_h d(\omega(t+h), \omega(t))/|h| \leq m(t)$ and proves $|\omega'| = m$ a.e.

We now need to prove the equality between the length of ω and the integral of
$|\omega'|$. The inequality (1.3), applied to arbitrary subdivisions $0 \leq t_0 < t_1 < \cdots <
t_N \leq T$ in the definition of the length, now that we know $m = |\omega'|$, provides
$\text{Length}(\omega) \leq \int_0^T |\omega'|(t) \, dt$.

In order to prove the opposite inequality, let us take $h = T/N > 0$ and, for
every $t \in (0, h)$, let us consider the subdivision $t_0 < t_1 < \cdots < t_{N-1}$ given
by $t_k := t + kh$, for $k = 0, \ldots, N - 1$. By definition we have $\text{Length}(\omega) \geq
\sum_{k=0}^{N-2} d(\omega(t_k), \omega(t_{k+1}))$ and hence, taking the average on $t \in (0, h)$,

$$\text{Length}(\omega) \geq \frac{1}{h} \int_0^h \sum_{k=0}^{N-2} d(\omega(t+kh), \omega(t+(k+1)h)) \, dt$$

$$= \int_0^{(N-1)h} \frac{d(\omega(s), \omega(s+h))}{h} \, ds,$$

where the last equality is based on a change of variable $s = t + kh$, inverting the sum and the integral, and putting the denominator h inside the integral. If we take the liminf in $h \to 0$ and apply Fatou's lemma, by pointwise a.e. convergence we obtain $\text{Length}(\omega) \geq \int_0^T |\omega'|(t)\,dt$ and hence the desired equality.   □

An easy corollary of the previous fact concerns the characterization of Lipschitz curves among absolutely continuous ones.

Corollary 1.11 *Let* $\omega : [0, T] \to X$ *be an absolutely continuous curve. Then* $\omega \in \text{Lip}_L$ *if and only if* $|\omega'| \leq L$ *a.e.*

Proof It is clear that $\omega \in \text{Lip}_L$ implies $\frac{d(\omega(t+h),\omega(t))}{h} \leq L$ and hence $|\omega'|(t) \leq L$ for every t where the metric derivative exists.

On the other hand, if $|\omega'| \leq L$ a.e., the inequality (1.3), together with $m = |\omega'|$, provides $d(\omega(t_0), \omega(t_1)) \leq \int_{t_0}^{t_1} |\omega'|(t)\,dt \leq L|t_1 - t_0|$ and hence $\omega \in \text{Lip}_L$.   □

Another important fact about absolutely continuous curves is that they can be reparameterized in time (through a monotone-increasing reparameterization) and become Lipschitz continuous. More precisely, we have the following.

Proposition 1.12 *Given an arbitrary absolutely continuous curve* $\omega : [0, 1] \to X$ *and a number* $\varepsilon > 0$, *there exists a reparameterized curve* $\tilde{\omega} : [0, 1] \to X$ *of the form* $\tilde{\omega} = \omega \circ \phi$ *for a strictly increasing homeomorphism* $\phi : [0, 1] \to [0, 1]$ *such that* $\tilde{\omega} \in \text{Lip}_L$ *where* $L := \text{Length}(\omega) + \varepsilon$.

Proof Define $G(t) := \int_0^t |\omega'|(s)\,ds$, then set $S(t) = \varepsilon t + G(t)$. The function S is continuous and strictly increasing, and valued in the interval $[0, L]$, where $L := \text{Length}(\omega) + \varepsilon$. Then for $t \in [0, L]$, set $\tilde{\omega}(t) = \omega(S^{-1}(t))$. It is easy to check that $\tilde{\omega} \in \text{Lip}_1$, since we have, for $t_0 < t_1$,

$$d(\omega(t_0), \omega(t_1)) \leq \int_{t_0}^{t_1} |\omega'|(s)\,ds = G(t_1) - G(t_0) \leq S(t_1) - S(t_0)$$

and then

$$d(\tilde{\omega}(t_0), \tilde{\omega}(t_1)) = d(\omega(S^{-1}(t_0)), \omega(S^{-1}(t_1))) \leq S(S^{-1}(t_1)) - S(S^{-1}(t_0)) = t_1 - t_0.$$

In order to have a parameterization defined on the same interval $[0, 1]$, we just need to rescale by a factor L.   □

In the above proposition we note that, if the interval $[0, 1]$ is replaced by another interval $[0, T]$, then we have to set $L := (\text{Length}(\omega) + \varepsilon)/T$. We also observe that, if the curve ω is injective (and actually it is enough that it is not constant on any non-trivial interval, so that the function G is indeed strictly increasing), the result can be strengthened, as follows.

Proposition 1.13 *Given an absolutely continuous curve* $\omega : [0, 1] \to X$, *assume that there is no subinterval* $(t_0, t_1) \subset [0, 1]$ *on which* ω *is constant. Then there* *exists a reparameterized curve* $\widetilde{\omega} : [0, 1] \rightarrow X$ *of the form* $\widetilde{\omega} = \omega \circ \phi$ *for a strictly increasing homeomorphism* $\phi : [0, 1] \rightarrow [0, 1]$ *such that* $\widetilde{\omega} \in \mathrm{Lip}_L$ *where* $L := \mathrm{Length}(\omega)$ *and* $|\widetilde{\omega}'| = L$ *a.e.*

Proof We use again the function $G(t) := \int_0^t |\omega'|(s)\, ds$ and we observe that it is continuous and strictly increasing. Indeed, if $G(t_0) = G(t_1)$ then we obtain $|\omega'| = 0$ on (t_0, t_1) and Proposition 1.10 shows that ω is constant on (t_0, t_1), which is a contradiction. Then for $t \in [0, L]$ (where $L := \mathrm{Length}(\omega)$), set $\widetilde{\omega}(t) = \omega(G^{-1}(t))$. We obtain again $\widetilde{\omega} \in \mathrm{Lip}_1$ as in Proposition 1.12. If we scale by a factor L we obtain a curve, which we call again $\widetilde{\omega}$, which is Lip_L and defined on $[0, 1]$. Since L also equals its length, we have

$$L = \mathrm{Length}(\omega) = \mathrm{Length}(\widetilde{\omega}) = \int_0^1 |\widetilde{\omega}'|(t)\, dt \leq \int_0^1 L\, dt = L,$$

which shows that we must have $|\widetilde{\omega}'|(t) = L$ for a.e. t. $\qquad\square$

Remark 1.14 We observe that in Proposition 1.13 the reparameterized curve $\widetilde{\omega}$ which is built also satisfies $\mathrm{Lip}(\widetilde{\omega}) \leq \mathrm{Lip}(\omega)$, since $\mathrm{Lip}(\widetilde{\omega}) = L = \mathrm{Length}(\omega) = \int_0^1 |\omega'|(t)\, dt \leq \mathrm{Lip}(\omega)$.

1.4.2 Geodesics in Proper Metric Spaces

In a metric space (X, d) we define the following quantity geod (called *geodesic distance*) by minimizing the length of all curves γ connecting two points $x_0, x_1 \in X$:

$$\mathrm{geod}(x_0, x_1) := \inf \left\{ \mathrm{Length}(\gamma) \ : \ \begin{array}{l} \gamma \in AC([0, 1]; X), \\ \gamma(0) = x_0, \gamma(1) = x_1 \end{array} \right\} \in [0, +\infty]. \quad (1.4)$$

This quantity is clearly a distance when it is finite. When (X, d) is a Euclidean space, geod $= d$ and the infimum value in (1.4) is achieved by the segment $[x_0, x_1]$. In general, a metric space such that geod $= d$ is called a *length space*.

The minimal length problem consists in finding a curve γ with $\gamma(0) = x_0, \gamma(1) = x_1$ and such that $\mathrm{Length}(\gamma) = \mathrm{geod}(x_0, x_1)$. The existence of such a curve, called a *geodesic*, is given by the following classical theorem, which requires the metric space to be proper.

Definition 1.15 A metric space (X, d) is said to be proper if every bounded closed subset of (X, d) is compact.

Theorem 1.16 *Assume that* (X, d) *is proper. Then, for any two points* x_0, x_1 *such that* $\mathrm{geod}(x_1, x_0) < +\infty$, *there exists a minimizing geodesic joining* x_0 *and* x_1.

Proof Let us take a minimizing sequence for the geodesic problem, i.e. a sequence ω_n of absolutely continuous curves defined on $[0, 1]$, with $\omega_n(0) = x_0$, $\omega_n(1) = x_1$, and Length$(\omega_n) \to \text{geod}(x_0, x_1)$. We take a sequence $\varepsilon_n > 0$, with $\varepsilon_n \to 0$ and, using Proposition 1.12, we define a new sequence of curves $\tilde{\omega}_n$ such that $\tilde{\omega}_n(0) = x_0$, $\tilde{\omega}_n(1) = x_1$, and $\tilde{\omega}_n \in \text{Lip}_{L_n}$, where $L_n := \text{Length}(\omega_n) + \varepsilon_n \to \text{geod}(x_0, x_1)$. Because of this Lipschitz bound, for any n and any t we have $d(\tilde{\omega}_n(t), x_0) \le L_n t \le R$, where R is a constant independent of n and t. Hence, the images of all the curves $\tilde{\omega}_n$ are contained in the closed ball $B(x_0, R)$, which is compact. The sequence of curves $\tilde{\omega}_n$ is then equicontinuous and valued in some compact set, and the Ascoli–Arzelà theorem allows us to extract a uniformly convergent subsequence $\tilde{\omega}_{n_k} \to \omega$. We have

$$d(\omega(t_0), \omega(t_1)) = \lim_{k \to \infty} d(\tilde{\omega}_{n_k}(t_0), \tilde{\omega}_{n_k}(t_1)) \le L_{n_k} |t_0 - t_1| \to \text{geod}(x_0, x_1)|t_0 - t_1|.$$

This shows that ω is Lipschitz continuous and its Lipschitz constant is at most geod(x_0, x_1). As a consequence we have $|\omega'| \le \text{geod}(x_0, x_1)|$ and, integrating on $[0, 1]$, we also have Length$(\omega) \le \text{geod}(x_0, x_1)$, so that ω is indeed a geodesic. Moreover, we can also observe that we necessarily have $|\omega'|(t) = \text{geod}(x_0, x_1)$ for a.e. t.   □

Box 1.5 Memo—*Compactness for the Uniform Convergence*

We recall that a family $\mathcal{F}$ of functions $f : X \to Y$ is said to be equicontinuous if for every $\varepsilon > 0$ there exists a common $\delta > 0$ such that $d(f(x), f(y)) < \varepsilon$ for all pairs x, y with $d(x, y) < \delta$ and for all $f \in \mathcal{F}$. Equivalently, this means that there exists a continuous function $\omega : \mathbb{R}_+ \to \mathbb{R}_+$ such that $\omega(0) = 0$ with $|f(x) - f(y)| \le \omega(d(x, y))$ for all x, y, f. This function ω is called the *modulus of continuity*.

Theorem (Ascoli–Arzelà) *If X is a compact metric space and $f_n : X \to \mathbb{R}$ are equi-continuous and equi-bounded (i.e. there is a common constant C with $|f_n(x)| \le C$ for all $x \in X$ and all n), then the sequence (f_n) admits a subsequence f_{n_k} uniformly converging to a continuous function $f : X \to \mathbb{R}$.*

Conversely, a subset of $C(X)$ is relatively compact for the uniform convergence (if and) only if its elements are equi-continuous and equi-bounded. In particular, the functions of a uniformly convergent sequence are always equi-continuous and admit the same modulus of continuity.

The same results are true if the target space $\mathbb{R}$ and the equiboundedness assumption are replaced with a target space which is a compact metric space.

We note that in the above Theorem 1.16 we proved the existence of a *constant-speed geodesic*, i.e. a curve ω which, besides being a geodesic, also satisfies $|\omega'| = const$ (and this constant can only be equal to Length$(\omega) = \text{geod}(x^-, x^+)$ if the interval on which the curve is parameterized is $[0, 1]$). Constant-speed geodesics admit other characterizations, as we can see in the following propositions.

Proposition 1.17 *If (X, d) is a length space then $\omega : [0, 1] \to X$ is a constant-speed geodesic between $\omega(0)$ and $\omega(1) \in X$ if and only if it satisfies*

$$d(\omega(t), \omega(s)) = |t - s| d(\omega(0), \omega(1)) \quad \text{for all } t, s \in [0, 1]. \tag{1.5}$$

Proof Assume that ω satisfies (1.5) (and in particular it is Lipschitz continuous). Then, from the formula for the length as a sup over partitions, we have Length$(\omega) = d(\omega(0), \omega(1))$. Since no curve connecting $\omega(0)$ to $\omega(1)$ can have a strictly smaller length, then ω is a geodesic. Moreover, from (1.5), dividing by $|t - s|$ and sending $s \to t$, we obtain $|\omega'|(t) = d(\omega(0), \omega(1))$ for every t, which shows that the speed is constant.

We now assume that ω is a constant-speed geodesic. We then have $|\omega'(t)| = \ell := \text{Length}(\omega) = \text{geod}(\omega(0), \omega(1)) = d(\omega(0), \omega(1))$ (the last equality coming from the assumption that (X, d) is a length space). In particular we have $\omega \in \text{Lip}_\ell$. We then have

$$d(\omega(0), \omega(s)) \leq \ell s; \quad d(\omega(s), \omega(t)) \leq \ell(t - s); \quad d(\omega(t), \omega(1)) \leq \ell(1 - t).$$

If we sum the above three inequalities and use the triangle inequality we get

$$d(\omega(0), \omega(1)) \leq d(\omega(0), \omega(s)) + d(\omega(s), \omega(t)) + d(\omega(t), \omega(1)) \leq d(\omega(0), \omega(1)),$$

so that all the previous inequalities are indeed equalities and we have the equality $d(\omega(s), \omega(t)) = \ell(t - s)$ with $\ell = d(\omega(0), \omega(1))$, i.e. (1.5). $\qquad \square$

Proposition 1.18 *Given a metric space (X, d), a strictly convex, continuous, and strictly increasing function $h : \mathbb{R}_+ \to \mathbb{R}_+$, and two points $x_0, x_1 \in X$, the following are equivalent*

1. $\gamma : [0, 1] \to X$ is a constant-speed geodesic connecting x_0 to x_1;
2. $\gamma : [0, 1] \to X$ solves

$$\min \left\{ \int_0^1 h(|\omega'|(t)) \, dt \, : \, \omega \in AC([0, 1]; X), \, \omega(0) = x_0, \omega(1) = x_1 \right\}.$$

Proof Given an arbitrary $\omega : [0, 1] \to X$ with $\omega(0) = x_0, \omega(1) = x_1$ we have the inequalities

$$\int_0^1 h(|\omega'|(t)) \, dt \geq h \left(\int_0^1 |\omega'|(t) \, dt \right) \geq h(\text{geod}(x_0, x_1)), \tag{1.6}$$

where the first inequality comes from Jensen's inequality and the second from the definition of geod, together with the increasing behavior of h.

Let us assume that γ is a constant-speed geodesic with $|\gamma'| = \text{geod}(x_0, x_1)$. Then (1.6) implies

$$\int_0^1 h(|\omega'|(t)) \, dt \geq h(\text{geod}(x_0, x_1)) = \int_0^1 h(|\gamma'|(t)) \, dt$$

i.e. (1) implies (2).

For the reverse implication, take γ which minimizes $\omega \mapsto \int_0^1 h(|\omega'|(t)) \, dt$ and an absolutely continuous competitor ω with $\text{Length}(\omega) = L$. For $\varepsilon > 0$, we choose a reparameterization $\tilde{\omega}$ of ω with $|\tilde{\omega}'| \leq L + \varepsilon$ and the optimality of γ provides

$$h \left(\int_0^1 |\gamma'|(t) \, dt \right) \leq \int_0^1 h(|\gamma'|(t)) \, dt \leq \int_0^1 h(|\tilde{\omega}'|(t)) \, dt \leq h(L + \varepsilon).$$

We obtain $\text{Length}(\gamma) = \int_0^1 |\gamma'|(t) \, dt \leq L + \varepsilon$ and, taking $\varepsilon \to 0$, we deduce $\text{Length}(\gamma) \leq L$, i.e. γ minimizes the length and is a geodesic. Moreover, taking $\omega = \gamma$ we necessarily have equality in the inequality $h \left(\int_0^1 |\gamma'|(t) \, dt \right) \leq \int_0^1 h(|\gamma'|(t)) \, dt$ and the equality case in Jensen's inequality for strictly convex h implies that $|\gamma'|$ is constant. $\square$

1.4.3 Geodesics and Connectedness in Thin Spaces

The existence result of a geodesic curve connecting x_0 to x_1 in a proper space X, presented in the previous sub-section, required the assumption $\text{geod}(x_1, x_0) < +\infty$. A completely different question is whether a curve with finite length, or an AC curve, connecting x_0 to x_1 exists. This is clearly related to the notion of connectedness of the space X. Yet, it is well-known that connectedness has a topological definition which does not involve curves at all, and that the existence of connecting curves is a stronger condition.

Box 1.6 Memo—*Connectedness*

A topological space X is said to be connected if for any subset $A \subset X$ which is at the same time open and closed we necessarily have either $A = \emptyset$ or $A = X$.

   A topological space X is said to be arcwise-connected if for any two points $x_0, x_1 \in X$ there exists a continuous curve $\gamma : [0, 1] \to X$ such that $\gamma(t) = x_t$ for $t = 0, 1$.

(continued)

> **Box 1.6** (continued)
>   A metric space X is said to be AC-connected if for any two points $x_0, x_1 \in X$ there exists an absolutely continuous curve $\gamma : [0, 1] \to X$ such that $\gamma(t) = x_t$ for $t = 0, 1$.
>   Of course AC-connectedness implies arcwise-connectedness, and arcwise-connectedness implies connectedness.

We want to consider here a class of metric spaces, that we call *thin spaces*, where connectedness implies AC-connectedness. This result is essentially taken from [15], where it is presented as a preliminary result for a notion which will be discussed in Chap. 6.

Definition 1.19 A metric space (X, d) is said to be thin if it has the following property: there exists a constant C such that if $\{B(x_i, r) : i = 1, \ldots, N\}$ is a finite family of disjoint balls in X, then $N \le \frac{C}{r}$.

The reason for the name "thin" lies in the fact that this property is typical of 1D sets. It is not satisfied by subsets of $\mathbb{R}^d$ which have a non-empty interior.
   First, we clarify that this notion implies compactness.

> **Box 1.7 Memo—*Totally Bounded and Precompact Sets***
>
> In a metric space (X, d) we say that a subset $E \subset X$ is pre-compact if its closure in X is compact. This means that from any sequence in E we can extract a convergent subsequence, but the limit does not necessarily lie in E.
>   We say that a subset $E \subset X$ is totally bounded if for every $\varepsilon > 0$ we can cover E with a finite number of balls of radius ε centered at points of E. The centers of these balls are called an ε-net.
> **Theorem** *If (X, d) is complete, then a subset E is precompact if and only if it is totally bounded. A metric space (X, d) is compact if and only if it is itself complete and totally bounded.*

Proposition 1.20 *If (X, d) is thin and complete then it is compact.*

Proof It is enough to prove that (X, d) is totally bounded (see Box 1.7). Given a number $\varepsilon > 0$, define a sequence of points $x_k \in X$ as follows: we start from an arbitrary x_0, then we take $x_k \in X \setminus \bigcup_{i=1}^{k-1} B(x_i, \varepsilon)$ if $\bigcup_{i=1}^{k-1} B(x_i, \varepsilon) \neq X$. As long as the construction proceeds, the balls $B(x_i, \varepsilon/2)$ are disjoint, so by the definition of thin space the construction must stop after $k = 2C/\varepsilon$ steps at most. Hence, X is contained in a finite union of balls of radius ε, which means that it is totally bounded, and hence compact because we also assumed completeness.   □

We now state the main theorem of this section.

Theorem 1.21 *If (X, d) is thin, complete, and connected, then it is AC-connected and for every pair of points $x_0, x_1 \in X$ there exists a geodesic connecting them.*

Proof Let us fix $\varepsilon > 0$ and define an equivalence relation on X: we say that two points $x, y \in X$ are ε-connected if there exists a finite chain of points $x_0, x_1, \ldots, x_N$ such that $x_0 = x$, $x_N = y$, and $d(x_i, x_{i+1}) \leq \varepsilon$ for every i. It is easy to see that this is an equivalence relation. For every point x_0, we denote by $A(x_0)$ its equivalence class. It is clear by definition that the equivalence classes are open sets, since we have $B(x, \varepsilon) \subset A(x_0)$ for every $x \in A(x_0)$. Since the complement of an equivalence class is the union of all the others, then every equivalence class is also closed as the complement of an open set. By connectedness, we have $A(x_0) = X$ for every x_0. Hence, for every pair $x, y \in X$ and every $\varepsilon > 0$ there exists a finite chain of points $x_0, x_1, \ldots, x_N$ such that $x_0 = x$, $x_N = y$, and $d(x_i, x_{i+1}) \leq \varepsilon$. Let us consider then a minimal chain connecting x to y (i.e. a chain for which N is minimal: such a chain exists since we are minimizing an integer-valued quantity). By minimality, $d(x_i, x_{i+2}) > \varepsilon$ (otherwise we could remove the point x_{i+1}). Hence the balls $B(x_i, \varepsilon/2)$ for i odd are disjoint, and this gives a bound on the number N, of the form $N \leq C/\varepsilon$, because X is thin.

Since we know that X is compact, we can isometrically embed X into ℓ^∞, and we can define a curve $\gamma_\varepsilon : [0, 1] \to \ell^\infty$ taking $\gamma(i/N) = x_i$ (with $N = N(\varepsilon)$) and connecting two subsequent points x_i with a constant-speed segment in ℓ^∞ (this is the reason for embedding into ℓ^∞, we need a vector space). For $t \in [i/N, (i+1)/N]$ we have $|\gamma_\varepsilon'|(t) = \frac{d(x_i, x_{i+1})}{1/N} = N d(x_i, x_{i+1}) \leq N\varepsilon \leq C$, so that these curves are equi-Lipschitz continuous. We want to apply the Ascoli–Arzelà theorem to extract a convergent subsequence, but the ambient space ℓ^∞ is not compact. Yet, if we take a sequence $\varepsilon_n \to 0$ and the corresponding curves γ_{ε_n}, the set $\bigcup_n \gamma_{\varepsilon_n}([0, 1])$ is totally bounded, and hence contained in a compact set. To see this, we fix $\varepsilon > 0$ and we note that $\bigcup_{n:\varepsilon_n \geq \varepsilon/2} \gamma_{\varepsilon_n}([0, 1])$ is compact since it is a finite union of compact sets (and hence can be covered by finitely many balls of radius ε), while $\bigcup_{n:\varepsilon_n < \varepsilon/2} \gamma_{\varepsilon_n}([0, 1])$ is contained in an $\varepsilon/2$-neighborhood of $X \subset \ell^\infty$. By taking finitely many balls of radius $\varepsilon/2$ which cover X we then cover this set as well with finitely many balls of radius ε.

Hence, we can extract a subsequence which uniformly converges to a curve γ which will also be Lipschitz continuous with the same Lipschitz constant C. This curve will also satisfy $\gamma(0) = x$ and $\gamma(1) = y$ as the approximating curves did, and since the image $\gamma_{\varepsilon_n}([0, 1])$ was contained in an ε_n-neighborhood of X, at the limit we have $\gamma(t) \in X$ for every t.

This provides the existence of at least an AC curve connecting arbitrary points $x, y \in X$. The existence of a minimal one follows from Theorem 1.16. $\square$

1.4.4 The Minimal Length Problem in Weighted Metric Spaces

Let (X, d) be a metric space and $K : X \to \mathbb{R}^+$ be a nonnegative function called the *weight function*. We define the K-length L_K of an absolutely continuous curve γ via the formula

$$L_K(\gamma) := \int_I K(\gamma(t)) \, |\gamma'|(t) \, dt$$

and we consider the minimization of L_K among curves with fixed endpoints. This problem is very natural and appears in many different instances both in applied and fundamental mathematics. For example, it can be considered as a particular case of the problem of looking for geodesics in Riemannian manifolds, since in a Riemannian manifold the length of a curve γ is computed by applying to the tangent vector $\gamma'(t)$ a norm induced by a scalar product on the tangent space depending on the point $\gamma(t)$: if we choose for (X, d) the Euclidean space with the Euclidean distance, the length L_K can be seen as a particular Riemannian length, when the scalar product is a positive scalar multiple of the standard Euclidean one (in practice, we do not use general positive symmetric matrices to define a Riemannian metric, but only a positive multiple of the identity: we obtain in this case a *conformal* metric). From the applied point of view, minimizing weighted lengths is natural for instance in traffic problems (see Sect. 4.6), where the cost of passing through a given point is not uniform, but depends on the amount of traffic at that point, as well as in optics, where the coefficient K depends on the refraction index of the material.

We first analyze the properties of the weighted length L_K. For future use, we define for every $r > 0$ and every $x \in X$ the following two quantities:

$$K_r(x) := \inf\{K(y) \,:\, d(x, y) \le r\}, \quad K^r(x) := \sup\{K(y) \,:\, d(x, y) \le r\}. \tag{1.7}$$

Lemma 1.22 *Assume that $K : X \to \mathbb{R}_+$ is l.s.c. If $\gamma : [0, 1] \to X$ has modulus of continuity ω then we have*

$$L_K(\gamma) = \sup\left\{ \sum_{i=0}^{N-1} K_r(\gamma(t_i)) d(\gamma(t_i), \gamma(t_{i+1})) \,:\, \begin{array}{l} \delta > 0, r > \omega(\delta), \\ 0 \le t_0 < \cdots < t_N \le 1, \\ |t_{i+1} - t_i| \le \delta \end{array} \right\}.$$

As a consequence, L_K is lower semicontinuous for the uniform convergence, i.e. $\gamma_n \to \gamma$ uniformly implies $L_K(\gamma) \le \liminf_n L_K(\gamma_n)$.

Proof If we take an absolutely continuous γ with modulus of continuity ω, and two instants $t_i < t_{i+1}$ with $|t_{i+1} - t_i| \leq \delta$, we then have $|\gamma(t) - \gamma(t_i)| \leq \omega(\delta) < r$ and hence $K(\gamma(t)) \geq K_r(\gamma(t_i))$ for all $t \in [t_i, t_{i+1}]$. We then have

$$\int_{t_i}^{t_{i+1}} K(\gamma(t))|\gamma'|(t)\,dt \geq K_r(\gamma(t_i)) \int_{t_i}^{t_{i+1}} |\gamma'|(t)\,dt \geq K_r(\gamma(t_i))d(\gamma(t_i), \gamma(t_{i+1})).$$

This shows that $L_K(\gamma)$ is larger than the sup in the first part of the claim. For the opposite inequality we first take, for every N, a number $r_N > 0$ so that $\lim_N r_N = 0$, and a partition with $t_i = i/N$ and define a function $k_N(t) := K_{r_N}(\gamma(t_i))$ for $t \in [t_i, t_{i+1})$. By the continuity of γ and the semicontinuity of K we have $\liminf_N k_N(t) \geq K(\gamma(t))$ for every t. Hence, by Fatou's lemma, we have

$$\liminf_N \int k_N(t)|\gamma'|(t)\,dt \geq L_K(\gamma),$$

so that for every ε there exists an N with $\int k_N(t)|\gamma'|(t)\,dt \geq L_K(\gamma)(1-\varepsilon)$. We now look at $\int_{t_i}^{t_{i+1}} k_N(t)|\gamma'|(t)\,dt = k_N(t_i)\int_{t_i}^{t_{i+1}} |\gamma'|(t)\,dt$. By definition of the length as a sup it is possible to sub-partition the interval $[t_i, t_{i+1}]$ by means of some $t_{i,j}$ with $\sum_j d(\gamma(t_{i,j}), \gamma(t_{i,j+1})) \geq (1 - \varepsilon) \int_{t_i}^{t_{i+1}} |\gamma'|(t)\,dt$.

We now have

$$\sum_{i,j} K_{r_N}(\gamma(t_i))d(\gamma(t_{i,j}), \gamma(t_{i,j+1})) \geq (1 - \varepsilon)^2 L_K(\gamma).$$

We need to replace the coefficients $K_{r_N}(\gamma(t_i))$ with some coefficients of the form $K_r(\gamma(t_{i,j}))$, with $r > \omega(\max_{i,j}|t_{i,j} - t_{i,j+1}|)$, yet satisfying $K_{r_N}(\gamma(t_i)) \leq K_r(\gamma(t_{i,j}))$. Since we have $B(\gamma(t_{i,j}), r) \subset B(\gamma(t_i), r + \omega(1/N))$, it is enough to choose $r = \omega(1/N)$ and $r_N = 2\omega(1/N)$. This shows that the partition $\{t_{i,j}\}$ with $r = \omega(1/N)$ satisfies

$$\sum_{i,j} K_r(\gamma(t_{i,j}))d(\gamma(t_{i,j}), \gamma(t_{i,j+1})) \geq (1 - \varepsilon)^2 L_K(\gamma)$$

and shows that the sup in the claim is larger than $L_K(\gamma)$.

Once we have established this representation formula, the proof of the lower semicontinuity is easy. We take a sequence $\gamma_n \to \gamma$ uniformly and we denote by ω a common modulus of continuity (see Box 1.5) for the curves γ_n and γ. On all these curves we can use the first part of the statement to write L_K as a sup, and use the fact that lower semicontinuity is stable under sup (see Chap. 3). Then, for fixed $\delta, r, t_0, t_1, \ldots, t_N$ the quantity

$$\gamma \mapsto \sum_{i=0}^{N-1} K_r(\gamma(t_i))d(\gamma(t_i), \gamma(t_{i+1}))$$

is clearly continuous for the uniform convergence (and actually pointwise convergence would be enough), which shows the semicontinuity of L_K by taking the sup (see Proposition 3.3). □

We can now prove an existence theorem for curves minimizing the weighted length L_K.

Theorem 1.23 *Given a proper metric space (X, d), an l.s.c. function $K : X \rightarrow \mathbb{R}$ which is strictly positive everywhere, and two points $x_0, x_1 \in X$, consider the minimization problem*

$$\inf\{L_K(\gamma) : \gamma \in \mathrm{AC}([0, 1]; X), \gamma(0) = x_0, \gamma(1) = x_1\}.$$

Then, there exists a curve solving this minimization problem whenever the above inf is finite and we have $\int_0^\infty (\inf_{B(x_0,r)} K) \, dr = +\infty$.

Proof We start from a minimizing sequence γ_n, so that we have $L_K(\gamma_n) \leq C$. First of all, we want to prove that all curves γ_n are contained in a ball $B(x_0, R)$ with R independent of n. Let us denote by $R_n := \max\{d(\gamma_n(t), x_0), t \in [0, 1]\}$ the maximal distance from x_0 attained by γ_n. This maximum exists as we look at the continuous function $t \mapsto d(\gamma_n(t), x_0)$ on a compact interval. For every integer $k < R_n$ we can define $t_{k,n} := \min\{t : d(\gamma_n(t), x_0) \geq k\}$. All these numbers $t_{k,n}$ are well-defined by continuity of $t \mapsto d(\gamma_n(t), x_0)$ and we have $t_{1,n} < t_{2,n} < t_{3,n} \dots$. We then have

$$\int_{t_{k,n}}^{t_{k+1,n}} K(\gamma(t))|\gamma'|(t) \, dt \geq \left(\inf_{B(x_0,k+1)} K \right) \int_{t_{k,n}}^{t_{k+1,n}} |\gamma'|(t) \, dt$$

$$\geq \left(\inf_{B(x_0,k+1)} K \right) d(\gamma(t_{k,n}), \gamma(t_{k+1,n})).$$

Using $d(\gamma(t_{k,n}), \gamma(t_{k+1,n})) \geq |d(\gamma(t_{k,n}), x_0) - d(x_0, \gamma(t_{k+1,n}))| = 1$ we get

$$L_K(\gamma) \geq \sum_{k=0}^{\lfloor R_n \rfloor} \inf_{B(x_0,k+1)} K.$$

Using the comparison between series and integral for non-decreasing functions, the condition $\int_0^\infty (\inf_{B(x_0,r)} K) \, dr = +\infty$ is equivalent to $\sum_{k=0}^{\infty} (\inf_{B(x_0,k+1)} K) = +\infty$, and this shows an upper bound on $\lfloor R_n \rfloor$.

We now know that all curves γ_n are contained in a bounded, and hence compact, set. On this set K is bounded from below by a strictly positive number, so that we also have a uniform bound on $\mathrm{Length}(\gamma_n) \leq L$. We then reparameterize these curves into some curves $\tilde{\gamma}_n$ with $|\tilde{\gamma}_n'| \leq L + \varepsilon$ and this does not change the value of the weighted length L_K, which is also invariant under reparameterization. We can now apply the Ascoli–Arzelà theorem to this new minimizing sequence, extract a uniformly convergent subsequence, apply Lemma 1.22 which gives the lower semicontinuity of L_K, and obtain the existence of a minimizer.   □

The condition $\int_0^\infty (\inf_{B(x_0,r)} K)\, dr = +\infty$, introduced for instance in [43] and also used in [90, 150, 151], is crucial in order to obtain a bound on the minimizing sequence. We now want to provide an example where the existence of weighted geodesics fails if it is not satisfied. We will consider a Euclidean space, say $\mathbb{R}^2$, endowed with a positive weight $K > 0$ which does not satisfy the above non-integrability condition.

Theorem 1.24 *Given an arbitrary continuous and strictly positive function g : $\mathbb{R}_+ \to (0, \infty)$ such that $\int_0^\infty g(s)\, ds < +\infty$, there exists a continuous weight $K : \mathbb{R}^2 \to \mathbb{R}_+$ and two points $x_0, x_1 \in \mathbb{R}^2$ such that*

- *the weight $K(x)$ coincides with $g(d(x, x_0))$ on an unbounded set;*
- *there is no curve connecting x_0 to x_1 and minimizing L_K.*

Proof First we define $G(t) = \int_0^t g(s)\, ds$ and $G(\infty) := \int_0^\infty g(s)\, ds \in \mathbb{R}$. Also take a smooth function $h : \mathbb{R} \to [0, 1]$ such that $h(0) = 1$ and $h(\pm 1) = 0$ (which implies $h'(0) = h'(\pm 1) = 0$), and choose it so that h' only vanishes at $-1, 0, 1$. Build a function $f : \mathbb{R}^2 \to \mathbb{R}_+$ via

$$f(x, y) = h(y)(2G(\infty) - G(|x|)) + (1 - h(y))G(|x|).$$

Note that we have $f \in C^1(\mathbb{R}^2 \setminus \{x = 0\})$. At $x = 0$, the partial derivative of f w.r.t. x is discontinuous and changes its sign, because of the modulus in the definition. However, $|\nabla f|$ is well-defined, and we take $K = |\nabla f|$. This means that we are defining

$$K(x, y) = \sqrt{|1 - 2h(y)|^2 g(|x|)^2 + 4h'(y)^2(G(\infty) - G(|x|))^2}.$$

It is easy to see that we have $K(x, 0) = K(x, \pm 1) = g(|x|)$ and $K > 0$ on $\mathbb{R}^2$.

Consider now the two points $x_0 = (0, 1)$ and $x_1 = (0, -1)$.

Any continuous curve $\gamma : [0, 1] \to \mathbb{R}^2$ connecting these two points must cross the line $\{y = 0\}$. Let us call $(\ell, 0)$ a point where γ crosses such a line, i.e. $\gamma(t_0) = (\ell, 0)$ for $t_0 \in [0, 1]$. We can estimate the weighted length of γ via

$$\int_0^1 K(\gamma(t))|\gamma'(t)|\, dt = \int_0^{t_0} |\nabla f|(\gamma(t))|\gamma'(t)|\, dt + \int_{t_0}^1 |\nabla f|(\gamma(t))|\gamma'(t)|\, dt$$

$$\geq \int_0^{t_0} \nabla f(\gamma(t)) \cdot \gamma'(t)\, dt - \int_{t_0}^1 \nabla f(\gamma(t)) \cdot \gamma'(t)\, dt$$

$$= 2f(\gamma(t_0)) - f(x_0) - f(x_1) = 2f(\ell, 0) - f(0, 1) - f(0, -1)$$

$$= 2(2G(\infty) - G(|\ell|)) > 2G(\infty).$$

This shows that no curve joining these two points can have a weighted length less than or equal to $2G(\infty)$. If we are able to construct a sequence of curves approaching this value, we have shown that the minimum of the weighted length does not exist. To do so, take a sequence $x_n \to \infty$ with $g(x_n) \to 0$ (which is possible because of the condition $\int_0^\infty g(s) \, ds < +\infty$). Consider a curve γ_n defined on the interval $[0, 2x_n + 2]$ (and possibly reparameterized on $[0, 1]$) in the following way

$$\gamma_n(t) = \begin{cases} (t, 1) & \text{if } t \in [0, x_n], \\ (x_n, 1 - t + x_n) & \text{if } t \in [x_n, x_n + 2], \\ (2x_n + 2 - t, -1) & \text{if } t \in [x_n + 2, 2x_n + 2]. \end{cases}$$

It is easy to see that we have

$$\int_0^{2x_n+2} K(\gamma(t))|\gamma'(t)| \, dt = 2 \int_0^{x_n} g(t) \, dt$$

$$+ \int_{-1}^1 \sqrt{|1 - 2h(y)|^2 g(|x_n|)^2 + 4h'(y)^2 (G(\infty) - G(|x_n|))^2} \, dy.$$

Using

$$\sqrt{|1 - 2h(y)|^2 g(|x_n|)^2 + 4h'(y)^2 (G(\infty) - G(x_n))^2}$$

$$\leq C(g(|x_n|) + G(\infty) - G(x_n)),$$

we can also obtain

$$\int_0^{2x_n+2} K(\gamma(t))|\gamma'(t)| \, dt = 2G(x_n) + O(g(|x_n|) + G(\infty) - G(x_n)) \xrightarrow[n\to\infty]{} 2G(\infty),$$

which concludes the example.   □

We are now interested in studying the minimization of L_K when K may vanish. This case is of interest, for instance, in the applications to the existence of heteroclinic connections presented in [150, 151] and in Sect. 1.4.5. If K vanishes on a connected set this creates a free region where movement has no cost: this can be interesting (see for instance [54] and Sect. 7.5 for an application in the modeling of transport networks) but goes beyond the scope of this section. We will only consider the case where K vanishes on a finite set Σ. This already creates a certain number of difficulties. In particular, K will no longer be bounded from below by positive constants on bounded sets, and it will be impossible to obtain bounds on the length (in the sense of the length for the original distance d, and not of the weighted length) for minimizing sequences. It will then be hard to apply Lemma 1.22 to obtain the semicontinuity, even if this lemma does not require $K > 0$.

Moreover, it will not be natural to work with curves γ which are absolutely continuous anymore, since this assumption corresponds to $\int |\gamma'| \, dt < +\infty$ and we cannot enforce bounds on this integral, but only on the weighted integral with the (possibly vanishing) factor K.

In the sequel we will make the following assumptions on (X, d, K):

(H1) (X, d) is a proper length metric space.
(H2) K is continuous.
(H3) We have $K(x) \geq \kappa(d(x, \Sigma))$ for a function $\kappa : \mathbb{R}_+ \to \mathbb{R}_+$ such that $\int_0^\infty \kappa(r) \, dr = +\infty$.

For a given interval I we now consider the class $\mathrm{AC}_{ploc}(I, X)$ of curves $\gamma \in C(I; X)$ which are piecewise locally absolutely continuous, i.e. there exists a finite partition $(t_0, t_1, \ldots, t_N)$ of I such that $\gamma \in \mathrm{AC}([t_i + \varepsilon, t_{i+1} - \varepsilon]; X)$ for every i and every $\varepsilon > 0$ such that $2\varepsilon < t_{i+1} - t_i$.

Our aim is to investigate the existence of a curve $\gamma \in \mathrm{AC}_{ploc}([0, 1], X)$ minimizing the K-length, defined by

$$L_K(\gamma) := \int_0^1 K(\gamma(t)) \, |\gamma'(t)| \, dt.$$

Namely, we want to find a curve $\gamma \in \mathrm{AC}_{ploc}([0, 1]; X)$ which minimizes the K-length among curves connecting two given points x_0, x_1. We define

$$d_K(x_0, x_1) := \inf\{L_K(\gamma) \; : \; \gamma \in \mathrm{AC}_{ploc}([0, 1]; X), \; \gamma(0) = x_0, \gamma(1) = x_1\}.$$

We are going to prove that d_K is a metric on X s.t. (X, d_K) is proper and such that L_K coincides with the length induced by the distance d_K, which we will denote by Length_{d_K}, thus implying the existence of a geodesic between two joinable points, in view of Theorem 1.16 (see Theorem 1.26 below).

Proposition 1.25 *The quantity d_K is a metric on X. Moreover (X, d_K) enjoys the following properties*

1. *d_K and d are equivalent (i.e. they induce the same topology) on all d-compact subsets of X.*
2. *Any d_K-bounded set is also d-bounded.*
3. *(X, d_K) is a proper metric space.*
4. *Any locally AC curve $\gamma : I \to X$ is also d_K-locally AC and the metric derivative of γ in (X, d_K), denoted by $|\gamma'|_K$, is given by $|\gamma'|_K(t) = K(\gamma(t)) \, |\gamma'|(t)$ a.e.*
5. *We have $L_K(\gamma) = \mathrm{Length}_{d_K}(\gamma)$ for all $\gamma \in \mathrm{AC}_{ploc}(I, X)$.*

Theorem 1.26 *For any $x_0, x_1 \in X$, there exists a $\gamma \in \mathrm{AC}_{ploc}([0, 1]; X)$ s.t. $L_K(\gamma) = d_K(x_0, x_1)$, $\gamma(0) = x_0$ and $\gamma(1) = x_1$.*

Proof Let us see how Proposition 1.25 implies Theorem 1.26. As (X, d_K) is a proper metric space, Theorem 1.16 ensures the existence of a Length_{d_K}-minimizing curve γ connecting x_0 to x_1. This curve is valued in a d_K-bounded set, which is then d-bounded, and hence contained in a d-compact set, where d_K and d are equivalent. This implies that γ is also d-continuous. Then, up to renormalization, one can assume that γ is parameterized by Length_{d_K}-arc length. By minimality, we also know that γ is injective: if $\gamma(t) = \gamma(s)$, then γ restricted to the interval (t, s) is reduced to a point, and $t = s$ as γ is parameterized by arc length. In particular, γ meets the finite set $\Sigma = \{K = 0\}$ at finitely many instants $t_1 < \cdots < t_N$. As K is bounded from below by some positive constant on each compact subinterval of (t_i, t_{i+1}) for $i \in \{1, \ldots, N\}$, Lemma 1.27 below implies that γ is piecewise locally d-Lipschitz, hence it belongs to $\text{AC}_{ploc}([0, 1]; X)$. Finally, thanks to the last statement in Proposition 1.25, the fact that γ minimizes L_{d_K} means that it also minimizes L_K among AC_{ploc} curves connecting x_0 to x_1. □

In order to prove Proposition 1.25, we will need the following estimates on d_K.

Lemma 1.27 *For all $x, y \in X$, one has*

$$K_{d(x,y)}(x)\, d(x, y) \le d_K(x, y) \le K^{d(x,y)}(x)\, d(x, y),$$

where $K_r(x)$ and $K^r(x)$ are defined in (1.7).

Proof Set $r := d(x, y)$. Since any curve γ connecting x to y has to get out of the open ball $B := B_d(x, r)$, it is clear that we have

$$L_K(\gamma) = \int_I K(\gamma(t))\, |\gamma'|(t)\, dt \ge r \inf_B K = r K_r(x).$$

Taking the infimum over the set of curves $\gamma \in \text{AC}_{ploc}$ joining x and y, one gets the first inequality.

For the second inequality, let us fix $\varepsilon > 0$. By definition, there exists a Lipschitz curve γ connecting x to y, that one can assume to be parameterized by arc-length, s.t. $\text{Length}_d(\gamma) \le r + \varepsilon$. In particular, the image of γ is included in the ball $B_d(x, r + \varepsilon)$. Thus, one has

$$d_K(x, y) \le L_K(\gamma) \le (r + \varepsilon)\, K^{r+\varepsilon}(x)$$

and the second inequality follows by sending $\varepsilon \to 0$. Indeed, the mapping $r \to K^r(x)$ is continuous on $\mathbb{R}_+$ since K is uniformly continuous on compact sets and since bounded closed subsets of X are compact (assumption (H1)). □

Proof *(of Proposition 1.25)* The proof is divided into six steps.

Step 1: d_K is a metric. First note that d_K is finite on $X \times X$. Indeed, given two points $x, y \in X$, just take a Lipschitz curve connecting them, and use $L_K(\gamma) \le L_d(\gamma) \sup_{\text{Im}(\gamma)} K < +\infty$. The triangle inequality for d_K is a consequence of the stability of the set AC_{ploc} under concatenation. The fact that $d_K(x, y) = 0$ implies $x = y$ is an easy consequence of the finiteness of the set $\{K = 0\}$. Indeed, if $x \ne y$, then any curve $\gamma : x \mapsto y$ has to connect $B_d(x, \varepsilon)$ to $B_d{}^c(x, 2\varepsilon)$ for all $\varepsilon > 0$ small enough. This implies that $L_K(\gamma) \ge \varepsilon \inf_C K$, where $C = \{y : \varepsilon \le d(x, y) \le 2\varepsilon\}$.

But for ε small enough, C does not intersect the set $\{K = 0\}$ so that $\inf_C K > 0$. In particular, $d_K(x, y) \geq \varepsilon \inf_C K > 0$.

Step 2: d_K and d are equivalent on d-compact sets. Take a compact set $Y \subset X$, and assume $Y \subset B_d(x_0, R)$ just to fix the ideas. Consider the identity map from (Y, d) to (Y, d_K). It is an injective map between metric spaces. Moreover, it is continuous, since, as a consequence of Lemma 1.27, we have $d_K \leq Cd$ on $Y \times Y$, where $C = \sup_{B_d(x_0, 3R)} K < +\infty$ (note that the closed ball $B_d(x_0, 3R)$ is d-compact, and that we assumed $d = \text{geo}d$ since (X, d) is a length space). Hence, as every injective continuous map defined on a compact space is a homeomorphism, d and d_K are equivalent (on Y).

Step 3: every ball in (X, d_K) is d-bounded. This is a consequence of assumptions (H1) and (H3). Let us take $x_0, x \in X$ with $d_K(x, x_0) \leq R$. By definition, there exists a $\gamma \in AC_{ploc}(I, X)$ s.t. $\gamma : x_0 \mapsto x$ and $L_K(\gamma) \leq d_K(x_0, x) + 1$. Now, set $\phi(t) := d(\gamma(t), \Sigma)$: since the function $x \mapsto d(x, \Sigma)$ has Lipschitz constant equal to 1, we have $\phi \in AC_{ploc}(I, \mathbb{R})$ and $|\phi'(t)| \leq |\gamma'(t)|$ a.e. Take the antiderivative $h : \mathbb{R}^+ \to \mathbb{R}^+$ of the function κ (the one in assumption (H3)), i.e. $h' = \kappa$ with $h(0) = 0$, and compute $[h(\phi(t))]' = \kappa(\phi(t))\phi'(t)$. Hence,

$$|[h(\phi(t))]'| = \kappa(\phi(t))|\phi'(t)| \leq K(\gamma(t))|\gamma'(t)|$$

and $h(d(\gamma(t), \Sigma)) \leq h(d(x_0, \Sigma)) + L_K(\gamma) \leq h(d(x_0, \Sigma)) + R + 1$. Since $\lim_{s \to \infty} h(s) = +\infty$, this provides a bound on $d(x, \Sigma)$, which means that the ball $B_{d_K}(x_0, R)$ is d-bounded.

Step 4: every closed ball in (X, d_K) is d_K-compact. Now that we know that closed balls in (X, d_K) are d-bounded, since (X, d) is proper, we know that they are contained in d-compact sets. But on these sets d and d_K are equivalent, hence these balls are also d_K-precompact and, since they are closed, d_K-compact.

Step 5: proof of Statement 4. Let $\gamma : I \to X$ be a d-locally Lipschitz curve valued in X. Thanks to the second inequality in Lemma 1.27, γ is also d_K-locally Lipschitz. Now, Lemma 1.27 provides

$$K_r(\gamma(t)) \frac{d(\gamma(t), \gamma(s))}{|t - s|} \leq \frac{d_K(\gamma(t), \gamma(s))}{|t - s|} \leq K^r(\gamma(t)) \frac{d(\gamma(t), \gamma(s))}{|t - s|}$$

with $r := d(\gamma(t), \gamma(s))$. In the limit $s \to t$ we get

$$K(\gamma(t))|\gamma'|(t) \leq |\gamma'|_K(t) \leq K(\gamma(t))|\gamma'|(t) \quad \text{a.e.,}$$

where the continuity of $r \to K^r(x)$ and $r \to K_r(x)$ on $\mathbb{R}_+$ have been used.

Last step: proof of Statement 5. This is an easy consequence of Statement 4. Indeed, by additivity of L_K and Length_{d_K} we can check the validity of the statement for curves which are locally AC (and not only piecewise locally AC), and the statement comes from the integration in time of the last equality in Statement 4. $\square$

1.4.5 An Application to Heteroclinic Connections

In this section we move to a different variational problem, i.e. our aim is to investigate the existence of a global minimizer of the energy

$$E_W(\gamma) = \int_{\mathbb{R}} \left(\frac{1}{2} |\gamma'(t)|^2 + W(\gamma(t)) \right) dt,$$

defined over locally Lipschitz curves $\gamma : \mathbb{R} \to X$ such that $\gamma(\pm\infty) = x^{\pm}$ (in the sense that they admit limits at $\pm\infty$). Here $W : X \to \mathbb{R}^+$ is a continuous function, called the *potential* in the sequel, and $x^{\pm} \in X$ are two wells, i.e. $W(x^{\pm}) = 0$. Note that $W(x^{\pm}) = 0$ is a necessary condition for the energy of γ to be finite. We will see how to attack this problem, whose main difficulty comes from the fact that the interval where curves are defined is not bounded and compactness is lacking, via the techniques concerning weighted geodesics that we developed here before. This problem, which allows us to find solutions to the equation $\gamma'' = W'(\gamma)$ with given boundary conditions at $\pm\infty$, is very classical and it has been studied, for instance, in [7, 18, 90, 184, 188]. The approach which is presented here is essentially taken from [150] and also appears in [201]. The results of [150] have been extended in [151] to the infinite-dimensional case with applications to the PDE $\Delta u = W'(u)$.

The main result of this section is the following:

Theorem 1.28 *Let us take (X, d) a metric space, $W : X \to \mathbb{R}_+$ a continuous function and $x^{\pm} \in X$ such that:*

(H) *(X, d, K) satisfies hypotheses (H1), (H2), (H3) of the previous section, where $K := \sqrt{2W}$.*

(STI) *$W(x^-) = W(x^+) = 0$ and d_K (defined above) satisfies the following strict triangle inequality on the set $\{W = 0\}$: for all $x \in X \setminus \{x^-, x^+\}$ s.t. $W(x) = 0$, $d_K(x^-, x^+) < d_K(x^-, x) + d_K(x, x^+)$.*

Then, there exists a heteroclinic connection between x^- and x^+, i.e. $\gamma \in \mathrm{Lip}(\mathbb{R}; X)$ such that

$$E_W(\gamma) = \inf\{E_W(\sigma) : \sigma \in \mathrm{AC}_{ploc}(\mathbb{R}, X), \sigma(\pm\infty) = x^{\pm}\}.$$

Moreover, $E_W(\gamma) = d_K(x^-, x^+)$.

Proof This theorem is a consequence of Theorem 1.26 and the following consequence of Young's inequality:

$$\text{for all } \gamma \in \mathrm{AC}_{ploc}(\mathbb{R}, X), \ E_W(\gamma) \geq L_K(\gamma), \tag{1.8}$$

where $K := \sqrt{2W}$. Indeed, thanks to assumption (H), Theorem 1.26 provides an L_K-minimizing curve $\gamma_0 : I \to X$, that one can assume to be parameterized by L_K-arc length (and injective by minimality), connecting x^- to x^+. Thanks to assumption

(STI), it is clear that the curve γ_0 cannot meet the set $\{W = 0\}$ at a third point $x \neq x^{\pm}$: in other words $K(\gamma(t)) > 0$ on the interior of I. Thus, γ_0 is also d-locally Lipschitz on I (and not only piecewise locally Lipschitz). In particular, one can reparameterize the curve γ_0 by L_d-arc length, so that $|\gamma_0'| = 1$ a.e. in I. Note that, since d and d_K are equivalent on $\mathrm{Im}(\gamma_0)$, γ_0 remains d_K-continuous, and γ_0 is even d_K-Lipschitz as $d_K(\gamma(t), \gamma(s)) \leq L_K(\gamma_{(t,s)}) \leq \sup K \circ \gamma \times L_d(\gamma_{(t,s)}) \leq C|t - s|$. In particular, γ_0 is still L_K-minimizing.

Then, in view of (1.8), it is enough to prove that γ_0 can be reparameterized into a locally Lipschitz curve γ satisfying the equipartition condition $|\gamma'| = K \circ \gamma$ a.e., for which (1.8) becomes an equality.

Namely, we look for an admissible curve $\gamma : \mathbb{R} \to X$ of the form $\gamma(t) = \gamma_0(\varphi(t))$, where $\varphi : \mathbb{R} \to I$ is C^1, increasing and surjective. For γ to satisfy $|\gamma'|(t) = K(\gamma(t))$ a.e., we need φ to solve the ODE

$$\varphi'(t) = F(\varphi(t)), \tag{1.9}$$

where $F : \overline{I} \to \mathbb{R}$ is the continuous function defined by $F = K \circ \gamma_0$ on I and $F \equiv 0$ outside I. Thanks to the Peano theorem, (1.9) admits at least one maximal solution $\varphi_0 : J = (t^-, t^+) \to \mathbb{R}$ such that $0 \in J$ and $\varphi_0(0)$ is any point inside I. Since F vanishes outside I, we know that $\mathrm{Im}(\varphi_0) \subset \overline{I}$. Moreover, since φ_0 is non-decreasing on I, it converges to two distinct stationary points of the preceding ODE. As $F > 0$ inside I, one has $\lim_{t \to t^+} \varphi_0(t) = \sup I$ and $\lim_{t \to t^-} \varphi_0(t) = \inf I$. We deduce that φ_0 is an entire solution of the preceding ODE, i.e. $I = \mathbb{R}$. Indeed, if $I \neq \mathbb{R}$, say $t^+ < +\infty$, then one could extend φ_0 by setting $\varphi_0(t) = \sup I$ for $t > t^+$. Finally, the curve $\gamma := \gamma_0 \circ \varphi_0$ satisfies $\gamma(\pm\infty) = x^{\pm}$, it is locally Lipschitz (by composition), and $|\gamma'|(t) = K(\gamma(t))$ a.e. Thus,

$$E_W(\gamma) = L_K(\gamma) = L_{d_K}(\gamma) = L_{d_K}(\gamma_0) = d_K(x^-, x^+)$$

$$\leq \inf\{E_W(\sigma) : \sigma \in \mathrm{AC}_{ploc}(\mathbb{R}; X), \sigma(\pm\infty) = x^{\pm}\}.$$

Namely, γ minimizes E_W over all admissible connections between x^- and x^+. □

- It is easy to see that the equipartition of the energy, that is, the identity $|\gamma'|^2(t) = 2W(\gamma(t))$, is a necessary condition for critical points of E_W.
- The assumption (STI) is not optimal but cannot be removed, and is quite standard in the literature. Without this assumption, a geodesic γ could meet the set $\{W = 0\}$ at a third point $x \neq x^{\pm}$. In this case, it is not always possible to parameterize γ in such a way that $|\gamma'|(t) = K(\gamma(t))$.
- However, if $K = \sqrt{2W}$ is not Lipschitz, it is possible that there exists a heteroclinic connection from x^- to x^+ meeting $\{W = 0\}$ at a third point $x \neq x^{\pm}$. Indeed, if $\lim \inf_{y \to x} K(y)/d(y, x) > 0$, then, there exists a heteroclinic connection γ^- from x^- to x which reaches x in finite time (say, $\gamma^-(t) = x$ for $t \geq 0$). Similarly, there exists a heteroclinic connection γ^+ from x to x^+ such

that $\gamma^+(t) = z$ for $t \leq 0$. Thus, there exists a heteroclinic connection between x^- and x^+ obtained by matching γ^- and γ^+.

1.5 The Ramsey Model for Optimal Growth

In this section we propose a detailed and rigorous treatment of the Ramsey model that we introduced in Sect. 1.1, proving the existence of solutions and studying them using a natural and original approach taken from [180] and based on the use of the BV space.

Box 1.8 Memo—*Bounded Variation Functions in One Variable*

BV functions are defined as L^1 functions whose distributional derivatives are measures (see Chap. 6). In dimension one this has a lot of consequences. In particular these functions coincide a.e. with functions which have bounded total variation in a pointwise sense. For each $f : [a, b] \to \mathbb{R}$ define

$$TV(f; [a, b]) := \sup \left\{ \begin{array}{l} \sum_{i=0}^{N-1} |f(t_{i+1}) - f(t_i)| : \\ a = t_0 < t_1 < t_2 < \cdots < t_N = b \end{array} \right\}.$$

Note that the definition of the total variation TV is essentially the same as Length, the only difference is that we do not apply it only to continuous curves. Functions of bounded variation (BV) are defined as those f such that $TV(f; [a, b]) < \infty$. It is easy to check that BV functions form a vector space, and that bounded monotone functions are BV (if f is monotone we have $TV(f; [a, b]) = |f(b) - f(a)|$). Lipschitz functions are also BV and $TV(f; [a, b]) \leq \mathrm{Lip}(f)(b - a)$. On the other hand, continuous functions are not necessarily BV (but absolutely continuous functions are BV).

BV functions have several properties:

Properties of BV functions in $\mathbb{R}$ – If $TV(f; [a, b]) < \infty$ then f is the difference of two monotone functions (in particular we can write $f(x) = TV(f; [a, x]) - (TV(f; [a, x]) - f(x))$, both terms being non-decreasing functions); it is a bounded function and $\sup f - \inf f \leq TV(f; [a, b])$; it has the same continuity and differentiability properties of monotone functions (it admits left and right limits at every point, it is continuous up to a countable set of points and differentiable a.e.).

In particular, in 1D, we have BV $\subset L^\infty$.

We also have a very important compactness result: given a sequence f_n of BV functions on an interval I such that $TV(f_n; I)$ is bounded, there exists a subsequence and a limit function f which is also BV such that $f_n(t) \to f(t)$

(continued)

Box 1.8 (continued)

for all $t \in I$ which are continuity points for f, i.e. for all but a countable set of points t.

We finish by stressing the connections with measures: for every positive measure μ on $[a, b]$ we can build a monotone function by taking its cumulative distribution function, i.e. $F_\mu(x) = \mu([a, x])$ and the distributional derivative of this function is exactly the measure μ. Conversely, every monotone non-decreasing function on a compact interval is the cumulative distribution function of a (unique) positive measure, and every BV function is the cumulative distribution function of a (unique) signed measure.

We come back to the problem we already described in Sect. 1.1, i.e.

$$\max_c \int_0^\infty e^{-rt} U(c(t)) \, dt \text{ subject to}$$

$$k'(t) = f(k(t), t) - c(t) \tag{1.10}$$

$$k(t) \geq 0, \ c(t) \geq 0 \text{ a.e. } k(0) = k_0 \text{ fixed.}$$

We assume that U is concave, non-negative and increasing, with $U'(\infty) = 0$, and that $f(\cdot, t)$ is also concave for every t.

We will now give a precise functional setting for Problem (1.10). The choice is motivated by the uniform bounds that we obtain below, in Lemma 1.29. Let us define, for fixed $k_0 \in \mathbb{R}_+$

$$\mathcal{A} := \left\{ (k, c) \ : \ k \in BV_{loc}(\mathbb{R}_+), \ c \in \mathcal{M}_{loc}(\mathbb{R}_+), \ \begin{array}{l} k \geq 0, \ k(0) = k_0, \ c \geq 0, \\ k' + c \leq f(k, t) \end{array} \right\},$$

where the last inequality is to be intended as an inequality between measures (the right-hand side being the measure on $\mathbb{R}_+$ with density $f(k(\cdot), \cdot)$). On the set of pairs (k, c) (and in particular on $\mathcal{A}$), we consider the following notion of (weak) convergence: we say $(k_n, c_n) \rightharpoonup (k, c)$ if c_n and k'_n weakly-* converge to c and k', respectively, as measures on every compact interval $I \subset \mathbb{R}_+$. It is important to observe that, thanks to the fact that the initial value $k(0)$ is prescribed in $\mathcal{A}$, the weak convergence as measures $k'_n \rightharpoonup k'$ also implies $k_n \to k$ a.e., as we have

$$k_n(t) = k_0 + \int_0^t dk'_n \to k_0 + \int_0^t dk' = k(t)$$

for every t which is not an atom for k' (i.e. every continuity point t for k). This condition is satisfied by all but a countable set of points t.

Lemma 1.29 *Consider any pair* $(k, c) \in \mathcal{A}$. *Then,* $k \in L^\infty(\mathbb{R}_+)$, *and for any compact subset* $I \subset \mathbb{R}_+$, *we have* $\|c\|_{\mathcal{M}(I)} \leq C$ *and* $\|k'\|_{\mathcal{M}(I)} \leq C$, *with the constant* C *depending on the choice of* I *but not on* (k, c). *In particular if* $I = [0, t]$, *we can choose* $C = C(t) = \bar{C}_0 + \bar{C}_1 t$ *for suitable* $\bar{C}_0, \bar{C}_1$.

Proof Consider first the constraint

$$k' + c \leq f(k, t), \qquad (1.11)$$

where the inequality is an inequality between measures. Since c is non-negative and f is negative for large k (say for $k \geq \bar{k}$), k is bounded from above by $\max\{k(0), \bar{k}\}$. Hence, $k \in L^\infty(\mathbb{R}_+)$.

We return to (1.11), and note that, since the right-hand side is positive and bounded above by C_1, we have $k'(t) \leq C_1$, for any t. This provides upper L^∞ bounds for k', but we also need lower bounds if we want to bound k' in $\in \mathcal{M}_{loc}(\mathbb{R}_+)$. We break up k' into its positive and negative parts as $k' = (k')^+ - (k')^-$. The positive part is bounded, by (1.11), and the upper bound is C_1. We proceed to obtain a bound for $(k')^-$. Clearly,

$$k(t) - k(0) = \int_0^t dk' = \int_0^t (k')^+(s)\,ds - \int_0^t d(k')^-,$$

which, using $k(t) \geq 0$ and the upper bound $(k')^+ \leq C_1$, leads to the estimate

$$- k(0) \leq k(t) - k(0) = \int_0^t (k')^+(s)\,ds - \int_0^t d(k')^- \leq C_1 t - \int_0^t (k')^-,$$

i.e.

$$\int_0^t (k')^- \leq k_0 + C_1 t. \qquad (1.12)$$

This allows us to estimate the total mass of $|k'| = (k')^+ + (k')^-$ as

$$\int_0^t |k'| = \int_0^t (k')^+ + \int_0^t (k')^- \leq C_1 t + k_0 + C_1 t = k_0 + 2C_1 t, \qquad (1.13)$$

where we used (1.12). This estimate implies $\|k'\|_{\mathcal{M}(I)} < C$ for any compact $I = [0, t]$, where C depends on t.

Once we have the desired bound for k' we return to (1.11), which yields $0 \leq c \leq C_1 - k'$, from which the desired bound on c follows. $\qquad \square$

The above lemma is the one which justifies the choice of the space BV as a functional setting. Indeed, Lemma 1.29 provides suitable compactness in this space, while the same could not be obtained in stronger spaces such as $W^{1,p}$.

We now define the quantity which is optimized, in this functional setting. Define

$$J(c) := \int_0^\infty e^{-rt} U(c^{ac}(t)) \, dt,$$

where c^{ac} is the absolutely continuous part of the measure c.

Box 1.9 Memo—*Radon–Nikodym Decomposition*

Given a finite measure μ and a positive σ-finite measure $\nu \geq 0$ on the same space X, we say that μ is absolutely continuous w.r.t. ν (and we write $\mu \ll \nu$) if $\nu(A) = 0$ implies $\mu(A) = 0$ for every measurable set A. We say that μ is orthogonal to ν if μ is concentrated on a set A such that $\nu(A) = 0$.

Two important facts can be proven:

- we have $\mu \ll \nu$ if and only if there exists a function $f \in L^1(\nu)$ such that $\mu = f \cdot \nu$, i.e. we have $\mu(A) = \int_A f \, d\nu$; such a function is called a density of μ w.r.t. ν;
- for any pair (μ, ν) we can decompose μ into two parts, $\mu = \mu^{ac} + \mu^{sing}$ where $\mu^{ac} \ll \nu$ and μ^{sing} is singular, i.e. orthogonal w.r.t. ν.

Combining the two statements we obtain the well-known *Radon–Nikodym* theorem saying that for every μ, ν there exists an $f \in L^1(\nu)$ such that $\mu = f \cdot \nu + \mu^{sing}$. Moreover, we often identify, by abuse of language, the measure μ^{ac} and its density f, so that μ^{ac} can be considered as a function belonging to $L^1(\nu)$. In most cases, when X is a subspace of Euclidean space, we use for ν the Lebesgue measure on X.

For notational purposes, we also introduce, for $T \in \mathbb{R}_+ \cup \{+\infty\}$, the quantity $J_T : \mathcal{M}_{loc}(\mathbb{R}_+) \to [0, +\infty]$:

$$J_T(c) := \int_0^T e^{-rt} U(c^{ac}(t)) \, dt,$$

and we will use the notations J and J_∞ interchangeably.

Lemma 1.30 *The following hold:*

1. *For any $\epsilon > 0$, there exists a $T = T(\epsilon) < \infty$ such that $J_\infty(c) \leq J_T(c) + \epsilon$ for any $(k, c) \in \mathcal{A}$ (in particular, $T(\epsilon)$ does not depend on c).*
2. *Whenever $(k_n, c_n) \rightharpoonup (k, c)$ in $\mathcal{A}$ we have $\limsup J_\infty(c_n) \leq J_\infty(c)$, i.e. J_∞ is u.s.c.*

Proof We treat the two points in order

1. We write

$$J_\infty(c) = J_T(c) + \int_T^\infty e^{-rt} U(c^{ac}(t)) \, dt$$

and we use the fact that U, as it is concave, is bounded above by an affine function $U(c) \leq A + Bc$. The integral $\int_T^\infty e^{-rt} A \, dt$ can be made small as soon as T is chosen large enough. For the integral $\int_T^\infty e^{-rt} Bc^{ac}(t) \, dt$ we use the bound in Lemma 1.29, applied to the intervals $[T + j, T + j + 1]$. We have $\int_{T+j}^{T+j+1} c^{ac}(t) \, dt \leq \bar{C}_0 + \bar{C}_1(T + j + 1)$ and hence

$$\int_{T+j}^{T+j+1} e^{-rt} Bc^{ac}(t) \, dt \leq B(\bar{C}_0 + \bar{C}_1(T + j + 1))e^{-j}e^{-T}.$$

The series of these quantities as j ranges from 0 to ∞ converges for every T, and can be also made small just by choosing T large enough.

2. Consider such a sequence (k_n, c_n). By (i), for any $\epsilon > 0$, there exists a $T = T(\epsilon)$ such that

$$J_\infty(c_n) \leq J_T(c_n) + \epsilon, \quad \forall n \in \mathbb{N}. \tag{1.14}$$

Furthermore, using $U \geq 0$,

$$J_T(c) \leq J_\infty(c), \quad \forall c \in \mathcal{M}_{loc}(\mathbb{R}_+). \tag{1.15}$$

The functional J_T is upper semicontinuous with respect to weak$-*$ convergence in $\mathcal{M}_{loc}(\mathbb{R}_+)$. Indeed, since U is concave, $-U$ is a convex function, and we can apply the semicontinuity results of Sect. 3.3.

Then, taking the limsup on both sides of (1.14) yields

$$\limsup_n J_\infty(c_n) \leq \limsup_n J_T(c_n) + \epsilon \leq J_T(c) + \epsilon \leq J_\infty(c) + \epsilon,$$

where we used first the upper semicontinuity of J_T and then (1.15). Taking the limit as $\epsilon \to 0^+$ we obtain the upper semicontinuity of J_∞.

$\square$

Proposition 1.31 *The optimization problem* (1.10) *admits a maximizer.*

Proof Consider a maximizing sequence $(\bar{k}_n, c_n)$ for (1.10). Since $(k_n, c_n) \in \mathcal{A}$ for any interval $I = [0, T]$, we conclude by Lemma 1.29 that k_n is bounded in $BV_{loc}(\mathbb{R}_+)$ and c_n in $\mathcal{M}_{loc}(\mathbb{R}_+)$ (in the sense that we have bounds on the mass of k_n' and c_n on any compact interval of $\mathbb{R}_+$). Using the weak$-*$ compactness of bounded sets in the space of measures on each compact interval, together with a suitable diagonal extraction procedure, we obtain the existence of $k \in \mathrm{BV}_{loc}(\mathbb{R}_+)$ and $\zeta, c \in \mathcal{M}_{loc}(\mathbb{R}_+)$ such that up to subsequences

$$k_n \to k \qquad a.e.$$

$$k_n' \overset{*}{\rightharpoonup} \zeta \ \text{in} \ \mathcal{M}_{loc}(\mathbb{R}_+),$$

$$c_n \overset{*}{\rightharpoonup} c \ \text{in} \ \mathcal{M}_{loc}(\mathbb{R}_+).$$

By looking at the distributional derivatives we can see that we necessarily have $\zeta = k'$.

Our first task is to show admissibility of the limit (k, c). It is straightforward to check $c \geq 0$ and $k \geq 0$. It remains to check $k' + c \leq f(k, t)$. In order to see this, note first that since $(k_n, c_n) \in \mathcal{A}$ for every $n \in \mathbb{N}$, we have

$$k_n' + c_n \leq f(k_n, t), \quad \forall n \in \mathbb{N}. \tag{1.16}$$

We need to pass to the limit as $n \to \infty$ (along the converging subsequence) in the above. The left-hand side is linear and $k_n' + c_n \overset{*}{\rightharpoonup} k' + c$. Some care must be taken for the nonlinear right-hand side. Since f is continuous and bounded from above, it is sufficient to use the pointwise convergence $k_n(t) \to k(t)$ a.e. in $t \in [0, T]$, guaranteed by the local BV bound on k_n.

To show that (k, c) is a maximizer for Problem (1.10), we use the upper semicontinuity result, $\limsup_n J_\infty(c_n) \leq J_\infty(c)$ contained in Lemma 1.30.   □

The above result guarantees the existence of a maximizer (k, c) where both k' and c are measures. In order to come back to a more standard functional setting we will show that, actually, such measures have no singular part. In terms of the state function k, this means $k \in W^{1,1}_{loc}$. This result is crucial both for the mathematical analysis of the optimization problem and for its economic meaning.

Lemma 1.32 *Any optimizer (k, c) satisfies $c \in L^1_{loc}(\mathbb{R}_+)$.*

Proof Let us recall that, in general, any c such that $(k, c) \in \mathcal{A}$ consists of an absolutely continuous and a singular part with respect to Lebesgue measure $c = c^{ac} + c^{sing}$.

Assume that (k, c) is optimal. Given an interval $I = [a, b] \subset \mathbb{R}_+$, we build a new competitor $(\tilde{k}, \tilde{c})$ in the following way: outside I, we take $\tilde{k} = k$ and $\tilde{c} = c$; then we choose $\tilde{k}$ to be affine on $[a, b]$, connecting the values $k(a^-)$ to $k(b^+)$; $\tilde{c}$ restricted to $[a, b]$ is an absolutely continuous measure, with constant density. We have to choose the constant value for $\tilde{c}$ for the inequality $\tilde{k}' + \tilde{c} \leq f(\tilde{k}(t), t)$ to be satisfied. Notice that on $[a, b]$ we have $\tilde{k}' = k'([a, b])/(b - a)$. If we set $\hat{c} = c([a, b])/(b - a)$ then we have $\tilde{k}' + \hat{c} \leq \int_a^b f(k(s), s)\, ds$. Since we can assume f to be bounded (k only takes values in a bounded set, and hence so does $\tilde{k}$), we have $\int_a^b f(k(s), s)\, ds \leq f(\tilde{k}(t), t) + M$. Thus, it is sufficient to set $\tilde{c} = \hat{c} - M$, provided $\hat{c} \geq M$.

Now, suppose that c is not absolutely continuous. Then there exists an $I = [a, b]$ such that $c^{sing}([a, b]) > M(b - a) + c^{ac}([a, b])$ (otherwise $c^{sing} \leq M + c^{ac}$ and c^{sing} would be absolutely continuous). The interval $[a, b]$ may be taken as small as we want, which means that we can assume $(b - a)e^{-ra} \leq 2\varepsilon_0$, where we set $\varepsilon_0 := \int_a^b e^{-rs} \, ds$. We choose such an interval and we build $\tilde{k}$ and $\tilde{c}$ as above, since we now have $\hat{c} \geq M$. We now claim we have a contradiction.

Let us set $m_0 = c^{ac}([b - a])/(b - a)$ and $m_1 = c^{sing}([b - a])/(b - a)$. The optimality of (k, c) gives

$$\int_a^b e^{-rt} U(c^{ac}(t)) \, dt \geq \int_a^b e^{-rt} U(\tilde{c}(t)) \, dt.$$

Using Jensen's inequality, we get

$$\int_a^b e^{-rt} U(c^{ac}(t)) \, dt \leq \varepsilon_0 U \left(\frac{1}{\varepsilon_0} \int_a^b c^{ac}(t) e^{-rt} \, dt \right)$$

$$\leq \varepsilon_0 U \left(\frac{e^{-ra}}{\varepsilon_0} \int_a^b c^{ac}(t) \, dt \right) \leq \varepsilon_0 U(2m_0),$$

and

$$\int_a^b e^{-rt} U(\tilde{c}(t)) \, dt = \varepsilon_0 U(m_0 + m_1 - M).$$

We deduce

$$U(m_0 + m_1 - M) \leq U(2m_0),$$

but the interval $[a, b]$ was chosen so that $m_1 > m_0 + M$, which gives a contradiction because of the strict monotonicity of U.   □

We can use the above result to prove some useful properties of our maximization problem.

Corollary 1.33 *For any optimizer (k, c), we have $k \in W^{1,1}_{loc}(\mathbb{R}_+)$ and saturation of the constraint $k' + c = f(k, t)$. Moreover, the optimal pair (k, c) is unique.*

Proof First, note that, if $k \notin W^{1,1}_{loc}(\mathbb{R}_+)$, then this means that k' has a negative singular part $((k')^{sing})_-$. Yet the inequality $k' + c \leq f(k, t)$ stays true if we replace c with $c + ((k')^{sing})_-$, which would be another optimizer. But this contradicts Lemma 1.32, since in every optimizer c should be absolutely continuous.

As soon as we know that both k' and c are absolutely continuous, it is clear that we must have $k' + c = f(k, t)$. Otherwise, we can replace c with $f(k, t) - k'$ and get a better result (we needed to prove the absolute continuity of k', since adding a singular part to c does not improve the functional).

Concerning uniqueness, we first stress that the functional we are maximizing is not strictly concave, for two reasons: on the one hand we did not assume U to be strictly concave, and on the other hand, anyway, there would be an issue as far as the singular part of c is involved. Yet, now that we know that optimizers are absolutely continuous and saturate the differential inequality constraint, uniqueness follows from the strict concavity of f.

Indeed, suppose that (k_1, c_1) and (k_2, c_2) are two minimizers. Then, setting $c_3 = (c_1 + c_2)/2$ and $k_3 = (k_1 + k_2)/2$, the pair (k_3, c_3) is also admissible (since f is concave in k) and optimal (since the cost is concave in c). Yet, strict concavity of f implies that (k_3, c_3) does not saturate the constraint, which is a contradiction with the first part of the statement, unless $k_1 = k_2$. As a consequence, we obtain uniqueness of k in the optimal pairs (k, c). But this, together with the relation $k' + c = f(k, t)$, which is now saturated, also provides uniqueness of c. $\square$

In [180] other bounds on the optimal c are proven (in particular upper and lower bounds on any bounded interval), but we will not insist on this here.

1.6   Non-autonomous Optimization in BV

In Sect. 1.2, as a very classical example for existence, we discussed problems of the form $\min \int_0^1 |u'|^p + F(t, u)$, in particular for $p = 2$, but we noted that $p > 1$ does not add substantial difficulties. In this section we discuss the case $p = 1$, which naturally requires the use of the space BV. As we are only dealing with one-variable BV functions, the theory that we need is that described in Box 1.8.

The reader will observe that, in the case $F = 0$, this amounts to the geodesic problem studied in Sect. 1.4, where there was no need to use the space BV, and we found a solution as a Lipschitz curve. The reason lies in the possibility of reparameterizing the curves without changing their length. In the presence of an extra running cost F this is no longer possible, since the term $\int F(t, u(t))\, dt$ is not invariant under reparameterization.

The problem that we will consider will then be of the form

$$\min \left\{ TV(\gamma; [0, 1]) + \int_0^1 F(t, \gamma(t))\, dt + \psi_0(\gamma(0)) + \psi_1(\gamma(1)) \; : \; \gamma \in BV([0, 1]) \right\},$$
$$(1.17)$$

where $TV(\gamma; [0, 1])$ stands for the total variation of γ on the interval $[0, 1]$, see Box 1.8, and the space $BV([0, 1])$ is the space of BV functions valued in $\mathbb{R}^d$ The functions $\psi_0, \psi_1 : \mathbb{R}^d \to [0, +\infty]$ are just assumed to be l.s.c., and among possible choices we mention those which impose Dirichlet boundary conditions, i.e.

$$\psi_i(x) = \begin{cases} 0 & \text{if } x = x_i, \\ +\infty & \text{if not.} \end{cases}$$

We stress the fact that functions in BV spaces are not continuous and can have jumps; even if we consider that BV functions of one variable are defined pointwisely, it is possible to change their value at a point very easily. In particular, Dirichlet boundary conditions have a very particular meaning: a curve which takes the value x_0 at $t = 0$ but immediately jumps at another point at $t = 0^+$ is considered to satisfy the condition $\gamma(0) = x_0$. In particular, it is possible to freely choose a curve γ on $(0, 1)$ and then add a jump to x_0 or x_1 at the boundary in order to satisfy the corresponding Dirichlet boundary condition, of course adding a price $|\gamma(0^+) - x_0|$ or $|\gamma(1^-) - x_1|$ to the total variation. In this way, we could decide to identify the values of γ at $t = 0$ or $t = 1$ with their right or left limits at these points, respectively, and replace Dirichlet boundary conditions with a boundary penalization. This could also be done for more general penalizations ψ_0, ψ_1, for which it is useful to define the relaxed functions

$$\tilde{\psi}_i(x) := \inf_y |y - x| + \psi_i(x).$$

It is important to observe that the functions $\tilde{\psi}_i$ are automatically 1-Lipschitz continuous, as an inf of Lip_1 functions of the variable x, indexed by the parameter y.

Box 1.10 Memo—*Continuity of Functions Defined as an* inf *or* sup

Proposition *Let* $(f_\alpha)_\alpha$ *be a family (finite, infinite, countable, uncountable...) of functions all satisfying the same condition*

$$|f_\alpha(x) - f_\alpha(x')| \leq \omega(d(x, x')).$$

Consider f *defined by* $f(x) := \inf_\alpha f_\alpha(x)$. *Then* f *also satisfies the same estimate.*

This can be easily seen from $f_\alpha(x) \leq f_\alpha(x') + \omega(d(x, x'))$, which implies $f(x) \leq f_\alpha(x') + \omega(d(x, x'))$ since $f \leq f_\alpha$. Then, taking the infimum over α on the r.h.s. one gets $f(x) \leq f(x') + \omega(d(x, x'))$. Interchanging x and x' one obtains

$$|f(x) - f(x')| \leq \omega(d(x, x')).$$

In particular, if the function $\omega : \mathbb{R}_+ \to \mathbb{R}_+$ satisfies $\lim_{t \to 0} \omega(t) = 0$ (which means that the family $(f_\alpha)_\alpha$ is equicontinuous), then f has the same modulus of continuity (i.e. the same function ω) as the functions f_α. The same idea obviously works for the supremum instead of the infimum.

In this way the problem (1.17) becomes

$$\min\left\{ TV(\gamma;[0,1]) + \int_0^1 F(t,\gamma(t))\,dt + \tilde{\psi}_0(\gamma(0^+)) + \tilde{\psi}_1(\gamma(1^-)) : \gamma \in BV([0,1]) \right\},$$

or, equivalently, we can replace $\tilde{\psi}_0(\gamma(0^+)) + \tilde{\psi}_1(\gamma(1^-))$ with $\tilde{\psi}_0(\gamma(0)) + \tilde{\psi}_1(\gamma(1))$ and impose continuity of γ at $t = 0, 1$.

First, we note that in the case where F does not depend explicitly on time the problem becomes trivial and it is completely equivalent to the following static problem

$$\min\left\{ F(x) + \tilde{\psi}_0(x) + \tilde{\psi}_1(x) \right\}.$$

Indeed, setting $J(\gamma) := TV(\gamma;[0,1]) + \int_0^1 F(\gamma(t))\,dt + \tilde{\psi}_0(\gamma(0^+)) + \tilde{\psi}_1(\gamma(1^-))$, for every curve γ we have, for every $t_0 \in (0,1)$

$$J(\gamma) \geq \tilde{\psi}_0(\gamma(0^+)) + \tilde{\psi}_1(\gamma(1^-)) + |\gamma(0^+) - \gamma(t_0)| + |\gamma(1^-) - \gamma(t_0)| + \inf_t F(\gamma(t)).$$

Using the 1-Lipschitz behavior of $\tilde{\psi}_0$ and $\tilde{\psi}_1$ we also obtain

$$J(\gamma) \geq \tilde{\psi}_0(\gamma(t_0)) + \tilde{\psi}_1(\gamma(t_0)) + \inf_t F(\gamma(t))$$

and, fixing $\varepsilon > 0$ and choosing a point t_0 such that $F(\gamma(t_0)) < \inf_t F(\gamma(t)) + \varepsilon$, we finally obtain

$$J(\gamma) \geq \tilde{\psi}_0(\gamma(t_0)) + \tilde{\psi}_1(\gamma(t_0) + F(\gamma(t_0)) - \varepsilon.$$

This shows that we have $\inf_\gamma J(\gamma) \geq \inf_x F(x) + \tilde{\psi}_0(x) + \tilde{\psi}_1(x)$. The opposite inequality is obvious since we can use constant curves $\gamma(t) = x$ in the minimization of J.

Note that, when saying that two optimization problems are equivalent, we do not only want to say that their minimal values are the same, but also have a rule on how to build a minimizer of one of them from a minimizer of the other. In this case, the procedure is the following.

Given an optimal x one can build an optimal γ by taking $\gamma(t) = x$ for every t.

Given an optimal γ one can build an optimal x by considering the set

$$X = \overline{\{\gamma(t) : t \in [0,1]\}}$$

and choosing $x \in \operatorname{argmin}_X F + \tilde{\psi}_0 + \tilde{\psi}_1$ (the reason for choosing the closed set X is that this guarantees, if F is l.s.c., that a point x attaining this minimum exists, thus making a clear rule on how to choose at least a point x out of γ).

In the above discussion we saw that the problem becomes somehow trivial when F does not explicitly depend on time, and we will now turn to the case where it does. This discussion is taken from [78]. We will start from an approximated problem.

Lemma 1.34 *Let $L : \mathbb{R}^d \to \mathbb{R}$ be a smooth and uniformly convex function which is assumed to be radial: $L(v) := \ell(|v|)$ for a convex and non-decreasing function $\ell : \mathbb{R}^+ \to \mathbb{R}$. Let $F : [0, 1] \times \mathbb{R}^d \to \mathbb{R}$ be a C^2 time-dependent potential satisfying $D_{xx}^2 F(t, x) \geq c_0 \mathbf{I}$ for a certain constant $c_0 > 0$ and $|\partial_t \nabla_x F(t, x)| \leq C_0$, and $\psi_0, \psi_1 : \mathbb{R}^d \to \mathbb{R}$ two Lipschitz continuous functions. Consider a solution γ of*

$$\min \left\{ \int_0^1 (L(\gamma'(t)) + F(t, \gamma(t))) \, dt + \psi_0(\gamma(0)) + \psi_1(\gamma(1)) \; : \; \gamma \in H^1([0, 1]) \right\}.$$

Then γ is Lipschitz continuous and satisfies $|\gamma'| \leq C$, where C is defined by

$$C := \max \left\{ \frac{C_0}{c_0}, (\ell')^{-1}(\operatorname{Lip} \psi_0), (\ell')^{-1}(\operatorname{Lip} \psi_1) \right\}$$

(the function ℓ' is strictly increasing as a consequence of the uniform convexity of L).

Proof Let us start from the Euler–Lagrange system of the above optimization problem. We have

$$\begin{cases} (\nabla L(\gamma'))' = \nabla_x F(t, \gamma(t)) \\ \nabla L(\gamma'(0)) = \nabla \psi_0(\gamma(0)) \\ \nabla L(\gamma'(1)) = -\nabla \psi_1(\gamma(1)). \end{cases}$$

First we observe that $\gamma \in C^0$ and $F \in C^1$ imply that the right-hand side in the first equation is a continuous function, so that we have $\nabla L(\gamma') \in C^1$. Inverting the injective function ∇L we obtain $\gamma \in C^2$ and, since $F \in C^2$, we obtain $\gamma \in C^3$.

Then, the transversality conditions show $|\nabla L(\gamma'(i))| \leq \operatorname{Lip} \psi_i$ for $i = 0, 1$. Using $|\nabla L(v)| = \ell'(|v|)$ we see that at $t = 0, 1$ the condition $|\gamma'| \leq C$ is satisfied.

Let us now consider the maximal value of $|\gamma'(t)|$. This maximum exists on $[0, 1]$ since $\gamma \in C^1$ and if it is attained on the boundary $t = 0, 1$ the desired Lipschitz bound $|\gamma'| \leq C$ is satisfied. We can now assume that it is attained in $(0, 1)$. Since ℓ' is an increasing and non-negative function, the maximum points of $|\gamma'|$ and of $|\nabla L(\gamma')|^2$ are the same. We can then write the optimality condition differentiating once and twice in t: we have

$$\nabla L(\gamma') \cdot (\nabla L(\gamma'))' = 0; \quad \nabla L(\gamma') \cdot (\nabla L(\gamma'))'' + |(\nabla L(\gamma'))'|^2 \leq 0.$$

In the last condition we can ignore the non-negative term $|(\nabla L(\gamma'))'|^2$ and observe that, since $\nabla L(\gamma')$ and γ' are vectors with the same orientation ($\nabla L(\gamma') = \frac{\ell'(|\gamma'|)}{|\gamma'|}\gamma'$), we have $\gamma' \cdot (\nabla L(\gamma'))'' \leq 0$.

We now differentiate the Euler–Lagrange equation in time and take the scalar product with γ', obtaining

$$0 \geq \gamma'(t) \cdot (\nabla L(\gamma'(t)))'' = (\nabla_x F(t, \gamma(t)))' \cdot \gamma'(t)$$

$$= \partial_t \nabla_x F(t, \gamma) \cdot \gamma'(t) + \gamma'(t) \cdot D^2_{xx} F(t, \gamma(t))\gamma'(t).$$

We deduce

$$c_0|\gamma'(t)|^2 \leq |\gamma'(t)||\partial_t \nabla_x F(t, \gamma)|,$$

which implies $|\gamma'(t)| \leq \frac{C_0}{c_0} \leq C$ and concludes the proof.   □

We now use the above result on an approximation of the original problem in BV.

Proposition 1.35 *Consider*

$$\min\left\{ TV(\gamma; [0, 1]) + \int_0^1 F(t, \gamma(t))\,dt + \psi_0(\gamma(0^+)) + \psi_1(\gamma(1^-)) : \gamma \in \mathrm{BV}([0, 1]) \right\}$$
$$(1.18)$$

where $F : [0, 1] \times \mathbb{R}^d \to \mathbb{R}$ is a C^2 time-dependent potential satisfying $D^2_{xx} F(t, x) \geq c_0 I$ for a certain constant $c_0 > 0$ and $|\partial_t \nabla_x F(t, x)| \leq C_0$, and $\psi_0, \psi_1 \in \mathrm{Lip}_1(\mathbb{R}^d)$ are two given penalization functions.

Then a minimizer γ for the above problem exists, is unique, and is actually Lipschitz continuous with $|\gamma'| \leq \frac{C_0}{c_0}$.

Proof Given $\varepsilon > 0$, we define $\ell_\varepsilon : \mathbb{R}_+ \to \mathbb{R}_+$ via $\ell_\varepsilon(s) := \sqrt{\varepsilon^2 + s^2} + \varepsilon h(s)$, where $h : \mathbb{R}_+ \to \mathbb{R}_+$ is a smooth, convex, and increasing function, with $\liminf_{s \to \infty} h''(s) > 0$. We then define $L_\varepsilon : \mathbb{R}^d \to \mathbb{R}$ via $L_\varepsilon(v) = \ell_\varepsilon(|v|)$, so that L_ε is smooth, uniformly convex, and radial.[2] Since the second derivative of $s \mapsto \sqrt{\varepsilon^2 + s^2}$ is strictly positive, it is enough to guarantee this lower bound far from 0.

We also choose some numbers $\alpha_\varepsilon < 1$ in such a way that $\lim_{\varepsilon \to 0} \alpha_\varepsilon = 1$ and $\lim_{\varepsilon \to 0} \frac{\varepsilon^2}{1 - \alpha_\varepsilon^2} = 0$ (for instance $\alpha_\varepsilon = \sqrt{1 - \varepsilon}$).

---

[2] Note that the easiest choice for h is $h(s) = \frac{1}{2}s^2$, but other choices are possible and reasonable, and the only role of h is to guarantee a lower bound on the Hessian of L_ε (and in particular, to provide a quadratic behavior to ℓ_ε so that the problem is well-posed in H^1). Later on we will see the interest in other choices of h.

We consider the solution γ_ε of the variational problem

$$\min\left\{\int_0^1 (L_\varepsilon(\gamma'(t)) + F(t, \gamma(t)))\, dt + \alpha_\varepsilon(\psi_0(\gamma(0)) + \psi_1(\gamma(1))) : \gamma \in H^1([0, 1])\right\}.$$

The solution exists by applying the same methods as in Sect. 1.2, and is unique because of the convexity of the problem, the function F being strictly convex, by assumption, in the variable γ.

We want to apply Lemma 1.34 to this approximated optimization problem. We first compute

$$\ell'_\varepsilon(s) = \frac{s}{\sqrt{\varepsilon^2 + s^2}} + \varepsilon h'(s) \geq \frac{s}{\sqrt{\varepsilon^2 + s^2}}$$

and observe that we have $(\ell'_\varepsilon)^{-1}(r) \leq \frac{r\varepsilon}{\sqrt{1 - r^2}}$. Since $\mathrm{Lip}(\alpha_\varepsilon \psi_i) = \alpha_\varepsilon \mathrm{Lip}(\psi_i) \leq \alpha_\varepsilon$ we obtain from Lemma 1.34

$$|\gamma'_\varepsilon| \leq \max\left\{\frac{C_0}{c_0}, \frac{\alpha_\varepsilon \varepsilon}{\sqrt{1 - \alpha_\varepsilon^2}}\right\},$$

and we observe that our choice of α_ε implies that the second term in the max above tends to 0 as $\varepsilon \to 0$. This means that the Lipschitz constant of γ_ε is at most $\frac{C_0}{c_0}$ if ε is small enough.

By comparing γ_ε with the constant curve $\gamma = 0$ we obtain

$$\int_0^1 F(t, \gamma_\varepsilon(t))\, dt + \alpha_\varepsilon(\psi_0(\gamma_\varepsilon(0)) + \psi_1(\gamma_\varepsilon(1)))$$

$$\leq \int_0^1 F(t, 0)\, dt + \alpha_\varepsilon(\psi_0(0) + \psi_1(0)) \leq C.$$

Let us define $M_\varepsilon := \sup_t |\gamma_\varepsilon(t)|$ and $m_\varepsilon := \inf_t |\gamma_\varepsilon(t)|$. Using the growth condition on F and the Lipschitz behavior of ψ_0, ψ_1 we have

$$\int_0^1 F(t, \gamma_\varepsilon(t))\, dt + \alpha_\varepsilon(\psi_0(\gamma_\varepsilon(0)) + \psi_1(\gamma_\varepsilon(1))) \geq c_0 \frac{m_\varepsilon^2}{2} - CM_\varepsilon$$

for some large constant C. Moreover, the uniform Lipschitz condition on γ_ε shows $M_\varepsilon - m_\varepsilon \leq C$, so that we obtain an upper bound on m_ε and hence on M_ε, independent of ε. This shows that the sequence $(\gamma_\varepsilon)_\varepsilon$ is an equibounded and equicontinuous sequence of curves. We can then apply the Ascoli–Arzelà theorem to obtain a limit curve $\gamma_\varepsilon \to \gamma_0$. This curve γ_0 is of course $\frac{C_0}{c_0}$-Lipschitz continuous, and we want to prove that it solves Problem (1.18). This is a typical question in optimization, and a general theory, called Γ-convergence, is developed in Chap. 7.

The optimality of γ_ε, together with the inequality $L_\varepsilon(v) \geq |v|$, shows that we have

$$TV(\gamma_\varepsilon; [0, 1]) + \int_0^1 F(t, \gamma_\varepsilon(t))\, dt + \alpha_\varepsilon(\psi_0(\gamma_\varepsilon(0)) + \psi_1(\gamma_\varepsilon(1)))$$

$$\leq \int_0^1 L_\varepsilon(\gamma')\, dt + \int_0^1 F(t, \gamma(t))\, dt + \alpha_\varepsilon(\psi_0(\gamma(0)) + \psi_1(\gamma(1)))$$

for every $\gamma \in H^1$. If we send $\varepsilon \to 0$ and use the lower semicontinuity of TV for the uniform convergence we obtain

$$TV(\gamma_0; [0, 1]) + \int_0^1 F(t, \gamma_0(t))\, dt + \alpha_\varepsilon(\psi_0(\gamma_0(0)) + \psi_1(\gamma_0(1)))$$

$$\leq \int_0^1 |\gamma'|\, dt + \int_0^1 F(t, \gamma(t))\, dt + \psi_0(\gamma(0)) + \psi_1(\gamma(1)),$$

where we used the dominated convergence $L_\varepsilon(\gamma') \to |\gamma'|$ as $\varepsilon \to 0$, and $\alpha_\varepsilon \to 1$.

This shows the optimality of γ_0 compared to any H^1 curve. It is now enough to approximate any BV curve with H^1 curves. We take $\gamma \in BV([0, 1]; \mathbb{R}^d)$, we define it as equal to $\gamma(0^+)$ on $[-1, 0]$ and to $\gamma(1^-)$ on $[1, 2]$ and we convolve it with a smooth compactly supported kernel η_δ tending to the identity so as to smooth it, thus obtaining a sequence of curves $\gamma * \eta_\delta$ such that $TV(\gamma * \eta_\delta; [-1, 2]) = \int_{-1}^2 |(\gamma * \eta_\delta)'(t)|\, dt \leq TV(\gamma, (0, 1))$; moreover, $\gamma * \eta_\delta$ is uniformly bounded and converges to γ at all continuity points of γ, which means that the convergence holds a.e. and at the boundary point. This proves

$$\limsup_{\delta \to 0} \int_0^1 |(\gamma * \eta_\delta)'|\, dt + \int_0^1 F(t, \gamma * \eta_\delta(t))\, dt + \psi_0(\gamma * \eta_\delta(0)) + \psi_1(\gamma * \eta_\delta(1))$$

$$\leq TV(\gamma, (0, 1)) + \int_0^1 F(t, \gamma(t))\, dt + \psi_0(\gamma(0)) + \psi_1(\gamma(1))$$

and concludes the proof of the optimality of γ_0.

This proof contains a proof of the existence of an optimal curve for (1.18), and uniqueness comes from strict convexity (which is guaranteed by the term F, not by the penalization of the time-derivative).   □

In the case where the target set is $\mathbb{R}$, we observe some very interesting behavior.

Proposition 1.36 *When $d = 1$, i.e. the target space of the curves in Problem (1.18) is one-dimensional, the minimizer γ satisfies $|\gamma'(t)||\nabla_x F(t, \gamma(t))| = 0$ a.e., i.e. at* *each instant of time either γ does not move or it is already located at the optimal point for F(t, ·).*

Proof We consider the same approximation as in Proposition 1.35, using the function $h(s) = (s - M)_+^2$ for a very large M. Since the uniform Lipschitz bound proven in Lemma 1.34 and Proposition 1.35 makes the behavior of ℓ_ε for large values of s irrelevant, we can write the Euler–Lagrange equation for the minimizer γ_ε in the form

$$\varepsilon^2(\varepsilon^2 + |\gamma_\varepsilon'|^2)^{-3/2}\gamma_\varepsilon'' = \left(L_\varepsilon'(\gamma_\varepsilon')\right)' = \nabla_x F(t, \gamma_\varepsilon),$$

where we explicitly computed[3] the second derivative of L_ε ignoring the term in h.

We write this as $\varepsilon^2\gamma_\varepsilon'' = (\varepsilon^2 + |\gamma_\varepsilon'|^2)^{3/2}\nabla_x F(t, \gamma_\varepsilon)$ and we differentiate it in time, thus obtaining

$$\varepsilon^2\gamma_\varepsilon''' = 3(\varepsilon^2 + |\gamma_\varepsilon'|^2)^{1/2}\gamma_\varepsilon' \cdot \gamma_\varepsilon''\nabla_x F(t, \gamma_\varepsilon) + (\varepsilon^2 + |\gamma_\varepsilon'|^2)^{3/2}\left(\nabla_x F(t, \gamma_\varepsilon)\right)'.$$

Re-using the Euler–Lagrange equation we have

$$\varepsilon^2\gamma_\varepsilon''' = 3\varepsilon^2(\varepsilon^2 + |\gamma_\varepsilon'|^2)^{-1}(\gamma_\varepsilon' \cdot \gamma_\varepsilon'')\gamma_\varepsilon'' + (\varepsilon^2 + |\gamma_\varepsilon'|^2)^{3/2}\left(\nabla_x F(t, \gamma_\varepsilon)\right)'.$$

We observe that the last term $(\varepsilon^2 + |\gamma_\varepsilon'|^2)^{3/2}\left(\nabla_x F(t, \gamma_\varepsilon)\right)'$ is bounded thanks to the Lipschitz bound on γ_ε and the regularity of F. We multiply by γ_ε' and obtain

$$\varepsilon^2\gamma_\varepsilon''' \cdot \gamma_\varepsilon' = 3\varepsilon^2(\varepsilon^2 + |\gamma_\varepsilon'|^2)^{-1}(\gamma_\varepsilon' \cdot \gamma_\varepsilon'')^2 + O(1),$$

where $O(1)$ stands for a quantity bounded independently of ε. We then compute

$$\varepsilon^2\int_0^1 |\gamma_\varepsilon''|^2\,dt = -\varepsilon^2\int_0^1 \gamma_\varepsilon''' \cdot \gamma_\varepsilon'\,dt + \left[\varepsilon^2\gamma_\varepsilon' \cdot \gamma_\varepsilon''\right]_0^1 \le C.$$

The last inequality is justified by the sign of the first term in the product $\varepsilon^2\gamma_\varepsilon''' \cdot \gamma_\varepsilon'$, the bound on the other term, and the fact that the boundary term is the product of two bounded quantities: γ_ε' and $\varepsilon^2\gamma_\varepsilon'' = (\varepsilon^2 + |\gamma_\varepsilon'|^2)^{3/2}\nabla_x F(t, \gamma_\varepsilon)$.

---

[3] Note that this computation is based on a 1D cancellation effect, since in higher dimensions we have instead

$$D^2 L_\varepsilon(v) = \frac{(\varepsilon^2 + |v|^2)I - v \otimes v}{(\varepsilon^2 + |v|^2)^{3/2}}$$

and the matrices $|v|^2 I$ and $v \otimes v$ do not cancel out.

Coming back to the equality $\varepsilon^2\gamma_\varepsilon'' = (\varepsilon^2 + |\gamma_\varepsilon'|^2)^{3/2}\nabla_x F(t, \gamma_\varepsilon)$ we take the L^2 norms of both sides, thus obtaining

$$\int_0^1 |\gamma_\varepsilon'|^6 |\nabla_x F(t, \gamma_\varepsilon)|^2 dt \le \int_0^1 (\varepsilon^2 + |\gamma_\varepsilon'|^2)^3 |\nabla_x F(t, \gamma_\varepsilon)|^2 dt = \int_0^1 \varepsilon^4 |\gamma_\varepsilon''|^2 dt \le C\varepsilon^2.$$

We deduce $\int_0^1 |\gamma_\varepsilon'|^6 |\nabla_x F(t, \gamma_\varepsilon)|^2 dt \to 0$ and, using the uniform convergence of γ_ε to the limit optimizer γ as well as the boundedness of $|\gamma_\varepsilon'|$ we can also write $\int_0^1 |\gamma_\varepsilon'|^6 |\nabla_x F(t, \gamma)|^2 dt \to 0$. The functions γ_ε' weakly converge in L^6 to γ' (actually, the convergence is weak-* in L^∞) and multiplying by the fixed function $|\nabla_x F(t, \gamma)|^{1/3} \in L^\infty$ we preserve the L^6 weak convergence, which implies, thanks to the semicontinuity of a norm w.r.t. the weak convergence in the same space

$$||\gamma'|\nabla_x F(t, \gamma)|^{1/3}||_{L^6} \le \liminf_{\varepsilon \to 0} ||\gamma_\varepsilon'|\nabla_x F(t, \gamma)|^{1/3}||_{L^6} = 0,$$

and proves the claim.   □

1.7 Discussion: Optimal Control and the Value Function

A mathematical theory which shares many common points with that of the 1D calculus of variations is that of optimal control problems. We will see that, depending on the definitions that we choose, we can either say that the 1D calculus of variations is a particular case of optimal control problems or that optimal control problems are a particular case of the 1D calculus of variations...

The problem reads as follows. Given a space X (for instance a metric space) and a time horizon $T > 0$, consider a measurable function $\alpha : [0, T] \to X$, a function $f \in C^0([0, T] \times \mathbb{R}^N \times X; \mathbb{R}^N)$ and a point $x_0 \in \mathbb{R}^N$. We then look at the Cauchy problem

$$y'(t) = f(t, y(t), \alpha(t)) \text{ in } [0, T] \text{ with } y(0) = x_0. \tag{1.19}$$

Equation (1.19) is a standard first-order ODE with an extra variable α, which is fixed so far. If f satisfies suitable conditions (for instance it is Lipschitz in its second variable), then there exists a unique solution, and this solution depends on x_0 and on α. This solution will be denoted by $y_{\alpha,0,x_0}$ and, more generally, y_{α,t_0,x_0} will stand for the solution of $y'(t) = f(t, y(t), \alpha(t))$ in $[t_0, T]$ with $y(t_0) = x_0$.

Given a running cost $= L(t, y, \alpha)$ and a final cost Ψ we can consider the problem of choosing the function α so as to solve

$$\min\left\{ J(\alpha) := \int_0^T L(t, y_{\alpha,0,x_0}(t), \alpha(t)) dt + \Psi(y_{\alpha,0,x_0}(T)) : \alpha : [0, T] \to X \right\}. \tag{1.20}$$

We can now see what we meant when we said that the standard 1D calculus of variations problems could be considered as particular cases of optimal control and conversely optimal control could be considered as particular cases of 1D calculus of variations problems. It is clear that, if one takes $X = \mathbb{R}^d$ and $f(t, x, \alpha) = \alpha$, then the problem (1.20) becomes precisely $\min \int_0^T L(t, y; y') \, dt + \psi(y(T))$, i.e. a standard Bolza problem, one of the most classical and general 1D calculus of variations problems. On the other hand, we can also consider that the calculus of variations includes all optimization problems where the unknown are functions, and in optimal control the unknown is the function α, even if the functional involves the solution of a differential equation where α is an ingredient (as occurs in many shape optimization problems, see for instance Chap. 6); with this in mind, we could instead say that optimal control is a particular case of the calculus of variations, and more precisely of the 1D calculus of variations, since α is a function of one variable.

This is just a matter of taste and does not change what we can say about optimal control problems and their interest. We do not want to discuss optimality conditions or existence proofs here, but concentrate on a crucial tool, which has mainly been developed in the framework of optimal control but is also useful for the calculus of variations, which is the *value function*.

The value function φ for Problem (1.20) is defined as a function of time and space

$$\varphi(t_0, x_0)$$

$$:= \min \left\{ \int_{t_0}^T L(t, y_{\alpha,t_0,x_0}(t), \alpha(t)) \, dt + \Psi(y_{\alpha,t_0,x_0}(T)) \; : \; \alpha : [t_0, T] \to X \right\}.$$
(1.21)

In practice, the value of φ at (t_0, x_0) stands for the best possible cost that one can achieve in the remaining time after t_0 if currently located at x_0.

Optimal control theory provides two important pieces of information about φ. The first is that it solves, in a suitable sense, a PDE called the Hamilton–Jacobi equation. The second is that the knowledge of $\nabla \varphi$ allows us to find the optimal α in the optimal control problems as a function of time and current position.

More precisely, let us define the Hamiltonian H through

$$H(t, x, p) := \sup \{ p \cdot f(t, x, \alpha) - L(t, x, \alpha) \; : \; \alpha \in X \}.$$

The function H is convex in p and coincides with the Legendre transform of L (see Sect. 4.2.1) when $f(t, x, \alpha) = \alpha$.

The equation solved by φ is then given by

$$(HJ) \qquad -\partial_t \varphi(t, x) + H(t, x, -\nabla \varphi(t, x)), \qquad \varphi(T, x) = \Psi(x).$$

This is a first-order equation, which is coupled with the final condition $\varphi(T, x) = \Psi(x)$. If the validity of this final condition can be easily seen (if $t_0 = T$ there is no integral term in the definition of φ and we obviously have $y_{\alpha,t_0,x_0}(T) = x_0$ since the initial and final time coincide), it is more delicate to see why the PDE should be satisfied. We will give a heuristic computation, based on the assumption that all involved functions are smooth.

Indeed, let us take (t_0, x_0) and a small parameter $\varepsilon > 0$. Fix some $\alpha_0 \in X$ and define x_ε as the value of y_{α,t_0,x_0} at time $t_0 + \varepsilon$ when the control α is chosen constant and equal to α_0. We have $x_\varepsilon = x_0 + \varepsilon f(t_0, x_0, \alpha_0) + o(\varepsilon)$. We can estimate $\varphi(t_0, x_0)$ in terms of $\varphi(t_0 + \varepsilon, x_\varepsilon)$, writing

$$\varphi(t_0, x_0) \leq \int_{t_0}^{t_0+\varepsilon} L(t, y_{\alpha,t_0,x_0}(t), \alpha_0)\, \mathrm{d}t + \varphi(t_0 + \varepsilon, x_\varepsilon)$$
$$= \varepsilon L(t_0, x_0, \alpha_0) + o(\varepsilon) + \varphi(t_0 + \varepsilon, x_\varepsilon),$$

and we can take the first-order Taylor expansion of φ, thus obtaining

$$\varphi(t_0, x_0) \leq \varepsilon L(t_0, x_0, \alpha_0) + \varphi(t_0, x_0) + \partial_t \varphi(t_0, x_0)\varepsilon$$
$$+ \nabla\varphi(t_0, x_0) \cdot \varepsilon f(t_0, x_0, \alpha_0) + o(\varepsilon).$$

If we now subtract $\varphi(t_0, x_0)$, divide by ε and send ε to 0 we obtain the inequality

$$- \partial_t \varphi(t_0, x_0) + (-\nabla\varphi(t_0, x_0)) \cdot f(t_0, x_0, \alpha_0) - L(t_0, x_0, \alpha_0) \leq 0$$

and, taking the sup over α_0,

$$- \partial_t \varphi(t_0, x_0) + H(t_0, x_0, -\nabla\varphi(t_0, x_0)) \leq 0.$$

We are now left to prove the opposite inequality. For this, we take the optimal control $\alpha : [t_0, T] \to X$ (supposing it exists) starting from (t_0, x_0) and the corresponding optimal trajectory y_{α,t_0,x_0}, and we write the equality

$$\varphi(t_0, x_0) = \int_{t_0}^{t_1} L(t, y_{\alpha,t_0,x_0}(t), \alpha)\, \mathrm{d}t + \varphi(t_1, y_{\alpha,t_0,x_0}(t_1)).$$

If we differentiate w.r.t. t_1 we find

$$0 = L(t_1, y_{\alpha,t_0,x_0}(t_1), \alpha(t_1)) + \partial_t \varphi(t_1, y_{\alpha,t_0,x_0}(t_1))$$
$$+ \nabla\varphi(t_1, y_{\alpha,t_0,x_0}(t_1)) \cdot f(t_1, y_{\alpha,t_0,x_0}(t_1), \alpha(t_1)),$$

where we used the equation $y'(t) = f(t, y(t), \alpha(t))$. We then compute everything at $t_1 = t_0$ and obtain, setting $\bar{\alpha} := \alpha(t_0)$,

$$0 = L(t_0, x_0, \bar{\alpha}) + \partial_t \varphi(t_0, x_0) + \nabla \varphi(t_0, x_0) \cdot f(t_0, x_0, \bar{\alpha})$$

$$\geq \partial_t \varphi(t_0, x_0) - H(t_0, x_0, -\nabla \varphi(t_0, x_0)).$$

This shows at the same time the equality

$$- \partial_t \varphi(t_0, x_0) + H(t_0, x_0, -\nabla \varphi(t_0, x_0)) = 0,$$

i.e. the Hamilton–Jacobi equation, but also that $\bar{\alpha}$ must be optimal in the definition of $H(t_0, x_0, -\nabla \varphi(t_0, x_0))$. Assume that (because of strict convexity or for other reasons) for every (t, x, p) there is a unique $\bar{\alpha}(t, x, p) = \mathrm{argmax}_{\alpha \in X} \, p \cdot f(t, x, \alpha) - L(t, x, \alpha)$. Define then $\bar{v}(t, x, p) := f(t, x, \bar{\alpha}(t, x, p))$. Then we obtain that optimal trajectories y should be solutions of

$$y'(t) = \bar{v}(t, y(t), -\nabla \varphi(t, y(t))), \tag{1.22}$$

which is a first-order equation which allows us to find the optimal trajectory as a solution of a Cauchy problem if we add the condition $y(t_0) = x_0$. Note that the envelope theorem clarifies that we have $\bar{v}(t, x, p) = \nabla_p H(t, x, p)$.

Box 1.11 Good to Know!—*Envelope Theorem*

Proposition *Let $(f_\alpha)_\alpha$ be a family (finite, infinite, countable, uncountable...) of functions of a variable $x \in \mathbb{R}^d$ and assume that each f_α is differentiable. Consider f defined through $f(x) := \sup_\alpha f_\alpha(x)$. Then, if f is differentiable at x_0 and α_0 is such that $f(x_0) = f_{\alpha_0}(x_0)$ we necessarily have $\nabla f(x_0) = \nabla_x f_{\alpha_0}(x_0)$.*

This simple statement is a crucial tool in mathematical economics, and is known as the *envelope theorem* in that community. We refer for instance to [189]. It can be easily proven using the fact that we have $f - f_{\alpha_0} \geq 0$ everywhere, with equality at $x = x_0$, so that x_0 is a minimum point of $f - f_{\alpha_0}$. Roughly speaking, this says that differentiating a function defined as a sup only requires us to differentiate the functions f_α and choose the derivative of the one corresponding to the optimal parameter $\alpha = \alpha_0$. The above statement is rigorous if we already assume the differentiability of f. Under suitable continuity assumptions on f_α and $\nabla_x f_\alpha$, in case of uniqueness of the optimal α_0, it is also possible to prove the differentiability of f. Uniqueness is not strictly necessary for the differentiability of f, but it is necessary that all maximizers α provide the same value for $\nabla_x f_\alpha(x_0)$.

(continued)

Box 1.11 (continued)

An example of non-differentiability is given by $f_\alpha(x) = \alpha x$ with $\alpha \in [-1, 1]$. We get $f(x) = |x|$, which is non-differentiable at $x = 0$, where any α is optimal.

For more precise mathematical details we refer to [62, Section 5.3].

Comparing this to the Euler–Lagrange equation for 1D variational problems, the interest here is to obtain a Cauchy problem, instead of having a second-order equation with initial and final transversality conditions. On the other hand, one should first find φ and hope that it is differentiable.

The problem of the differentiability (or regularity) of φ also arises when proving the validity of the Hamilton–Jacobi equation: we proved it supposing $\varphi \in C^1$, but the definition (1.21) need not provide a smooth function. In general, the function φ defined by (1.21) is not a classical solution of (HJ) but is only a solution in the viscosity sense, and it is possible to prove that the equation (HJ) together with its final condition does indeed admit a unique viscosity solution.

Box 1.12 Important Notion—*Viscosity Solution for First-Order PDEs*

Given a function $F : \Omega \times \mathbb{R} \times \mathbb{R}^d \to \mathbb{R}$ and $\Omega \subset \mathbb{R}^d$, we say that a function $u \in C^0(\Omega)$ is a viscosity solution of $F(x, u, \nabla u) = 0$ if it satisfies the following two properties

- for every $x_0 \in \Omega$ and $\varphi \in C^1(\Omega)$ such that $\varphi \geq u$ but $\varphi(x_0) = u(x_0)$ we have $F(x_0, \varphi(x_0), \nabla\varphi(x_0)) \leq 0$;
- for every $x_0 \in \Omega$ and $\varphi \in C^1(\Omega)$ such that $\varphi \leq u$ but $\varphi(x_0) = u(x_0)$ we have $F(x_0, \varphi(x_0), \nabla\varphi(x_0)) \geq 0$.

An interesting example is given by the Eikonal equation $|\nabla u| - 1 = 0$. In 1D, many saw-shaped functions solve $|u'| = 1$ a.e. but only those which only have singularities shaped like $-|x|$ and not like $|x|$ satisfy the above conditions. For instance, with $\Omega = (-1, 1)$ and imposing $u = 0$ on $\partial\Omega$ the only viscosity solution of $|u'| - 1 = 0$ is $u(x) = 1 - |x|$. More generally, the only viscosity solution of $|\nabla u| - 1 = 0$ with $u = 0$ on $\partial\Omega$ is, even in higher dimensions and in more general domains, $u(x) = d(x, \partial\Omega)$. Note that the solutions would change if the same equation was written $-|\nabla u| + 1 = 0$, changing the sign. The Eikonal equation $|\nabla u| - K = 0$ characterizes, instead, the distance function induced by the weighted geodesic problem of Sect. 1.4.4.

As a reference for first-order viscosity theory, see for instance [83, Chapter 10], or [21, 59] for connections with optimal control.

The use of the value function also allows a very informal approach to another classical issue in optimal control, that of optimality conditions in the form of the *Pontryagin Maximum Principle*. This principle can be roughly stated as follows: if α is a solution to Problem (1.20) and y is the associated trajectory, calling H the Hamiltonian which we already introduced, then there exists a curve $p : [0, T] \to \mathbb{R}^d$ such that

- for every t the value $\alpha(t)$ is a maximizer in the definition of $H(t, y(t), p(t))$;
- y and p together solve the Hamiltonian system

$$\begin{cases} y'(t) = \nabla_p H(t, y(t), p(t)) & \text{in } (0, T), \\ p'(t) = -\nabla_x H(t, y(t), p(t)) & \text{in } (0, T), \\ y(0) = x_0, \ p(T) = -\nabla \psi(y(T)). \end{cases}$$

Indeed, using the considerations that we presented above on the value function φ, it is enough to choose $p(t) = -\nabla \varphi(t, y(t))$. We already explained that the optimal α solves the maximization problem defining $H(t, y(t), -\nabla \varphi(t, y(t)))$ and the combination of (1.22) with $\bar{v}(t, x, p) = \nabla_p H(t, x, p)$ provides the first equation in the Hamiltonian system. Since the initial and final conditions for y and p are also satisfied with this choice of p (using the final condition for φ in the Hamilton–Jacobi system), we are only left to prove the validity of the second one. For this we compute

$$\frac{\mathrm{d}}{\mathrm{d}t} (-\nabla \varphi(t, y(t))) = -\partial_t \nabla \varphi(t, y(t)) - D^2 \varphi(t, y(t)) \cdot y'(t).$$

We then differentiate the Hamilton–Jacobi equation in space and obtain

$$- \partial_t \nabla \varphi(t, x) + \nabla_x H(t, x, -\nabla \varphi(t, x)) - \nabla_p H(t, x, -\nabla \varphi(t, x)) \cdot D^2 \varphi(t, x) = 0.$$

This allows us to obtain (the reader can easily check that all vector-matrix products provide the desired vector in these equations, using the symmetry of the Hessian $D^2 \varphi$)

$$\frac{\mathrm{d}}{\mathrm{d}t} (-\nabla \varphi(t, y(t))) = -\nabla_x H(t, x, -\nabla \varphi(t, x)).$$

Another very formal justification of the Pontryagin Maximum Principle can be obtained using methods inspired by convex duality, writing the problem as an optimization in (y, α) under the constraint $y' = f(t, y, \alpha)$ and transforming this constraint into a sup over test functions p, as proposed in Exercise 4.21.

1.8 Exercises

Exercise 1.1 Using the change of variable $u = \sqrt{H - z}$ prove that the brachistochrone problem can be reformulated as a convex optimization problem, and deduce the uniqueness of its solution.

Exercise 1.2 Prove that if z is a solution to the brachistochrone problem, then we have $(H - z(t))(1 + |z'(t)|^2) = const$. Verify that any portion of the cycloid curve, written in parametric form as $z(\theta) = H - k(1 - \cos\theta)$ and $t(\theta) = k(\theta - \sin\theta)$, satisfies this condition.

Exercise 1.3 Let $M \subset \mathbb{R}^d$ be a set defined via $M = \{x : h(x) = 0\}$ for a smooth function h such that $\nabla h \neq 0$ on M. Let $\gamma : [0, T] \rightarrow M$ be a smooth curve satisfying $\gamma''(t) = a(t)\nabla h(\gamma(t))$ for a scalar coefficient a. Prove that we necessarily have

$$a(t) = -\frac{D^2 h(\gamma(t))(\gamma'(t), \gamma'(t))}{|\nabla h(\gamma(t))|^2}$$

and that if T is small enough this curve is a geodesic on M.

Exercise 1.4 Solve the problem

$$\min\left\{ J(f) := \int_0^1 \left[\frac{1}{2}f'(t)^2 + tf(t) + \frac{1}{2}f(t)^2\right] dt \ : \ f \in \mathcal{A}\right\},$$

where $\mathcal{A} := \{f \in C^1([0, 1]) \ : \ f(0) = 0\}$. Find the minimal value of J on $\mathcal{A}$ and the function(s) f which attain(s) it, proving that they are actually minimizers.

Exercise 1.5 Consider the problem

$$\min\left\{ \int_0^T e^{-t}\left(u'(t)^2 + 5u(t)^2\right) dt \ : \ u \in C^1([0, T]), \ u(0) = 1\right\}.$$

Prove that it admits a minimizer and that it is unique. Find it, compute the value of the minimum, and the limit of the minimizer (in which sense?) and of the minimal value as $T \rightarrow +\infty$.

Exercise 1.6 Consider the problem

$$\min\left\{ J(u) := \int_0^1 \left[\frac{1}{2}u'(t)^2 + u(t)f(t)\right] dt \ : \ u \in W^{1,2}([0, 1])\right\}.$$

Find a necessary and sufficient condition on f so that this problem admits a solution.

Exercise 1.7 Let $L : \mathbb{R} \to \mathbb{R}$ be a strictly convex C^1 function, and consider

$$\min \left\{ \int_0^1 L(u'(t)) \, dt \; : \; u \in C^1([0, 1]), \, u(0) = a, \, u(1) = b \right\}.$$

Prove that the solution is $u(t) = (1 - t)a + tb$, whatever L is. What happens if L is not C^1, and if L is convex but not strictly convex?

Exercise 1.8 Prove that we have

$$\inf \left\{ \int_0^1 t|u'(t)|^2 \, dt \; : \; u \in C^1([0, 1]), \, u(0) = 1, \, u(1) = 0 \right\} = 0.$$

Is the infimum above attained? What about, instead

$$\inf \left\{ \int_0^1 \sqrt{t}|u'(t)|^2 \, dt \; : \; u \in C^1([0, 1]), \, u(0) = 1, \, u(1) = 0 \right\} \; ?$$

Exercise 1.9 Consider a minimization problem of the form

$$\min \left\{ F(u) := \int_0^1 L(t, u(t), u'(t)) \, dt \; : \; u \in W^{1,1}([0, 1]), \, u(0) = a, \, u(1) = b \right\},$$

where $L \in C^2([0, 1] \times \mathbb{R} \times \mathbb{R})$. We denote as usual by (t, x, v) the variables of L. Assume that $\bar{u}$ is a solution to the above problem. Prove that we have

$$\frac{\partial^2 L}{\partial v^2}(t, \bar{u}(t), \bar{u}'(t)) \geq 0 \text{ a.e.}$$

Exercise 1.10 Given $f \in C^2(\mathbb{R})$, consider the problem

$$\min \left\{ F(u) := \int_0^1 \left[(u'(t)^2 - 1)^2 + f(u(t)) \right] dt \; : \; \begin{array}{l} u \in C^1([0, 1]), \\ u(0) = a, \\ u(1) = b \end{array} \right\}.$$

Prove that the problem does not admit any solution if $|b - a| \leq \frac{1}{\sqrt{3}}$.

Exercise 1.11 Consider the problem

$$\min \left\{ \int_0^1 \left[|u'(t)|^2 + \arctan(u(t)) \right] dt \; : \; u \in C^1([0, 1]) \right\},$$

and prove that it has no solutions. Prove the existence of a solution if we add the boundary condition $u(0) = 0$. Write the optimality conditions and prove that any solution is C^∞

Exercise 1.12 Consider the functional $F : H^1([0, T]) \to \mathbb{R}$ defined by

$$F(u) = \int_0^T \left(u'(t)^2 + \arctan(u(t) - t) \right) dt.$$

Prove that

1. the problem $(P) := \min\{F(u) : u \in H^1([0, T])\}$ has no solution;
2. the problem $(P_a) := \min\{F(u) : u \in H^1([0, T]), u(0) = a\}$ admits a solution for every $a \in \mathbb{R}$;
3. we have $F(-|u|) \le F(u)$;
4. the solution of (P_a) is unique as soon as $a \le 0$;
5. there exists an $L_0 < +\infty$ such that for every $T \le L_0$ the solution of (P_a) is unique for every $a \in \mathbb{R}$
6. the minimizers of (P) and (P_a) are C^∞ functions.

Exercise 1.13 Given a function $V : X \to \mathbb{R}$ on a metric space (X, d) and the functional J defined via $J(\gamma) := \int_0^1 (\frac{1}{2}|\gamma'|^2 + V(\gamma)) dt$ prove the conservation of the energy $\frac{1}{2}|\gamma'|^2 - V(\gamma)$ for minimizers of J (among curves with fixed endpoints) without using the Euler–Lagrange equation (which has no meaning in a metric setting) but considering variations of the form $\gamma_\varepsilon(t) = \gamma(t + \varepsilon\eta(t))$ for $\eta \in C_c^\infty((0, 1))$.

Exercise 1.14 Given $p \in (1, \infty)$, set $J_p(\gamma) := \int_0^1 (\frac{1}{p}|\gamma'|^p + V(\gamma)) dt$, assume that γ minimizes J_p among curves with fixed endpoints. Prove the following p-variant of the conservation of the energy:

$$\frac{p-1}{p}|\gamma'|^p - V(\gamma) = const.$$

Also prove the same result by internal variations as in Exercise 1.13.

Exercise 1.15 Consider the problem

$$\min\left\{ \int_0^{2\pi} \frac{|u'(t)|^2}{2 + \sin t} dt \ : \ u \in C^1([0, 2\pi]), \ u(0) = 0, u(2\pi) = 1 \right\}.$$

Find the minimizer as well as its minimal value.

More generally, given a function $h \in C^0([a, b])$, $h > 0$, characterize the solution and find the minimal value of

$$\min\left\{ \int_a^b \frac{|u'(t)|^2}{h(t)} dt \ : \ u \in C^1([a, b]), \ u(a) = A, u(b) = B \right\}$$

in terms of h, A and B.

Exercise 1.16 Find an analogous Lipschitz regularity result to that of Proposition 1.35 for the minimizers of the infinite-horizon discounted variational problem

$$\min \left\{ \int_0^\infty e^{-\delta t}(|\gamma'|(t) + F(t, \gamma(t)))\,dt + \psi_0(\gamma_0) \; : \; \gamma \in BV_{loc}(\mathbb{R}_+; \mathbb{R}^d) \right\},$$

where the term $\int_0^\infty e^{-\delta t}|\gamma'|(t)\,dt$ can be defined, for $\gamma \in BV$ as a non-autonomous weighted-length equal to $\sup \sum_i e^{-\delta t_{i+1}}|\gamma(t_i) - \gamma(t_{i+1})|$, the sup being taken over all finite partitions of $\mathbb{R}_+$.

Exercise 1.17 Let $|| \cdot ||_p$ be the ℓ_p norm on $\mathbb{R}^d$, i.e. $||v||_p := (\sum_i |v_i|^p)^{1/p}$ and $TV_p(\gamma)$ be the total variation defined w.r.t. the ℓ_p norm. Consider the minimization problem of Proposition 1.35 but replace TV_2 with TV_1. Prove that we have an estimate of the form

$$c_1 \sup ||\gamma'||_2 \leq \sup ||\partial_t \nabla_x F||_2.$$

Exercise 1.18 Given a curve $\omega \in C^0(\mathbb{R}; \mathbb{R}^d)$ which is 2π-periodic, consider the minimization problem

$$\min \left\{ \int_0^{2\pi} \left(\sqrt{\varepsilon^2 + |u'(t)|^2} + \frac{\varepsilon}{2}|u'(t)|^2 + \frac{1}{2}|u(t) - \omega(t)|^2 \right) dt \; : \; u \in H^1_{per}(\mathbb{R}; \mathbb{R}^d) \right\},$$

where $H^1_{per}(\mathbb{R})$ denotes the set of functions $u \in H^1_{loc}(\mathbb{R})$ which are 2π-periodic.

1. Prove that this minimization problem admits a unique solution u_ε.
2. Prove that the sequence of minimizers is bounded, when $\varepsilon \to 0$, in $BV(I)$ for any interval $I \subset \mathbb{R}$.
3. Prove that u_ε converges (in which sense?) to the unique solution $\bar{u}$ of

$$\min \left\{ TV(u; [0, 2\pi)) + \int_0^{2\pi} \frac{1}{2}|u(t) - \omega(t)|^2\,dt \; : \; u \in BV_{per}(\mathbb{R}; \mathbb{R}^d) \right\},$$

where $BV_{per}(\mathbb{R})$ denotes the set of functions $u \in BV_{loc}(\mathbb{R})$ which are 2π-periodic and $TV(u; [0, 2\pi))$ is their total variation on one period.
4. Find $\bar{u}$ in the case $d = 2$ and $\omega(t) = (R \cos t, R \sin t)$ for $R > 1$.

Exercise 1.19 Solve the minimization problem

$$\min \left\{ \frac{\int_{-A}^A (u')^2(x)\,dx}{\int_{-A}^A u^2(x)\,dx} \; : \; u \in H^1_0((-A, A)) \right\},$$

i.e. prove that the solutions exist and find them, and deduce from this another way of finding the Poincaré constant of the interval $(-A, A)$.

Exercise 1.20 Prove that we have $\sum_{n\in\mathbb{Z}\setminus\{0\}} \frac{1}{4n^2-1} = 1$ and deduce $\left|\sum_{n\in\mathbb{Z}\setminus\{0\}} a_n\right|^2$
$\leq \sum_{n\in\mathbb{Z}\setminus\{0\}} (4n^2-1)|a_n|^2$ for any sequence $a_n \in \mathbb{C}$.

Prove that any 2π-periodic function $u \in H^1_{loc}(\mathbb{R})$ satisfies, when written as a Fourier series, $\sum_{n\in\mathbb{Z}} |a_n| < +\infty$ and that we have $\sum_{n\in\mathbb{Z}} a_n = 0$ if and only if $u(0) = u(2\pi) = 0$. Deduce the Poincaré inequality for functions in $H^1_0([0, 2\pi])$ and the case of equality.

Hints

Hint to Exercise 1.1 One obtains an integrand of the form $\sqrt{|u'|^2 + u^{-2}}$.

Hint to Exercise 1.2 Use the Beltrami formula.

Hint to Exercise 1.3 Differentiate the relation $h(\gamma(t)) = 0$ twice for every t in order to identify a. Prove the sufficiency of this condition for minimality in $\min \int_0^T |\gamma'(t)|^2 dt$ by adapting the computations on the sphere (where $h(x) = |x|^2 - 1$).

Hint to Exercise 1.4 It is enough to write the Euler–Lagrange equation, solve it, and prove by an explicit expansion of the squares that the solution of the equation is a minimizer.

Hint to Exercise 1.5 The Euler–Lagrange equation is again linear with constant coefficients. Its explicit expression allows us to obtain the limit, which is locally uniform in time.

Hint to Exercise 1.6 If f has zero average we can restrict to zero-mean competitors and use the Poincaré–Wirtinger inequality, otherwise the infimum is $-\infty$.

Hint to Exercise 1.7 If $L \in C^1$ the Euler–Lagrange equation provides $L(u') = const$, i.e. $u' = const$. If not, we can use Jensen's inequality with its equality case.

Hint to Exercise 1.8 In the second case the Hölder inequality $(\int |u'|) \leq \int \sqrt{t}|u'|^2 \int \frac{1}{\sqrt{t}}$ provides the lower bound; in the other case, one can draw inspiration from the same inequality and try to get as close as possible to equality.

Hint to Exercise 1.9 Consider a second-order expansion in ε for $J(u+\varepsilon\phi)$ and then use a sequence of very oscillating functions ϕ_n.

Hint to Exercise 1.10 Apply the necessary condition found in the previous exercise,

Hint to Exercise 1.11 Follows the strategy in Sects. 1.2.1 and 1.3.1. For non-existence, use $u = -n$.

Hint to Exercise 1.12 Follows the standard strategy. Uniqueness is obtained once we restrict the arctan function to a region where it is convex.

Hint to Exercise 1.13 Consider a first-order expansion in ε and use the arbitrariness of η to obtain an ODE.

Hint to Exercise 1.14 Use either the Euler–Lagrange equation or the same procedure as the previous exercise.

Hint to Exercise 1.15 Find $u' = ch$ and adjust the value of c so as to satisfy the boundary data.

Hint to Exercise 1.16 Follow the same strategy as in Sect. 1.6. There is an extra term due to the coefficient e^{-rt} but it vanishes due to the first-order condition.

Hint to Exercise 1.17 Find another quantity to be maximized over t, of the form $\sum_i h(\gamma_i'(t))$.

Hint to Exercise 1.18 Compare to the problem studied in Sect. 1.6.

Hint to Exercise 1.19 Take a minimizing sequence which, up to multiplication, can be assumed to be such that the denominator is equal to 1, and use its compactness. Then write the optimality conditions by considering the first-order expansion of $u + \varepsilon\phi$.

Hint to Exercise 1.20 Use the Hölder inequality, with its equality case.

Chapter 2
Multi-Dimensional Variational Problems

This short chapter establishes the bases of the multi-dimensional case, starting from the techniques which were described in the previous chapter for the 1D case and highlighting the necessary modifications. After discussing the existence of the minimizers and their properties in terms of a multi-dimensional Euler–Lagrange equation, which typically takes the form of an elliptic PDE, a section is devoted to the study of harmonic functions, which are the solutions to the simplest example of a variational problem, that of minimizing the Dirichlet energy $u \mapsto \int |\nabla u|^2$.

As a choice, throughout the chapter, the multidimensional case will only be treated when the target space is, instead, one-dimensional. Many notions can be easily extended to vector-valued maps without particular difficulties.

2.1 Existence in Sobolev Spaces

In this section we will discuss the classes of problems for which we are easily able to prove the existence of a solution in the Sobolev space $W^{1,p}(\Omega)$. In the sequel Ω will be a bounded (and, when needed, smooth) open set in $\mathbb{R}^d$. We will consider optimization problems in $W^{1,p}(\Omega)$ for $p > 1$.

Box 2.1 Memo—*Sobolev Spaces in Higher Dimensions*
 Given an open domain $\Omega \subset \mathbb{R}^d$ and an exponent $p \in [1, +\infty]$ we define the Sobolev space $W^{1,p}(\Omega)$ as

$$W^{1,p}(\Omega) := \{u \in L^p(\Omega) \ : \ \forall i \exists g_i \in L^p(\Omega) \text{ s.t. } \int u \partial_{x_i} \varphi = -\int g_i \varphi \text{ for all } \varphi \in C_c^\infty(\Omega)\}.$$

(continued)

© The Author(s), under exclusive license to Springer Nature Switzerland AG 2023
F. Santambrogio, *A Course in the Calculus of Variations*, Universitext,
https://doi.org/10.1007/978-3-031-45036-5_2

Box 2.1 (continued)

The functions g_i, if they exist, are unique, and together they provide a vector which will be denoted by ∇u since it plays the role of the derivative in the integration by parts. The space $W^{1,p}$ is endowed with the norm $||u||_{W^{1,p}} :=$ $||u||_{L^p} + ||\nabla u||_{L^p}$. With this norm $W^{1,p}$ is a Banach space, separable if $p <$ $+\infty$, and reflexive if $p \in (1, \infty)$.

The functions in $W^{1,p}(\Omega)$ can also be characterized as those functions $u \in L^p(\Omega)$ such that their translations u_h (defined via $u_h(x) := u(x + h)$) satisfy $||u - h_h||_{L^p(\Omega')} \leq C|h|$ for every subdomain Ω' compactly contained in Ω and every h such that $|h| < d(\Omega'\partial\Omega)$. The optimal constant C in this inequality equals $||\nabla u||_{L^p}$.

Sobolev Injections If $p < d$ all functions $u \in W^{1,p}$ are indeed more summable than just L^p and they actually belong to L^{p^*}, where $p^* := \frac{pd}{d-p} >$ p. The injection of $W^{1,p}$ into L^q is compact (i.e. it sends bounded sets into precompact sets) for any $q < p^*$ (including $q = p$), while the injection into L^{p^*} is continuous. If $p = d$ then $W^{1,p}$ injects compactly in all spaces L^q with $q < +\infty$, but does not inject into L^∞. If $p > d$ then all functions in $W^{1,p}$ admit a continuous representative, which is Hölder continuous of exponent $\alpha = 1 - \frac{d}{p} > 0$, and the injection from $W^{1,p}$ into $C^0(\Omega)$ is compact if Ω is bounded.

The space $W_0^{1,p}(\Omega)$ can be defined as the closure in $W^{k,p}(\Omega)$ of $C_c^\infty(\Omega)$ and the Poincaré inequality $||\varphi||_{L^p} \leq C(\Omega)||\nabla\varphi||_{L^p(\Omega)}$, which is valid for all $\varphi \in C_c^\infty(\Omega)$ if Ω is bounded (this is not a sharp condition, for instance bounded in one direction—a notion to be made precise—would be enough), allows us to use as a norm on $W_0^{1,p}(\Omega)$ the quantity $||u||_{W_0^{1,p}} := ||\nabla u||_{L^p}$.

If $p = 2$ the space $W^{1,p}$ can be given a Hilbert structure exactly as in 1D and is denoted by H^1. Higher-order Sobolev spaces $W^{k,p}$ and H^k can also be defined exactly as in dimension 1 (and $W_0^{k,p}$ is again defined as the closure of $C_c^\infty(\Omega)$ for the $W^{k,p}$ norm).

We can also define negative-order Sobolev spaces: the space $W^{-k,p'}$ is defined as the dual of the space $W_0^{k,p}$.

Theorem 2.1 *Given a function $g \in W^{1,p}(\Omega)$ and a measurable function $F : \Omega \times \mathbb{R} \to \mathbb{R}_+$ such that $F(x, \cdot)$ is lower semicontinuous for a.e. x, let us consider the problem*

$$\min\left\{ J(u) := \int_\Omega \left(F(x, u(x)) + |\nabla u(x)|^p \right) \, dx : u \in W^{1,p}(\Omega), \ u - g \in W_0^{1,p}(\Omega) \right\}.$$
(2.1)

This minimization problem admits a solution.

Proof Take a minimizing sequence u_n such that $J(u_n) \to \inf J$. From the bound on $J(u_n)$ we obtain an upper bound for $\|\nabla u_n\|_{L^p}$. Applying the Poincaré inequality for $W_0^{1,p}$ functions to $u_n - g$ we obtain

$$
\begin{aligned}
\|u_n\|_{L^p} &\leq \|u_n - g\|_{L^p} + \|g\|_{L^p} \\
&\leq C\|\nabla(u_n - g)\|_{L^p} + \|g\|_{L^p} \\
&\leq C\|\nabla u_n\|_{L^p} + C\|\nabla g\|_{L^p} + \|g\|_{L^p},
\end{aligned}
$$

which allows us to bound $\|u_n\|_{L^p}$.

Hence, $(u_n)_n$ is a bounded sequence in $W^{1,p}$ and we can extract a subsequence which weakly converges in $W^{1,p}$ to a function u (see Boxes 2.2 and 2.3). This convergence implies the strong convergence $u_n \to u$ in $L^p(\Omega)$ and, up to a subsequence, a.e. convergence $u_n(x) \to u(x)$. Moreover, we also have $\nabla u_n \rightharpoonup \nabla u$ in $L^p(\Omega; \mathbb{R}^d)$ by applying the continuous linear mapping $W^{1,p}(\Omega) \ni v \mapsto \nabla v \in L^p$.

The space $W_0^{1,p}(\Omega)$ is a closed subspace (by definition, as it is defined as the closure of C_c^∞ in $W^{1,p}$) of $W^{1,p}(\Omega)$ and, since it is a vector space and hence a convex set, it is also weakly closed (see Proposition 3.7). Then, from $u_n \rightharpoonup u$ we deduce $u_n - g \rightharpoonup u - g$ and $u - g \in W_0^{1,p}(\Omega)$. Hence, the limit u is also admissible in the optimization problem.

We now need to show that J is l.s.c. for the weak convergence in $W^{1,p}(\Omega)$. The semicontinuity of the L^p norm for the weak convergence easily provides

$$
\int_\Omega |\nabla u|^p \, \mathrm{d}x \leq \liminf_n \int_\Omega |\nabla u_n|^p \, \mathrm{d}x.
$$

The a.e. pointwise convergence together with the lower semicontinuity of F and the use of Fatou's lemma provides

$$
\int_\Omega F(x, u(x)) \, \mathrm{d}x \leq \liminf_n \int_\Omega F(x, u_n(x)) \, \mathrm{d}x.
$$

This shows that u is a minimizer. □

Box 2.2 Memo—*Weak Convergence and Compactness*

On a normed vector space X we say that a sequence x_n weakly converges to x if for every $\xi \in X'$ we have $\langle \xi, x_n \rangle \to \langle \xi, x \rangle$, and we write $x_n \rightharpoonup x$. Of course, if $x_n \to x$ (in the sense of $\|x_n - x\| \to 0$), then we have $x_n \rightharpoonup x$. On the dual space X' we say that a sequence ξ_n weakly-* converges to ξ if for

(continued)

Box 2.2 (continued)

every $x \in \mathcal{X}$ we have $\langle \xi_n, x \rangle \to \langle \xi, x \rangle$, and we write $\xi_n \overset{*}{\rightharpoonup} \xi$. Note that this is a priori different than the notion of weak convergence on $\mathcal{X}'$, which involves the bidual $\mathcal{X}''$.

   Theorem (Banach–Alaoglu) – If $\mathcal{X}$ is a separable normed space and ξ_n is a bounded sequence in $\mathcal{X}'$, then there exists a subsequence ξ_{n_k} and a vector $\xi \in \mathcal{X}'$ such that $\xi_{n_k} \overset{*}{\rightharpoonup} \xi$.

Box 2.3 Memo—*Reflexive Banach Spaces*

   A normed vector space $\mathcal{X}$ can always be embedded in its bidual $\mathcal{X}''$ since we can associate with every $x \in \mathcal{X}$ the linear form ι_x on $\mathcal{X}'$ defined via $\langle \iota_x, \xi \rangle := \langle \xi, x \rangle$. In general not all elements of $\mathcal{X}''$ are of the form ι_x for $x \in \mathcal{X}$. When $\iota : \mathcal{X} \to \mathcal{X}''$ is surjective we say that $\mathcal{X}$ is reflexive. In this case it is isomorphic to its bidual which, as a dual space, is a Banach space.

   All L^p spaces for $p \in (1, \infty)$ are reflexive as their dual is $L^{p'}$ and their bidual is again L^p. In contrast, L^∞ is the dual of L^1 but L^1 is not the dual of L^∞. The space $C(X)$ of continuous functions on a compact metric space X is not reflexive either, since its dual is $\mathcal{M}(X)$, the space of measures, and the dual of $\mathcal{M}(X)$ is not $C(X)$. Of course all Hilbert spaces are reflexive, as they coincide with their own dual. Sobolev spaces $W^{k,p}$ for $p \in (1, \infty)$ are also reflexive.

Theorem (Eberlein–Smulian) *A Banach space X is reflexive if and only if any bounded sequence in X is weakly compact.*

   In optimization the most important part of this theorem is that it provides weak compactness of bounded sequences. If $\mathcal{X}$ is separable this is a consequence of the Banach–Alaoglu theorem, as this convergence coincides with the weak-* convergence on the dual of $\mathcal{X}'$. Yet, separability is not necessary since one can restrict to the closure of the vector space generated by the sequence: this space is separable, closed subspaces of reflexive spaces are reflexive, and a space is reflexive and separable if and only if its dual is reflexive and separable.

Many variants of the above result are possible. A first possibility is to replace the positivity assumption on F with a lower bound of the form $F(x, u) \geq -a(x) - C|u|^r$ for $a \in L^1(\Omega)$ and $r < p$. In this case, we cannot immediately deduce a bound on $\|\nabla u_n\|_{L^p}$ from the bound on $J(u_n)$. Yet, we can act as follows:

$$J(u_n) \leq C \Rightarrow \|\nabla u_n\|_{L^p}^p \leq C + C\|u_n\|_{L^r}^r \leq C + C\|u_n\|_{L^p}^r \leq C + C\|\nabla u_n\|_{L^p}^r$$

and the condition $r < p$ allows us to prove that we cannot have $||\nabla u_n||_{L^p} \to \infty$. This is enough to obtain the desired compactness but the lower semicontinuity of the functional J is still to be proven, since we cannot directly apply Fatou's lemma to F, which is no longer non-negative. The idea is then to define

$$\tilde{F}(x, u) := F(x, u) + C|u|^r$$

and write

$$J(u) = \tilde{J}(u) - C||u||^r_{L^r}, \quad \text{where } \tilde{J}(u) := \int_\Omega \left(\tilde{F}(x, u(x)) + |\nabla u(x)|^p \right) \, dx.$$

The semicontinuity of $\tilde{J}$ can be handled exactly as in Theorem 2.1, while on the L^r part we need, because of the negative part, continuity (upper semicontinuity of the norm would be enough, but we already know that norms are lower semicontinuous). This is easy to handle because of the compact injection $W^{1,p} \hookrightarrow L^r$ (an injection which is compact also for $r = p$ and even for some $r > p$).

The reader may be disappointed that we need to require $r < p$, while we know that $W^{1,p}$ implies a summability better than L^p and that the injection is still compact. Indeed, it would be possible to handle the case where the functional includes a positive part of the form $||\nabla u_n||^p_{L^p}$ and a negative part of the form $||u_n||^{r_2}_{L^{r_1}}$ for $r_1 < p^*$, but we would still need $r_2 < p$ if we want to prove that minimizing sequences are bounded. Since only $r_1 = r_2$ gives a classical integral function, we only consider $r_1 = r_2 = r < p$.

Another variant concerns the Dirichlet boundary data. In dimension one we considered the case where each endpoint was either penalized or fixed, independently. A first natural question is how to deal with a Dirichlet boundary condition on a part of $\partial\Omega$ and not on the full boundary. In order to deal with the values on the boundary we recall the notion of the trace of a Sobolev function.

Box 2.4 Important Notion—*Traces of Sobolev Functions*

If Ω is a smooth open domain in $\mathbb{R}^d$ (we assume that its boundary has Lipschitz regularity, i.e. Ω can locally be written as the epigraph of a Lipschitz function) and $p > 1$, there exists an operator $\mathrm{Tr} : W^{1,p}(\Omega) \to L^p(\partial\Omega)$ (where the boundary $\partial\Omega$ is endowed with the natural surface measure $\mathcal{H}^{d-1}$) with the following properties

- Tr is a linear continuous and compact operator when $W^{1,p}(\Omega)$ and $L^p(\partial\Omega)$ are endowed with their standard norms;
- for every Lipschitz function $u : \Omega \to \mathbb{R}$ we have $\mathrm{Tr}[u] = u_{|\partial\Omega}$;
- the kernel of Tr coincides with $W^{1,p}_0(\Omega)$, i.e. with the closure in $W^{1,p}(\Omega)$ of $C^\infty_c(\Omega)$;

(continued)

Box 2.4 (continued)
- for every Lipschitz function $h : \mathbb{R} \to \mathbb{R}$ we have $\mathrm{Tr}[h(u)] = h(\mathrm{Tr}[u])$;
- the operator Tr actually takes values in a space $L^q(\partial\Omega)$ with $q > p$ and is continuous when valued in this space: we can take $q = (d-1)p/(d-p)$ when $p < d$, $q = \infty$ when $p > d$ and any $q \in (p, \infty)$ when $p = d$; when $p > d$ the operator is also valued and continuous in $C^{0,\alpha}$ with $\alpha = 1 - d/p$.

For the theory of traces of Sobolev functions we refer to [84, Chapter 4].

We need the following Poincaré-like inequality, that we state in the most general form so as to be able to re-use it when dealing with boundary penalization instead of Dirichlet boundary data. We will summarize in the same statement both a result involving the trace and a result involving the behavior inside the domain.

Proposition 2.2 *Given a connected open domain Ω, for every $\varepsilon > 0$ there exist two constants C_1, C_2 such that we have*

1. *if $u \in W^{1,p}(\Omega)$ satisfies $\mathcal{H}^{d-1}(\{x \in \partial\Omega : \mathrm{Tr}[u](x) = 0\}) \geq \varepsilon$ then we have $||u||_{L^p} \leq C_1 ||\nabla u||_{L^p}$;*
2. *if $u \in W^{1,p}(\Omega)$ satisfies $|\{x \in \Omega : u(x) = 0\}| \geq \varepsilon$, then we have $||u||_{L^p} \leq C_2 ||\nabla u||_{L^p}$.*

The proof of the above proposition will be done by contradiction, which provides a useful technique to prove similar inequalities, even if the constants are in general neither explicit nor quantified.

Proof We suppose by contradiction that one of these constants, for a certain $\varepsilon > 0$, does not exist. This means that for every n we find a function $u_n \in W^{1,p}(\Omega)$ which violates the claimed inequality with $C = n$. Up to multiplying this function by a multiplicative constant we can assume $||u_n||_{L^p} = 1$ and $||\nabla u_n||_{L^p} < 1/n$. The sequence u_n is hence bounded in $W^{1,p}$ and a weakly convergent subsequence can be extracted. We call u the weak limit and the compact injection of $W^{1,p}$ into L^p provides $u_n \to u$ in L^p and hence $||u||_{L^p} = 1$ as well. We also have $||\nabla u||_{L^p} \leq \liminf_n ||\nabla u_n||_{L^p} = 0$, so that u is a constant function. The condition on its L^p norm implies $u = c$ with $c \neq 0$. Moreover, the convergence $u_n \to u = c$ is actually strong in $W^{1,p}$ since we also have $||\nabla u_n - \nabla u||_{L^p} = ||\nabla u_n||_{L^p} \to 0$.

We now distinguish the two cases. For the sake of the constant C_1 related to the behavior on the boundary, we observe that the strong convergence in $W^{1,p}$ implies the convergence of the trace, so that $\mathrm{Tr}[u_n]$ strongly converges to $\mathrm{Tr}[u]$ and we have $\mathrm{Tr}[u] = c$. Then we write

$$||\mathrm{Tr}[u_n] - \mathrm{Tr}[u]||^p_{L^p(\partial\Omega)} \geq c^p \mathcal{H}^{d-1}(\{x \in \partial\Omega : \mathrm{Tr}[u_n](x) = 0\}) \geq c^p \varepsilon,$$

which is a contradiction since the limit of the left-hand side is 0.

For the sake of the constant C_2 related to the behavior inside Ω, the situation is easier, since we already have strong convergence $u_n \to u$ in $L^p(\Omega)$. Yet, we have

$$||u_n - u||^p_{L^p(\partial\Omega)} \geq c^p |(\{x \in \Omega \,:\, u_n(x) = 0\})| \geq c^p \varepsilon,$$

and, again, this provides a contradiction when $n \to \infty$. $\square$

The idea behind the inequality proven above is that the L^p norm of a function is bounded by the L^p norm of its gradient as soon as we impose sufficient conditions which prevent the function from being a non-zero constant. We note that the very same technique could also be used to prove another inequality of this form, which is the one concerning zero-mean functions, and is known as *Poincaré–Wirtinger* inequality.

Proposition 2.3 *Given a connected, Lipschitz, and open domain Ω there exists a constant C such that we have $||u||_{L^p} \leq C||\nabla u||_{L^p}$ for all functions $u \in W^{1,p}(\Omega)$ satisfying $\fint_\Omega u = 0$.*

Proof We suppose by contradiction that such a constant does not exist. For every n we find a function $u_n \in W^{1,p}(\Omega)$ with $||u_n||_{L^p} = 1$ and $||\nabla u_n||_{L^p} < 1/n$. As before, we extract a weakly convergent subsequence and call u the weak limit. The compact injection of $W^{1,p}$ into L^p provides again $u_n \to u$ in L^p and hence $||u||_{L^p} = 1$ as well. We also have $||\nabla u||_{L^p} \leq \liminf_n ||\nabla u_n||_{L^p} = 0$, so that u is a constant function because Ω is connected. We also have $\fint u_n \to \fint u = 0$ so that $\fint u = 0$. We have a contradiction since u is a constant with zero mean (hence $u = 0$) but $||u||_{L^p} = 1$. $\square$

The above result on zero-mean functions was presented for the sake of completeness. We now come back to the notion of trace, which we are able to use in order to provide a more general existence result.

Theorem 2.4 *Given two measurable functions $\psi : \partial\Omega \times \mathbb{R} \to [0, +\infty]$ and $F : \Omega \times \mathbb{R} \to \mathbb{R}_+$ such that $\psi(x, \cdot)$ is lower semicontinuous for $\mathcal{H}^{d-1}$-a.e. x and $F(x, \cdot)$ is lower semicontinuous for a.e. x, let us consider the problem*

$$\min\left\{ J(u) \,:\, u \in W^{1,p}(\Omega) \right\}, \tag{2.2}$$

where

$$J(u) := \int_\Omega \left(F(x, u(x)) + |\nabla u(x)|^p \right) dx + \int_{\partial\Omega} \psi(x, \mathrm{Tr}[u](x)) \, d\mathcal{H}^{d-1}(x).$$

This minimization problem admits a solution provided one of the following conditions is satisfied for a function $g : \mathbb{R} \to \mathbb{R}_+$ with $\lim_{s \to \pm\infty} g(s) = +\infty$

1. *either there exists an $A \subset \partial\Omega$ with $\mathcal{H}^{d-1}(A) > 0$ and such that $\psi(x, s) \geq g(s)$ for all $x \in A$;*

2. *or there exists a $B \subset \Omega$ with positive Lebesgue measure $|B| > 0$ and such that $F(x, s) \geq g(s)$ for all $x \in B$.*

Proof Take a minimizing sequence u_n with $J(u_n) \leq C$.

We first analyze the case (1). From the positivity of all the terms in the optimization problem we can deduce that the set $\{x \in A : g(\mathrm{Tr}[u_n](x)) > \ell\}$ has measure at most $\mathcal{H}^{d-1}(A) - C/\ell$ so that, taking $\ell = 2C/\mathcal{H}^{d-1}(A)$, we have $g(\mathrm{Tr}[u_n](x)) \leq \ell$ on a set $A_n \subset A$ with $\mathcal{H}^{d-1}(A_n) \geq \mathcal{H}^{d-1}(A)/2$. Using the condition on g we deduce that there exists an M independent of n such that $|(\mathrm{Tr}[u_n](x)| \leq M$ on A_n. If instead, we are in the situation described in case 2), we see analogously that we have a set $B_n \subset B$ with $|B_n| \geq |B|/2$ and $|u| \leq M$ on B_n.

In both cases, defining the function $h : \mathbb{R} \to \mathbb{R}$ as

$$h(s) = \begin{cases} s - M & \text{if } s > M, \\ 0 & \text{if } |s| \leq M, \\ s + M & \text{if } s < -M, \end{cases}$$

we can apply the bounds provided by Proposition 2.2 to $v_n = h(u_n)$. Using $|h'| \leq 1$ the norm $\|\nabla v_n\|_{L^p}$ is bounded, and so is the L^p norm of v_n. Using $|v_n - u_n| \leq M$ this implies a bound on $\|u_n\|_{L^p}$ and then on $\|u_n\|_{W^{1,p}}$.

We can then extract a subsequence which weakly converges in $W^{1,p}$ to a function u. We have $u_n \to u$ in $L^p(\Omega)$ as well as $\mathrm{Tr}[u_n] \to \mathrm{Tr}[u]$ in $L^p(\partial\Omega)$ (since both the injection of $W^{1,p}$ into L^p and the trace operator are compact). This means, up to a subsequence, a.e. convergence $u_n(x) \to u(x)$ in Ω and $\mathrm{Tr}[u_n](x) \to \mathrm{Tr}[u](x)$ a.e. for the $\mathcal{H}^{d-1}$ measure on $\partial\Omega$. Fatou's lemma provides the semi-continuity of both the integral term $\int_\Omega F(x, u(x)) \, dx$ and the boundary integral $\int_{\partial\Omega} \psi(x, \mathrm{Tr}[u](x)) \, d\mathcal{H}^{d-1}(x)$, and the gradient term is treated by semicontinuity of the norm as usual. This shows that u is a minimizer. $\square$

It is important to observe that the assumption on the boundary penalization ψ in the above theorem allows us to consider the case

$$\psi(x, s) = \begin{cases} 0 & \text{if } x \notin A, \\ 0 & \text{if } x \in A, s = h(x), \\ +\infty & \text{if } x \in A, s \neq h(x), \end{cases}$$

which is equivalent to imposing the Dirichlet boundary condition $u = h$ on $A \subset \partial\Omega$. In particular, the function ψ satisfies $\psi(x, s) \geq g(s) := (s - M)_+$ for all $x \in A$ with $|h(x)| \leq M$, a set of points x which is of positive measure for large M.

2.2 The Multi-Dimensional Euler–Lagrange Equation

This section is just a translation to higher dimensions of Sect. 1.3. We consider an open domain $\Omega \subset \mathbb{R}^d$ and an integrand $L : \Omega \times \mathbb{R} \times \mathbb{R}^d \to \mathbb{R}$, where the variables will be called x (the position, which is now the independent variable, replacing time), s (the value of the function) and v (its variation, which here is a gradient vector). The function L will be assumed to be C^1 in (u, v) and its partial derivatives w.r.t. s and v_i will be denoted by $\partial_s L$ and $\partial_{v_i} L$, respectively (and the gradient vector with all the corresponding partial derivatives w.r.t. v, by $\nabla_v L$).

As we did in Chap. 1, we write the optimality conditions for multi-dimensional variational problems starting from the case where the boundary data are fixed. For simplicity, in particular in order to handle the boundary, we will stick to the case where the functional space X is a Sobolev space $W^{1,p}(\Omega)$. In 1D we presented a more abstract situation where X was a generic functional space contained in C^0 but in higher dimensions few spaces are at the same time a reasonable choice for the most classical optimization problems and are made of continuous functions, so we prefer to directly choose the Sobolev formalism, and the boundary data will be imposed using the space $W_0^{1,p}$. We then face

$$\min \left\{ J(u) := \int_\Omega L(x, u(x), \nabla u(x)) \, dx \; : \; u \in W^{1,p}(\Omega), \, u - g \in W_0^{1,p}(\Omega) \right\}.$$

We assume a lower bound of the form $L(x, u, v) \geq -a(x) - c(|u|^p + |v|^p)$ for $a \in L^1$, so that for every $u \in W^{1,p}(\Omega)$ the negative part of $L(\cdot, u, \nabla u)$ is integrable,[1] thus J is a well-defined functional from $W^{1,p}(\Omega)$ to $\mathbb{R} \cup \{+\infty\}$. We assume that u is a solution of such a minimization problem. Since all functions in $u + C_c^\infty(\Omega)$ are admissible competitors, for every $\varphi \in C_c^\infty(\Omega)$ we have $J(u) \leq J(u + \varepsilon\varphi)$ for small ε.

We now fix the minimizer u and a perturbation φ and consider the one-variable function

$$j(\varepsilon) := J(u_\varepsilon), \quad \text{where} \quad u_\varepsilon := u + \varepsilon\varphi,$$

which is defined in a neighborhood of $\varepsilon = 0$, and minimal at $\varepsilon = 0$. We compute exactly as in Chap. 1 the value of $j'(0)$.

The assumption to justify the differentiation under the integral sign will be the same as in Chap. 1: we assume that for every $u \in W^{1,p}(\Omega)$ with $J(u) < +\infty$ there exists a $\delta > 0$ such that we have

$$x \mapsto \sup \left\{ |\partial_s L(x, s, v)| + |\nabla_v L(x, s, v)| : |s - u(x)| < \delta, \, v \in B(\nabla u(x), \delta) \right\}$$

$$\in L^1(\Omega). \qquad (2.3)$$

---

[1] Actually, exploiting $W^{1,p} \subset L^{p^*}$, we could accept a weaker lower bound in terms of u, i.e. $L(x, u, v) \geq -a(x) - c(|u|^{p^*} + |v|^p)$.

Examples of sufficient conditions on L in order to guarantee (2.3) are the same as in 1D. Under these conditions, we can differentiate w.r.t. ε the function $\varepsilon \mapsto L(t, u_\varepsilon, \nabla u_\varepsilon)$, and obtain

$$\frac{d}{d\varepsilon} L(t, u_\varepsilon, \nabla u_\varepsilon) = \partial_s L(x, u_\varepsilon, \nabla u_\varepsilon)\varphi + \nabla_v L(x, u_\varepsilon, \nabla u_\varepsilon) \cdot \nabla\varphi.$$

Then, using the fact that both φ and $\nabla\varphi$ are bounded, we can apply the assumption in (2.3) in order to obtain domination in L^1 of the pointwise derivatives and thus, for small ε, we have

$$j'(\varepsilon) = \int_\Omega (\partial_s L(x, u_\varepsilon, \nabla u_\varepsilon)\varphi + \nabla_v L(x, u_\varepsilon, \nabla u_\varepsilon) \cdot \nabla\varphi) \, dx$$

and

$$j'(0) = \int_\Omega (\partial_s L(x, u, \nabla u)\varphi + \nabla_v L(x, u, \nabla u) \cdot \nabla\varphi) \, dx.$$

Imposing $j'(0) = 0$, which comes from the optimality of u, means precisely that we have, in the sense of distributions, the following partial differential equation, also known as the *Euler–Lagrange* equation:

$$\nabla \cdot (\nabla_v L(x, u, \nabla u)) = \partial_s L(x, u, \nabla u).$$

This second-order partial differential equation (PDE) is usually of elliptic type, because of the standard assumption (which is satisfied in all the examples that we saw for the existence of a solution and will be natural from the lower semicontinuity theory that will be developed in Chap. 3) that L is convex in v, so that $\nabla_v L(x, u, \nabla u)$ is a monotone function of ∇u. It is formally possible to expand the divergence and indeed obtain

$$\nabla \cdot (\nabla_v L(x, u, \nabla u)) = \sum_j \frac{d}{dx_j} \left(\partial_{v_j} L(x, u, \nabla u)\right)$$

$$= \sum_j \left(\partial^2_{x_j, v_j} L(x, u, \nabla u) + \partial^2_{u, v_j} L(x, u, \nabla u)u_j\right)$$

$$+ \sum_{j,k} \partial^2_{v_k, v_j} L(x, u, \nabla u)u_{j,k},$$

where u_j and $u_{j,k}$ stand for the various partial derivatives of the solution u. This computation shows that the second-order term in this PDE is ruled by the matrix $A^{j,k}(x) := \partial^2_{v_k, v_j} L(x, u(x), \nabla u(x))$, which is (semi-)positive-definite if L is convex in v.

It is important to stress the exact meaning of the Euler–Lagrange equation.

Box 2.5 Memo—*Distributions*

We call a distribution on an open domain Ω any linear operator T on $C_c^\infty(\Omega)$ with the following continuity property: for every compact subset $K \subset \Omega$ and every sequence $\phi_n \in C_c^\infty(\Omega)$ with $\operatorname{spt}\phi_n \subset K$ and such that all derivatives $D^\alpha\phi_n$ of any order of ϕ_n uniformly converge on K to the corresponding derivative $D^\alpha\phi$ we have $\langle T, \phi_n \rangle \to \langle T, \phi \rangle$.

On the set of distributions we define a notion of convergence: we say $T_n \to T$ if for every $\phi \in C_c^\infty(\Omega)$ we have $\langle T_n, \phi \rangle \to \langle T, \phi \rangle$.

L^1_{loc} functions are examples of distributions in the sense that for any $f \in L^1_{loc}(\Omega)$ we can define a distribution T_f by $\langle T_f \phi \rangle := \int_\Omega f\phi$. Locally finite measures are also distributions (with $\langle \mu, \phi \rangle := \int \phi \, d\mu$).

Some operations are possible with distributions: the derivative $\partial_{x_i} T$ is defined via the integration-by-parts formula $\langle \partial_{x_i} T, \phi \rangle := -\langle T, \partial_{x_i}\phi \rangle$; the product with a C^∞ function u by $\langle uT, \phi \rangle := \langle T, u\phi \rangle$, and the convolution (for simplicity with an even function η with compact support) by $\langle \eta * T, \phi \rangle := \langle T, \eta * \phi \rangle$. It is possible to show that, whenever $\eta \in C_c^\infty$, then $\eta * T$ is a distribution associated with a locally integrable function which is C^∞. If η is not C^∞ but less smooth, then the regularity of $\eta * T$ depends both on that of η and of T.

A distribution is said to be of order m on a compact set K whenever $|\langle T, \phi \rangle| \le C\|\phi\|_{C^m(K)}$ for all functions $\phi \in C_c^\infty(\Omega)$ with $\operatorname{spt}\phi \subset K$. When convolving a function $\eta \in C^k$ with a convolution T of order m we have $\eta * T \in C^{k-m}$ if $k \ge m$ while, if $k < m$, we obtain a distribution $\eta * T$ of order $m - k$.

Whenever a function $f \in L^1_{loc}$ belongs to a Sobolev space $W^{1,p}$ it is easy to see that the derivative in terms of distributions coincides with that in Sobolev spaces. The notion of convergence is also essentially the same: for every space $W^{k,p}$ with $1 < p < \infty$ a sequence $u_n \rightharpoonup u$ weakly converges in $W^{k,p}$ to u if and only if it is bounded in $W^{k,p}$ and converges in the sense of distributions to u (in the sense that the associated distributions converge).

It is easy to see that $T_n \to T$ in the sense of distributions implies the convergence of the derivatives $\partial_{x_i} T_n \to \partial_{x_i} T$. The convergence in the sense of distributions is essentially the weakest notion of convergence that is used in functional analysis and is implied by all other standard notions of convergence.

Indeed, in the expression that we obtained by expanding the divergence, the terms $\partial^2_{v_k, v_j} L(x, u, \nabla u) u_{j,k}$ are the product of a measurable function (whose regularity is the same as that of ∇u, which is just summable) and the Hessian of a Sobolev function, which is in general a distribution but not a function. Since distributions can only be multiplied by C^∞ functions, our Euler–Lagrange equation cannot be written in the distributional sense in this way. Indeed, the precise meaning of the PDE under the current assumptions is the following: the vector-valued function $\nabla_v L(x, u, \nabla u)$ is an L^1 function whose distributional divergence equals $\partial_s L(x, u, \nabla u)$ since it satisfies

$$\int_\Omega (\partial_s L(x, u, \nabla u)\varphi + \nabla_v L(x, u, \nabla u) \cdot \nabla\varphi) \, dx = 0$$

for every $\varphi \in C_c^\infty(\Omega)$. In dimension one, it was easy to deduce from this condition extra regularity for u, at least on some examples, since, for instance, having a C^0 distributional derivative is equivalent to being C^1. In higher dimensions we do not have an equation on each partial derivative of $\nabla_v L(x, u, \nabla u)$, but only on its divergence, which makes it difficult to deduce regularity.

A classical example is the case $L(x, u, v) = \frac{1}{2}|v|^2$, i.e. the minimization of

$$H^1(\Omega) \ni u \mapsto \int_\Omega |\nabla u|^2$$

with prescribed boundary conditions. In this case we have $\nabla_v L(x, u, \nabla u) = \nabla u$ and the equation is $\nabla \cdot \nabla u = 0$. This corresponds to the Laplace equation $\Delta u = 0$ which would mean, in the distributional sense, $\int_\Omega u\Delta\varphi = 0$ for all $\varphi \in C_c^\infty(\Omega)$. Here, instead, since we know $u \in H^1$ and hence ∇u is a well-defined integrable function, we require $\int_\Omega \nabla u \cdot \nabla\varphi = 0$. We will see anyway that these two conditions coincide and both imply $u \in C^\infty$ and $\Delta u = 0$ in the most classical sense (see Sect. 2.3, devoted to the study of harmonic functions and distributions). On the other hand, the situation could be trickier when the equation is non-linear, or if there is an explicit dependence on x and/or some low-regularity lower order terms, and in this case the regularity of the optimal solution u is usually obtained from the Euler–Lagrange equation by choosing suitable test functions φ. As we will see in Chap. 5, for instance, it is often useful to choose a function φ related to the solution u itself. This requires the weak formulation of the Euler–Lagrange equation to be extended to test functions which, instead of being C^∞, have the same regularity as u. This can be done via the following lemma.

Lemma 2.5 *Let $p > 1$ be a given exponent, with $p' = \frac{p}{p-1}$ its dual exponent, and let $z \in L^{p'}(\Omega; \mathbb{R}^d)$ be a vector field such that $\int_\Omega z \cdot \nabla\varphi = 0$ for all $\varphi \in C_c^\infty(\Omega)$. Then we also have $\int_\Omega z \cdot \nabla\varphi = 0$ for all $\varphi \in W_0^{1,p}(\Omega)$. This applies in particular to the case $z = \nabla_v L(x, u, \nabla u)$ whenever $u \in W^{1,p}$ and L satisfies $|\nabla_v L(x, u, v)| \leq C(1 + |v|^{p-1})$.*

Proof The condition $z \in L^{p'}$ implies that $L^p \ni v \mapsto \int z \cdot v$ is continuous in L^p and hence $W^{1,p} \ni \varphi \mapsto \int z \cdot \nabla\varphi$ is continuous in $W^{1,p}$. Thus, if this linear form vanishes on $C_c^\infty(\Omega)$, it also vanishes on its closure for the $W^{1,p}$ norm, i.e. on $W_0^{1,p}(\Omega)$.

For the second part of the claim, note that whenever $u \in W^{1,p}(\Omega)$ we have $|\nabla u|^{p-1} \in L^{\frac{p}{p-1}}$, which implies, under the growth assumption of the claim, $\nabla_v L(x, u, \nabla u) \in L^{p'}$. □

Boundary and Transversality Conditions We switch now to the case where no boundary value is prescribed. Exactly as in the 1D case we can consider $u_\varepsilon = u + \varepsilon \varphi$ for arbitrary $\varphi \in C^\infty$, without imposing a zero boundary value or compact support for φ. The same computations as before provide the necessary optimality condition

$$\int_\Omega (\partial_s L(x, u, \nabla u)\varphi + \nabla_v L(x, u, \nabla u) \cdot \nabla\varphi) \, dx = 0 \quad \text{for all } \varphi \in C^\infty(\Omega).$$

Distributionally, this corresponds to saying that the vector field $x \mapsto \nabla_v L(x, u, \nabla u)$ extended to 0 outside Ω has a distributional divergence equal to the scalar function $\partial_s L(x, u, \nabla u)$, also extended to 0 outside Ω. A formal integration by parts would instead give

$$0 = \int_\Omega (\partial_s L(x, u, \nabla u) - \nabla \cdot (\nabla_v L(x, u, \nabla u))) \, \varphi \, dx + \int_{\partial\Omega} \nabla_v L(x, u, \nabla u) \cdot \mathbf{n} \, \varphi \, d\mathcal{H}^{d-1},$$

where $\mathbf{n}$ is the normal vector to $\partial\Omega$, so that besides the differential condition inside Ω we also obtain a Neumann-like boundary condition on $\partial\Omega$, i.e. we obtain a weak version of the system

$$\begin{cases} \nabla \cdot (\nabla_v L(x, u, \nabla u)) = \partial_s L(x, u, \nabla u), & \text{in } \Omega, \\ \nabla_v L(x, u, \nabla u)) \cdot \mathbf{n} = 0 & \text{on } \partial\Omega. \end{cases}$$

In the particular case where $L(x, sv) = \frac{1}{2}|v|^2 + F(x, s)$ the boundary condition is exactly a homogeneous Neumann condition $\partial u / \partial \mathbf{n} = 0$.

It is of course possible to consider mixed cases where part of the boundary is subject to a Dirichlet boundary condition and part of the boundary is either free or penalized. We then consider the optimization problem

$$\min \left\{ \begin{array}{c} J(u) := \int_\Omega L(x, u(x), \nabla u(x)) \, dx + \int_{\partial\Omega} \psi(x, \text{Tr}[u]) \, d\mathcal{H}^{d-1} : \\ u \in W^{1,p}(\Omega), \text{Tr}[u] = g \text{ on } A \end{array} \right\}, \quad (2.4)$$

where $A \subset \partial\Omega$ is a prescribed part of the boundary (which is possibly empty) and $\psi : \partial\Omega \times \mathbb{R} \to \mathbb{R}$ is a prescribed penalization function, which could be zero for x in some parts of $\partial\Omega$ and is irrelevant for $x \in A$. In order to perform a differentiation under the integral sign we need to add to condition (2.3) a similar condition on the boundary, i.e.

$$x \mapsto \sup\{|\partial_s \psi(x,s)| \; : \; |s - \mathrm{Tr}[u](x)| < \delta\} \in L^1(\partial\Omega \setminus A) \qquad (2.5)$$

for every $u \in W^{1,p}$ with $J(u) < +\infty$. The conditions to guarantee this bound are similar to those for the corresponding bound on L. Condition (2.5) is in particular satisfied when $p \leq d$ every time that $\partial_s \psi(x,s)$ has a growth of order q in s (i.e. $|\partial_s \psi(x,s)| \leq C(1 + |s|)^q$) and $\mathrm{Tr}[u] \in L^q(\partial\Omega)$ (in particular, we can take $q = p$ since the trace operator is valued in $L^p(\partial\Omega)$, but also $q = (d-1)p/(d-p)$, if we use the last property of the trace operator in Box 2.4). In the case $p > d$ any local bound on $\partial_s \psi$ is enough since $\mathrm{Tr}[u]$ is bounded, in particular this can apply to the case where ψ is differentiable in s with $\partial_s \psi \in C^0(\partial\Omega \times \mathbb{R})$.

A computation which will be now standard for the reader provides the necessary optimality condition for the problem (2.4): u should satisfy

$$\int_\Omega (\partial_s L(x, u, \nabla u)\varphi + \nabla_v L(x, u, \nabla u) \cdot \nabla \varphi) \, dx + \int_{\partial\Omega} \partial_s \psi(x, \mathrm{Tr}[u])\varphi \, d\mathcal{H}^{d-1} = 0$$
$$(2.6)$$

for all $\varphi \in C_c^\infty(\overline{\Omega} \setminus A)$.

When $A = \emptyset$, distributionally this means that the vector field $x \mapsto \nabla_v L(x, u, \nabla u)$ extended to 0 outside Ω has a distributional divergence equal to the scalar function $\partial_s L(x, u, \nabla u)$, also extended to 0 outside Ω, plus a singular measure concentrated on $\partial\Omega$ with a density w.r.t. the surface measure $\mathcal{H}^{d-1}$, this density being equal to $\partial_s \psi(x, \mathrm{Tr}[u])$. When $A \neq \emptyset$, this distributional condition is satisfied on the open set $\mathbb{R}^d \setminus \overline{A}$.

A formal integration by parts provides the Euler–Lagrange system

$$\begin{cases} \nabla \cdot (\nabla_v L(x, u, \nabla u)) = \partial_s L(x, u, \nabla u), & \text{in } \Omega, \\ \nabla_v L(x, u, \nabla u)) \cdot \mathbf{n} = -\partial_s \psi(x, u) & \text{on } \partial\Omega \setminus A \\ u = g & \text{on } A, \end{cases} \qquad (2.7)$$

where we dared to write u instead of $\mathrm{Tr}[u]$ since, anyway, the condition is quite formal.

2.3 Harmonic Functions

This section will be devoted to the study of harmonic functions, which are the solutions of the simplest and most classical variational problem, i.e. the minimization of the Dirichlet energy $\int |\nabla u|^2$ among functions with prescribed values on the boundary of a given domain. However, we will not make use here of the variational properties of these functions, but only of the equation they solve. Recall that we denote by Δ the differential operator given by

$$\Delta := \sum_{i=1}^{d} \frac{\partial^2}{\partial x_i^2}.$$

We will start from the case of a C^2 function u satisfying $\Delta u = 0$ in a given open domain Ω. The first property that we consider is the mean value formula, which connects the value of u at a point x_0 to the averages of u on balls or spheres around x_0.

Proposition 2.6 *If $u \in C^2(\Omega)$ and $\Delta u = 0$ then, for every point x_0 and $R > 0$ such that $B(x_0, R) \subset \Omega$, we have the following properties*

- $(0, R) \ni r \mapsto \fint_{\partial B(x_0,r)} u \, d\mathcal{H}^{d-1}$ *is constant and equal to $u(x_0)$;*
- $(0, R) \ni r \mapsto \fint_{B(x_0,r)} u \, dx$ *is also constant and equal to $u(x_0)$.*

Proof We start from the function $h(r) := \fint_{\partial B(x_0,r)} u \, d\mathcal{H}^{d-1}$, which we can write as $h(r) = \fint_{\partial B(0,1)} u(x_0+re) \, d\mathcal{H}^{d-1}(e)$. The smoothness of u allows us to differentiate and obtain

$$h'(r) = \fint_{\partial B(0,1)} \nabla u(x_0 + re) \cdot e \, d\mathcal{H}^{d-1}(e) = \fint_{\partial B(0,1)} \nabla u(x_0 + re) \cdot \mathbf{n} \, d\mathcal{H}^{d-1}(e),$$

where we used that, for every point e on the boundary of the unit ball, the vector e coincides with the outward normal vector. The last integral can be then re-transformed into an integral on the sphere $\partial B(x_0, r)$, and it equals

$$\fint_{\partial B(0,r)} \nabla u \cdot \mathbf{n} \, d\mathcal{H}^{d-1}.$$

We can then integrate by parts

$$\int_{\partial B(0,r)} \nabla u \cdot \mathbf{n} \, d\mathcal{H}^{d-1} = \int_{B(0,r)} \Delta u \, dx = 0,$$

which proves that h is constant as soon as $\Delta u = 0$. Since u is continuous, we have $\lim_{r \to 0} h(r) = u(x_0)$, so that we have $h(r) = u(x_0)$ for every r, which proves the first part of the claim.

We then use the polar coordinate computation

$$\fint_{B(0,r)} u \, dx = \frac{\int_0^r s^{d-1} h(s) \, ds}{\int_0^r s^{d-1} \, ds} = u(x_0),$$

to obtain the second part of the claim. $\square$

Remark 2.7 It is clear from the above proof that if one replaces the assumption $\Delta u = 0$ with $\Delta u \geq 0$ then the functions $r \mapsto \fint_{\partial B(x_0,r)} u \, d\mathcal{H}^{d-1}$ and $r \mapsto \fint_{B(x_0,r)} u \, dx$ are no longer constant but non-decreasing. Functions with non-negative Laplacian are called *sub-harmonic*.

A key point about harmonic functions is that if u is harmonic and C^∞, then all derivatives of u are also harmonic. This allows us to prove the following bounds.

Proposition 2.8 *If $u \in C^\infty(\Omega)$ and $\Delta u = 0$ then, for every point x_0 and $R > 0$ such that $B(x_0, R) \subset \Omega \subset \mathbb{R}^d$, we have*

$$|\nabla u(x_0)| \leq \frac{d}{R} \|u\|_{L^\infty(B(x_0,R))}.$$

Proof We consider the vector function ∇u, which is also harmonic (each of its components is harmonic) and use

$$\nabla u(x_0) = \fint_{B(x_0,R)} \nabla u \, dx = \frac{1}{\omega_d R^d} \int_{B(x_0,R)} \nabla u \, dx = \frac{1}{\omega_d R^d} \int_{\partial B(x_0,R)} u \mathbf{n} \, d\mathcal{H}^{d-1},$$

where ω_d is the volume of the unit ball in $\mathbb{R}^d$. To establish the last equality we used an integration by parts: for every constant vector e we have $\nabla \cdot (ue) = \nabla u \cdot e$ and $\int_A \nabla \cdot (ue) = \int_{\partial A} ue \cdot \mathbf{n}$ for every open set A; the vector e being arbitrary, this provides $\int_A \nabla u = \int_{\partial A} u\mathbf{n}$.

We finish the estimate by bounding $\left| \int_{\partial B(x_0,R)} u\mathbf{n} \, d\mathcal{H}^{d-1} \right|$ by the maximal norm of $u\mathbf{n}$ times the measure of the surface, i.e. by $\|u\|_{L^\infty(B(x_0,R))} \mathcal{H}^{d-1}(\partial B(x_0, R))$ or, equivalently, by $\|u\|_{L^\infty(B(x_0,R))} d\omega_d R^{d-1}$. $\square$

In order to analyze higher-order derivatives of u we recall the notation with multi-indices: a multi-index $\alpha = (\alpha_1, \dots, \alpha_d) \in \mathbb{N}^d$ is a vector of natural numbers indicating how many times we differentiate w.r.t. each variable, and we write

$$D^\alpha u := \frac{\partial^{|\alpha|}}{\partial_{x_1}^{\alpha_1} \dots \partial_{x_d}^{\alpha_d}} u,$$

where $|\alpha| := \sum_{i=1}^d \alpha_i$ is the order of the multi-index α. A consequence of the previous estimate, if iteratively applied to the derivatives of u, is the following result.

Proposition 2.9 *If $u \in C^\infty(\Omega)$ and $\Delta u = 0$ then, for every point x_0 and $R > 0$ such that $B(x_0, R) \subset \Omega \subset \mathbb{R}^d$ and every multi-index α with $|\alpha| = m$, we have*

$$|D^\alpha u(x_0)| \leq \frac{d^m e^{m-1} m!}{R^m} \|u\|_{L^\infty(B(x_0,R))}. \tag{2.8}$$

Proof Given x_0 and r, assume $B(x_0, mr) \subset \Omega$. We can find a finite family of harmonic functions v_k, $k = 0, \dots, m$ such that each v_{k+1} is a derivative (with respect to a suitable variable x_i) of the previous v_k, with $v_0 = u$, and $v_m = D^\alpha u$. Applying iteratively Proposition 2.8 we obtain $|v_k(x)| \leq \frac{d}{r}\|v_{k-1}\|_{L^\infty(B(x,r))}$, so that we have

$$|D^\alpha u(x_0)| = |v_m(x_0)| \leq \frac{d}{r}\|v_{m-1}\|_{L^\infty(B(x,r))} \leq \frac{d^2}{r^2}\|v_{m-2}\|_{L^\infty(B(x,2r))} \leq \cdots$$

$$\cdots \leq \frac{d^k}{r^k}\|v_{m-k}\|_{L^\infty(B(x,kr))} \leq \frac{d^m}{r^m}\|u\|_{L^\infty(B(x,mr))}.$$

We then take $r = R/m$ and get

$$|D^\alpha u(x_0)| \leq \frac{m^m d^m}{R^m}\|u\|_{L^\infty(B(x,R))}.$$

We then apply the inequality $m^m \leq e^{m-1}m!$ and conclude.

To prove this inequality we observe that we have $m \log m = m \log m - 1 \log 1 = \int_1^m (\log s + 1)\, ds \leq \sum_{k=2}^m (\log k + 1) = \log(m!) + m - 1$, which gives exactly the desired inequality when taking the exponential on both sides.[2]   □

With this precise estimate in mind we can also prove the following fact.

Proposition 2.10 C^∞ *harmonic functions are analytic.*

We recall that analytic functions are those functions which locally coincide with their Taylor series around each point.

Proof For a C^∞ function u we can write its Taylor expansion as

$$u(x) = u(x_0) + \sum_{k=1}^m \sum_{\alpha:\,|\alpha|=k} \frac{1}{\alpha!} D^\alpha u(x_0)x^\alpha + R(m+1, x),$$

where the notation $\alpha!$ stands for $\alpha_1!\alpha_2!\ldots\alpha_d!$, the notation x^α for $x_1^{\alpha_1} x_2^{\alpha_2} \ldots x_d^{\alpha_d}$ and the remainder $R(m+1, x)$ has the form

$$R(m+1, x) = \sum_{\alpha:\,|\alpha|=m+1} \frac{1}{\alpha!} D^\alpha u(\xi)x^\alpha$$

for a certain point $\xi \in [x_0, x]$. In order to prove that the function u is analytic we need to prove $R(m, x) \to 0$ as $m \to \infty$ for x in a neighborhood of x_0. Assume $B(x_0, 2R) \subset \Omega$; if $x \in B(x_0, r)$ the point ξ also belongs to the same ball and if

---

[2] A refinement of this argument provides the well-known Stirling formula for the factorial $n! = \sqrt{2\pi n}(n/e)^n(1 + O(1/n))$.

$r \leq R$ we have $B(\xi, R) \subset B(x_0, 2R) \subset \Omega$. Then, we can estimate, thanks to (2.8),

$$|R(m, x)| \leq \sum_{\alpha : |\alpha|=m} \frac{1}{\alpha!} r^m \frac{d^m e^{m-1} m!}{R^m} ||u||_{L^\infty(B(x_0, 2R))}$$

$$= C \left(\frac{d \cdot e \cdot r}{R} \right)^m \sum_{\alpha : |\alpha|=m} \frac{m!}{\alpha!}.$$

We then use the well-known formula for multinomial coefficients

$$\sum_{\alpha : |\alpha|=m} \frac{m!}{\alpha!} = d^m$$

and obtain

$$|R(m, x)| \leq C \left(\frac{d^2 \cdot e \cdot r}{R} \right)^m.$$

The convergence is guaranteed as soon as $r < \frac{R}{d^2 e}$, which proves the analyticity of u.   □

Another consequence of Proposition 2.9 is the following regularity result.

Proposition 2.11 *If* $u \in L^1_{loc}(\Omega)$ *satisfies* $\Delta u = 0$ *in the distributional sense, then* $u \in C^\infty(\Omega)$.

Proof Given a sequence (with $\varepsilon \to 0$) of standard convolution kernels η_ε we consider $u_\varepsilon := u * \eta_\varepsilon$. We consider a ball $B(x_0, R)$ compactly contained in Ω; we have $u_\varepsilon \in C^\infty(B(x_0, R))$ for small ε and $\Delta u_\varepsilon = 0$ in the distributional, and hence classical, sense, in the same ball. Moreover, $||u_\varepsilon||_{L^1(B(x_0,R))}$ is bounded independently of ε because of $u \in L^1_{loc}(\Omega)$. We deduce a uniform bound, independent of ε, on $||u_\varepsilon||_{L^\infty(B(x_0,R/2))}$, as a consequence of the mean formula. Indeed, for each $x \in B(x_0, R/2)$ we have

$$u_\varepsilon(x) = \fint_{B(x,R/2)} u_\varepsilon(y) \, dy = \frac{1}{\omega_d (R/2)^d} \int_{B(x,R/2)} u_\varepsilon(y) \, dy$$

and using $B(x, R/2) \subset B(x_0, R)$ we obtain $|u_\varepsilon(x)| \leq C||u_\varepsilon||_{L^1(B(x_0,R))}$. Then, the estimate (2.8) allows us to bound all the norms $||D^\alpha u_\varepsilon||_{L^\infty(B(x_0,R/4))}$ by $||u_\varepsilon||_{L^\infty(B(x_0,R/2))}$. This shows that, on the ball $B(x_0, R/4)$, all derivatives of u_ε are bounded by a constant independent of ε. Using $u_\varepsilon \to u$ in the distributional sense, we obtain $u \in C^k(B(x_0, R/4))$ for every k and the same bounds on the derivatives are satisfied by u. This shows $u \in C^\infty(B(x_0, R/4))$ and, the point $x_0 \in \Omega$ being arbitrary, $u \in C^\infty(\Omega)$.   □

Box 2.6 Memo—*Convolutions and Approximation*

Given two functions u, v on $\mathbb{R}^d$ we define their convolution $u * v$ via

$$(u * v)(x) := \int_{\mathbb{R}^d} u(x - h)v(h)\,dh$$

when this integral makes sense, for instance when one function is L^1 and the other L^∞. Via a simple change of variables we see that we have $(u * v)(x) := \int u(x')v(x - x')$, so that the convolution is commutative and $u * v = v * u$. When $u, v \in L^1$ it is possible to prove that $u * v$ is well-defined for a.e. x and it is an L^1 function itself.

In the particular case where $v \geq 0$ and $\int v = 1$ we can write $u * v = \int u_h\,dv(h)$, where u_h is the translation of u defined via $u_h(x) := u(x - h)$ and we identify v with a probability measure. This means that $u*v$ is a suitable average of translations of u.

A very common procedure consists in using a function $\eta \in C_c^\infty$ with $\eta \geq 0$ and $\int \eta = 1$, and taking $v_n(x) := n^d \eta(nx)$, which is also a probability density, supported in a ball whose radius is $O(1/n)$. It is clear that when u is uniformly continuous we have $u * v_n \to u$ uniformly. Moreover, using the density of C_c^∞ in L^p, it is possible to prove that for any $u \in L^p$ we have $u * v_n \to u$ strongly in L^p.

Derivatives of convolutions of smooth functions can be computed very easily: we have $\partial_{x_i}(u * v) = (\partial_{x_i} u) * v$ as a consequence of the differentiation under the integral sign. Using $u * v = v * u$, the derivative can be taken on any of the two functions. If one of the functions is C^k and the other C^m the convolution is C^{k+m}, and if one is C^k and the other is only L^1 the convolution inherits the regularity of the best one. In particular, for $u \in L^1$ the sequence $u_n := u * v_n$ defined above is an approximation of u made of C^∞ functions.

A striking fact is the following: if F is any functional defined on a functional space which is convex and invariant under translations (which is the case, for instance, for all the L^p norms, but also for norms depending on the derivatives) and v is a probability measure, we necessarily have $F(u * v) \leq F(u)$.

Finally, it is also possible to define convolutions for distributions, thanks to the procedure in Box 2.5. When using the convolution kernel v_n above, for any distribution T we have $T * v_n \to T$ in the sense of distributions.

We want to go on with our analysis, proving that all harmonic distributions are actually analytic functions. In order to proceed, we first need to make a short digression about the equation $\Delta u = f$ and its fundamental solution. We introduce the fundamental solution of the Laplacian as the function Γ given by

$$\Gamma(x) = \begin{cases} \frac{1}{2\pi} \log|x| & \text{if } d = 2, \\ -\frac{1}{d(d-2)\omega_d}|x|^{2-d} & \text{if } d > 2. \end{cases}$$

We can see that we have

- $\Gamma \in L^1_{loc}, \nabla\Gamma \in L^1_{loc}$, but $D^2\Gamma \notin L^1_{loc}$.
- $\int_{\partial B(0,R)} \nabla\Gamma \cdot \mathbf{n} = 1$ for every R.
- $\Gamma \in C^\infty(\mathbb{R}^d \setminus \{0\})$ and $\Delta\Gamma = 0$ on $\mathbb{R}^d \setminus \{0\}$.
- $\Delta\Gamma = \delta_0$ in the sense of distributions.

As a consequence, for every $f \in C^\infty_c(\mathbb{R}^d)$, the function $u = \Gamma * f$ is a smooth function solving $\Delta u = f$ in the classical sense. It is of course not the unique solution to this equation, since we can add to u arbitrary harmonic functions.

For this solution, we have the following estimate

Proposition 2.12 *Given $f \in C^\infty_c(\mathbb{R}^d)$, let u be given by $u = \Gamma * f$. Then we have $\int_{\mathbb{R}^d} |D^2u|^2 = \int_{\mathbb{R}^d} |f|^2$, where $|D^2u|^2$ denotes the squared Frobenius norm of the Hessian matrix, given by $|A|^2 = \text{Tr}(A^t A) = \sum_{i,j} A^2_{ij}$.*

Proof We consider a ball $B(0, R)$ and obtain, by integration by parts,

$$\int_{B(0,R)} f^2 \, dx = \int_{B(0,R)} |\Delta u|^2 \, dx = \sum_{i,j} \int_{B(0,R)} u_{ii} u_{jj} \, dx$$

$$= -\sum_{i,j} \int_{B(0,R)} u_{iij} u_j \, dx + \sum_{i,j} \int_{\partial B(0,R)} u_{ii} u_j \mathbf{n}^j \, d\mathcal{H}^{d-1}.$$

If the support of f is compactly contained in $B(0, R)$ then the last boundary term vanishes since it equals $\int_{\partial B(0,R)} f \nabla u \cdot \mathbf{n}$. Going on with the integration by parts we have

$$\int_{B(0,R)} f^2 \, dx = \sum_{i,j} \int_{B(0,R)} u_{ij} u_{ij} \, dx - \sum_{i,j} \int_{\partial B(0,R)} u_{ij} u_j \mathbf{n}^i \, d\mathcal{H}^{d-1},$$

hence

$$\int_{B(0,R)} f^2 \, dx = \int_{B(0,R)} |D^2u|^2 \, dx - \int_{\partial B(0,R)} D^2 u \nabla u \cdot \mathbf{n} \, d\mathcal{H}^{d-1}.$$

We then note that, for $R \gg 1$, we have $|\nabla u(x)| \le C|x|^{1-d}$ and $|D^2 u(x)| \le C|x|^{-d}$, as a consequence of the shape of Γ and the compact support of f, so that we have

$$\left| \int_{\partial B(0,R)} D^2 u \nabla u \cdot \mathbf{n} \, d\mathcal{H}^{d-1} \right| \le C R^{d-1} \cdot R^{-d} \cdot R^{1-d} = C R^{-d} \to 0 \text{ as } R \to \infty,$$

which proves the claim. □

Using Γ, $\nabla\Gamma \in L^1_{loc}$ we see that $f \in L^2$ also implies that the L^2 norms of $\Gamma * f$ and $\nabla(\Gamma * f)$ are locally bounded by constants multiplied by the L^2 norm of f, so that we have local bounds on the full norm $||\Gamma * f||_{H^2}$ in terms of $||f||_{L^2}$.

Applying the same result to the derivatives of f we immediately obtain the following.

Corollary 2.13 *Given $f \in C_c^\infty(\Omega)$ for a bounded set Ω, let u be given by $u = \Gamma * f$. Then for every $k \ge 0$ we have*

$$||u||_{H^{k+2}(\Omega)} \le C(k, \Omega)||f||_{H^k(\Omega)}.$$

Lemma 2.14 *Assume that $u \in C^\infty(\Omega)$ is harmonic in Ω and take $x_0, r < R$ such that we have $\overline{B(x_0, R)} \subset \Omega$. Then, for every integer $k \ge 1$, we have*

$$||u||_{H^{1-k}(B(x_0,r))} \le C(k, r, R)||u||_{H^{-k}(B(x_0,R))}.$$

Proof We want to take $\varphi \in C^\infty$ with $\mathrm{spt}(\varphi) \subset B(x_0, r)$ and estimate $\int u\varphi$ in terms of $||\varphi||_{H^{k-1}}$ and $||u||_{H^{-k}(B(x_0,R))}$. To do this, we first consider $v = \Gamma * \varphi$ and a cutoff function $\eta \in C^\infty(\Omega)$ with $\eta = 1$ on $B(x_0, r)$ and $\eta = 0$ outside of $B_R := B(x_0, R)$. We write

$$0 = \int_{B_R} u \Delta(v\eta) \, dx = \int_{B_R} u\varphi\eta \, dx + \int_{B_R} uv\Delta\eta \, dx + 2\int_{B_R} u\nabla v \cdot \nabla\eta \, dx.$$

Using $\varphi\eta = \varphi$ (since $\eta = 1$ on $\mathrm{spt}(\varphi)$), we obtain

$$\int_{B_R} u\varphi \, dx = -\int_{B_R} uv\Delta\eta \, dx - 2\int_{B_R} u\nabla v \cdot \nabla\eta \, dx$$

$$\le ||u||_{H^{-k}(B(x_0,R))} \left(||v\Delta\eta||_{H^k(B(x_0,R))} + 2||\nabla v \cdot \nabla\eta||_{H^k(B(x_0,R))} \right).$$

Since η is smooth and fixed, and its norms only depend on r, R, we obtain

$$||v\Delta\eta||_{H^k(B(x_0,R))}, ||\nabla v \cdot \nabla\eta||_{H^k(B(x_0,R))} \le C(k, r, R)||\nabla v||_{H^k(B(x_0,R))}.$$

Applying the Corollary 2.13 we obtain $||\nabla v||_{H^k} \le ||v||_{H^{k+1}} \le C(k, r, R)||\varphi||_{H^{k-1}}$, which provides the desired result. □

We can then obtain

Proposition 2.15 *Assume that $u \in H_{loc}^{-k}(\Omega)$ is harmonic in Ω in the sense of distributions. Then u is an analytic function.*

Proof The proof is the same as in Proposition 2.11: we regularize by convolution and apply the bounds on the Sobolev norms. Lemma 2.14 allows us to pass from H^{-k} to H^{1-k} and, iterating, arrive at L^2. Once we know that u is in L_{loc}^2 we directly apply Proposition 2.11. $\square$

Finally, we have

Proposition 2.16 *Assume that u is a harmonic distribution in Ω. Then u is an analytic function.*

Proof We just need to show that u locally belongs to a space H^{-k} (said differently, every distribution is of finite order, when restricted to a compact set). This is a consequence of the definition of distributions. Indeed, we have the following: for every distribution u and every compact set $K \subset \Omega$ there exist n, C such that $\langle u, \varphi \rangle \leq C\|\varphi\|_{C^n}$ for every $\varphi \in C^\infty$ with $\text{spt}(\varphi) \subset K$. If we want to prove this we just need to proceed by contradiction: if it is not true, then there exists a distribution u and a compact set K such that for every n we can find a φ_n with

$$\langle u, \varphi_n \rangle = 1, \quad \|\varphi_n\|_{C^n} \leq \frac{1}{n}, \quad \text{spt}(\varphi_n) \subset K.$$

Note that we define the C^n norm as the sup of all derivatives up to order n; so that $\|\varphi\|_{C^{n+1}} \geq \|\varphi\|_{C^n}$. Yet, this is a contradiction since the sequence φ_n tends to 0 in the space of C_c^∞ functions and u should be continuous for this convergence. So we have the inequality $\langle u, \varphi \rangle \leq C\|\varphi\|_{C^n}$ which can be turned into $< u, \varphi > \leq C\|\varphi\|_{H^k}$ because of the continuous embedding of Sobolev spaces into C^n spaces (take $k > n + d/2$). $\square$

The techniques that we presented for the case $\Delta u = 0$ now allow us to discuss the regularity for the Poisson equation $\Delta u = f$ in Sobolev spaces H^k, and we can prove the following.

Proposition 2.17 *Assume that u is a distributional solution in Ω of $\Delta u = f$, where $f \in H_{loc}^{-k}(\Omega)$. Then $u \in H_{loc}^{2-k}(\Omega)$.*

Proof Let B be an open ball compactly contained in Ω and let us define a distribution $\tilde{f}$ of the form ηf, where $\eta \in C_c^\infty(\Omega)$ is such that $B \subset \{\eta = 1\}$. Since u and $\Gamma * \tilde{f}$ have the same Laplacian in B, they differ by a harmonic function, and we know that harmonic functions are locally smooth, so that we just need to prove $\Gamma * \tilde{f} \in H_{loc}^{2-k}(B)$. This result is proven in Corollary 2.13 if $k \leq 0$, so we can assume $k \geq 1$.

We take a test function $\phi \in C_c^\infty(B)$ and we use the properties of the convolution to write

$$\int_B (\Gamma * \tilde{f})\phi \, dx = \int_B \tilde{f}(\Gamma * \phi) \, dx = \int_B f\eta(\Gamma * \phi) \, dx \leq \|f\|_{H^{-k}(B)} \|\eta(\Gamma * \phi)\|_{H^k(B)}.$$

We then exploit the C^∞ regularity of η to obtain $\|\eta(\Gamma * \phi)\|_{H^k(B)} \leq C\|\Gamma * \phi\|_{H^k(B)}$ and, if $k \geq 2$, we have, using again Corollary 2.13, $\|\eta(\Gamma * \phi)\|_{H^k(B)} \leq C\|\phi\|_{H^{k-2}(B)}$. This proves $\int (\Gamma * \tilde{f})\phi \leq C\|\phi\|_{H^{k-2}(B)}$, i.e. $\Gamma * \tilde{f} \in H^{2-k}(B)$.

We are left with the case $k = 1$. Since this case is treated in a different way, we refer to a separate proposition for it. □

Proposition 2.18 Let $B_i = B(x_0, R_i)$, $i = 1, 2$, be two concentric balls with $R_1 < R_2$. Then, given $f \in H^{-1}(B_2)$, setting $u = \Gamma * f$, we have $\|u\|_{H^1(B_1)} \leq C(R_1, R_2)\|f\|_{H^{-1}(B_2)}$.

Proof By density, it is enough to prove this inequality if $f \in C^\infty$. We choose a smooth cut-off function χ with $0 \leq \chi \leq 1$, $\chi = 1$ on B_1 and spt $\chi \subset B_2$. We then write

$$\int_{B_2} fu\chi^2 \, dx = \int_{B_2} \Delta u(u\chi^2) \, dx = -\int_{B_2} \nabla u \nabla(u\chi^2) \, dx,$$

where in the integration by parts there are no boundary terms since χ is compactly supported. We then obtain

$$\int_{B_1} |\nabla u|^2 \, dx \leq \int_{B_2} |\nabla u|^2 \chi^2 \, dx$$

and

$$\int_{B_2} |\nabla u|^2 \chi^2 \, dx = -\int_{B_2} fu\chi^2 \, dx - 2\int_{B_2} \nabla u \cdot \nabla \chi u\chi \, dx.$$

The right-hand side above can be bounded by

$$\|f\|_{H^{-1}} \|\nabla(u\chi^2)\|_{L^2} + C\|\nabla u\chi\|_{L^2}\|u\nabla\chi\|_{L^2}.$$

Using $\|\nabla(u\chi^2)\|_{L^2} \leq C\|\nabla u\chi\|_{L^2} + C\|u\nabla\chi\|_{L^2}$ (an estimate where we use $|\chi| \leq 1$) and applying Young's inequality we obtain

$$\int_{B_2} |\nabla u|^2 \chi^2 \leq C\|f\|_{H^{-1}(B_2)}^2 + C\|f\|_{H^{-1}(B_2)}\|u\nabla\chi\|_{L^2(B_2)} + C\|u\nabla\chi\|_{L^2(B_2)}^2.$$

The same estimate as in Proposition 2.17, for $k = 2$, shows that we have $\|u\|_{L^2(B_2)} \leq C\|f\|_{H^{-2}(B_2)} \leq C\|f\|_{H^{-1}(B_1)}$, so that we can deduce from the previous inequality

$$\int_{B_2} |\nabla u|^2 \chi^2 \, dx \leq C(r, R)\|f\|_{H^{-1}(B_2)}^2$$

and then

$$\int_{B_1} |\nabla u|^2 \, dx \leq C(r, R)\|f\|_{H^{-1}(B_2)}^2,$$

which is the claim.   □

Box 2.7 Good to Know—*Manifold- and Metric-Valued Harmonic Maps*

All the regularity theory on harmonic functions which has been presented in this section relies on the linear structure of the target space (we presented the results for scalar harmonic maps, but working componentwise they are easily seen to be true for vector-valued harmonic maps).

Given a manifold $M \subset \mathbb{R}^N$, one could consider the problem

$$\min\{\int_\Omega |\nabla u|^2 \; : \; u \in H^1(\Omega; \mathbb{R}^N), \; u \in M \text{ a.e.}, u - g \in H_0^1(\Omega)\}$$

together with its optimality conditions written as a PDE. This PDE would not be $\Delta u = 0$ as it should take into account the constraint, and formally one finds $\Delta u = c(x)\mathbf{n}_M(u)$ (the Laplacian is pointwisely a vector normal to the manifold M at the point u). In this case, because of the non-convexity of the problem, the PDE would not be sufficient for the minimization, and because of the right-hand side in $\Delta u = f$ the regularity is not always guaranteed.

For instance, when $M = \mathbb{S}^2 \subset \mathbb{R}^3$ the equation is $\Delta u = -|\nabla u|^2 u$ (see Exercise 2.10) and $u(x) = x/|x|$, a function defined on $\Omega = B(0, 1) \subset \mathbb{R}^3$ which solves it, is H^1, but not continuous. This lack of regularity is not surprising considering that the right-hand side $-|\nabla u|^2 u$ is only L^1. It has been pointed out in [165] that the condition of being an H^1 distributional solution of $\Delta u = -|\nabla u|^2 u$ is so weak that for any boundary data there are infinitely many solutions. Yet, the function $u(x) = x/|x|$ is not only a solution of this PDE, but also a minimizer of the Dirichlet energy for M-valued maps which are the identity on $\partial B(0, 1)$. This nontrivial fact was first proven by Brezis, Coron and Lieb in [49] and then extended to other dimensions (and even other $W^{1,p}$ norms, for $p < d$) in [116] and [139].

(continued)

Box 2.7 (continued)

The situation is different when M has non-positive curvature, in which case it is possible to prove regularity of harmonic maps valued in M. These harmonic maps were first studied by Ells and Sampson in [80]. Then, Schoen and Uhlenbeck proved in [181], for maps minimizing the Dirichlet energy $\int |\nabla u|^2$ (and not only solving the Euler–Lagrange equation), some partial regularity results (smoothness out of a lower-dimensional set), guaranteeing in some cases that the singular set is empty. The interested reader can have a look at the book [122, Chapter 8] or at the survey by Hélein and Woods [112, Section 3.1].

Finally, it is possible to study the problem of minimizing the Dirichlet energy when replacing the Euclidean space or its submanifolds as a target space with a more abstract metric space. This requires us to give a definition of this energy, as the notion of gradient is not well-defined for maps valued in a metric space (consider that we also had to provide a suitable definition for the derivative of curves valued in a metric space, see Sect. 1.4.1). In this case the key observation to define a Dirichlet energy is the fact that, for the Euclidean case, we have for $u \in C^1_c(\mathbb{R}^d)$

$$\int_{\mathbb{R}^d} |\nabla u|^2 \, dx = c(d) \lim_{\varepsilon \to 0^+} \varepsilon^{-(d+2)} \int \int_{|x-y|<\varepsilon} |u(x) - u(y)|^2 \, dx \, dy,$$

where $c(d)$ is a dimensional constant. Then it is possible to replace $|u(x) - u(y)|$ with $d(u(x), u(y))$ and study the maps minimizing the limit energy. For this approach we refer to [127] and [122]. Besides the definition of the energy, these references provide a study of the harmonic maps valued in a metric space (X, d), provided it has negative curvature in a suitable sense.

2.4 Discussion: p-Harmonic Functions for $1 \le p \le \infty$

Section 2.3 provides a full description of the functions which minimize $\|\nabla u\|_{L^2}$ with given boundary conditions, which are the solutions of $\Delta u = 0$ (and actually more, since it also considers harmonic distributions, i.e. we discuss distributional solutions of $\Delta u = 0$ without requiring $u \in H^1$). A similar problem can be considered by minimizing $\|\nabla u\|_{L^p}$, for arbitrary p. Let us start with $p \in (1, \infty)$.

From the variational point of view, minimizing $\int_\Omega |\nabla u|^p \, dx$ with given boundary conditions $u - u_0 \in W^{1,p}_0(\Omega)$ is a very classical and simple problem: not only can we easily show that the minimizers exist thanks to the techniques presented in this chapter, but they are unique (because of strict convexity), and they are characterized by the p-Laplacian equation

$$\nabla \cdot (|\nabla u|^{p-2} \nabla u) = 0,$$

whose solutions are called *p-harmonic functions*. Moreover, the growth conditions on the integrand allow us to formulate the equation as $\int |\nabla u|^{p-2} \nabla u \cdot \nabla \varphi = 0$ for every $\varphi \in W_0^{1,p}(\Omega)$.

For $p = 2$ we have harmonic functions, and we proved that they are actually smooth and even analytic. What about p-harmonic functions? These functions have been the object of a huge number of works in elliptic PDEs: there are of course regularity results, but the situation is actually much trickier than the linear case $p = 2$. We will see in Chap. 4 some results—which are presented in this book through techniques coming from convex duality but could have been introduced using more classical PDE tools—about the Sobolev regularity of solutions of $\nabla \cdot (|\nabla u|^{p-2} \nabla u) = f$ (even for $f \neq 0$, but depending of course on the regularity of f). When we mention Sobolev regularity, considering that we already assume $u \in W^{1,p}$, we mean Sobolev regularity of the gradient, i.e. proving that ∇u, or some function of ∇u, belongs to H^1 or some other spaces H^s.

Another possible, and classical, direction of research could be the Hölder regularity of the gradient. This started with [192], which proved that, among other results, for $p \geq 2$, functions which are p-harmonic are $C^{1,\alpha}$. In dimension $d = 2$ the result can be made much stronger and sharp, as it is possible to prove (see [119]) $u \in C^{k,\alpha}$ with

$$k + \alpha = \frac{1}{6} \left(7 + \frac{1}{p-1} + \sqrt{1 + \frac{14}{p-1} + \frac{1}{(p-1)^2}} \right). \tag{2.9}$$

The literature on the p-Laplacian equation is huge (see, for instance, the classical references [30, 131]), and these considerations as well as those in Chap. 4 only cover a small part of the results.

It is then interesting to consider the limit cases $p = 1$ and $p = \infty$. Let us start from the latter. We can take two different approaches: either we look at the limit of the minimization problem, and we consider functions minimizing $\|\nabla u\|_{L^\infty}$ with prescribed boundary datum, or we look at the limit of the equation. The equation $\nabla \cdot (|\nabla u|^{p-2} \nabla u) = 0$ can be expanded and it can be written as

$$|\nabla u|^{p-2} \Delta u + (p-2)|\nabla u|^{p-4} \nabla u \cdot D^2 u \nabla u = 0.$$

We can factorize $(p-2)|\nabla u|^{p-4}$ and this becomes

$$\frac{\Delta u}{p-2} + \nabla u \cdot D^2 u \nabla u = 0.$$

In the limit $p \to \infty$, only the second terms remains, and we define ∞-harmonic functions to be those which solve, in a suitable sense, the non-linear equation

$$\nabla u \cdot D^2 u \nabla u = 0.$$

Since the equation is non-linear and is not in divergence form, we cannot use distributional or weak solutions. On the other hand, the function $(p, M) \mapsto p \cdot Mp$ is non-decreasing in M, which allows us to use the notion of viscosity solutions (see below). Thus, we define ∞-harmonic functions as those which solve the above equation in the viscosity sense. An important result (many proofs are available, but we suggest the reader to look at [19], which is very elementary) states that there exists a unique viscosity solution u of $\nabla u \cdot D^2 u \nabla u = 0$ with given boundary conditions. Solutions of this equation, that we also write as $\Delta_\infty u = 0$, are said to be ∞-harmonic.

Box 2.8 Important Notion—*Viscosity Solution of Second-Order PDEs*

Let us take a function $F : \Omega \times \mathbb{R} \times \mathbb{R}^d \times \mathbb{R}^{d \times d}_{sym} \to \mathbb{R}$, where $\mathbb{R}^{d \times d}_{sym}$ stands for the space of $d \times d$ symmetric matrices. Let us assume that F is non-decreasing in the last variable, in the sense that $F(x, s, p, M) \leq F(x, s, p, N)$ every time that $N - M$ is positive-definite. We then say that a function $u \in C^0(\Omega)$ is a viscosity solution of $F(x, u, \nabla u, D^2 u) = 0$ if it satisfies the following two properties

- for every $x_0 \in \Omega$ and $\varphi \in C^2(\Omega)$ such that $\varphi \geq u$ but $\varphi(x_0) = u(x_0)$ we have $F(x_0, \varphi(x_0), \nabla \varphi(x_0), D^2 \varphi(x_0)) \geq 0$;
- for every $x_0 \in \Omega$ and $\varphi \in C^2(\Omega)$ such that $\varphi \leq u$ but $\varphi(x_0) = u(x_0)$ we have $F(x_0, \varphi(x_0), \nabla \varphi(x_0), D^2 \varphi(x_0)) \leq 0$.

As a reference for the theory of viscosity solutions for second-order PDEs, see [58].

Another approach to the limit $p \to \infty$ concerns the limit of the optimization problem. Limits of optimization problems will be discussed in detail in Chap. 7, but in this case it is very easy, as one only needs to replace the L^p norm with the L^∞ norm. What can we say about minimizers of $u \mapsto \|\nabla u\|_{L^\infty}$ with given boundary conditions? On convex domains, this quantity coincides with the Lipschitz constant, which makes it easy to prove the existence of a solution using the Ascoli–Arzelà theorem. More than this, it is also possible to identify the optimal Lipschitz constant, which is equal, when speaking of real-valued functions, to the Lipschitz constant of the boundary datum. Indeed, given an L-Lipschitz function on a set A (not necessarily $\partial \Omega$), it is always possible to extend it to the whole space keeping L as a Lipschitz constant. For instance, given $u : A \to \mathbb{R}$, one can define $\underline{u}(x) :=$

$\sup\{L|x - y| + u(y) \, : \, y \in A\}$. But this is only a possible choice, since we can, for instance, also choose $\overline{u}(x) := \inf\{-L|x - y| + u(y) \, : \, y \in A\}$. These two functions do not coincide in general, and they are actually the smallest and the largest possible L-Lipschitz extension of u, in the sense that any $u \in \text{Lip}_L$ with prescribed value on A satisfies $\underline{u} \leq u \leq \overline{u}$.

As a consequence, we understand that many different functions can minimize $u \mapsto ||\nabla u||_{L^\infty}$ with given boundary conditions, but which is the one that we can find as the limit of the solutions u_p of $-\Delta_p u = 0$ with the same boundary conditions? (or, equivalently, which is the limit of the minimizers of $u \mapsto ||\nabla u||_{L^p}$?). We see now that the minimization problem of the L^∞ norm lacks a property which is obviously satisfied by the minimization of the L^p norm and, more generally, of all integral functionals: when we minimize on a domain Ω an integral cost of the form $\int_\Omega L(x, u, \nabla u)\,dx$, the solution u is such that, for every subdomain $\Omega' \subset \Omega$, it also minimizes $\int_{\Omega'} L(x, u, \nabla u)\,dx$ among all functions defined on Ω' sharing the same value with u on $\partial\Omega'$. This is just due to the additivity of the integral, so that if a better function existed on Ω' we could use it on Ω' instead of u, and strictly improve the value of the integral. When the functional is defined as a sup, if the sup is not realized in Ω' but in $\Omega \setminus \Omega'$, most likely modifying u on Ω' does not change the value of the functional. This means that minimizers of the L^∞ norm of the gradient can, in many situations, be almost arbitrary in large portions of the domain Ω and, unlike the case of the L^p norm, cannot be characterized by any PDE. This motivates the following definition.

Definition 2.19 A Lipschitz function $u \, : \, \Omega \to \mathbb{R}$ is said to be an *absolute minimizer* of the L^∞ norm of the gradient if it satisfies the following condition: for every $\Omega' \subset \Omega$ we have

$$||\nabla u||_{L^\infty(\Omega')} = \min\{||\nabla \tilde{u}||_{L^\infty(\Omega')} \, : \, \tilde{u} \in \text{Lip}(\overline{\Omega'}), \, \tilde{u} = u \text{ on } \partial\Omega'\}.$$

We then have the following results, which we will not prove.

Theorem 2.20 *A Lipschitz function $u \, : \, \Omega \to \mathbb{R}$ is an absolute minimizer if and only if it is a viscosity solution of $\Delta_\infty u = 0$.*

Given a Lipschitz boundary datum $g \, : \, \partial\Omega \to \mathbb{R}$, the functions u_p defined as the unique solution of $\Delta_p u_p = 0$ in Ω with $u_p = g$ on $\partial\Omega$ uniformly converge as $p \to \infty$ to the function u_∞ defined as the unique viscosity solution of $\Delta_\infty u = 0$ with the same boundary data.

We now switch to the other extreme case, i.e. $p = 1$. The first difficulty in solving $\min ||\nabla u||_{L^1}$ with given boundary data consists in the fact that the space L^1 is neither reflexive (as is the case for $1 < p < \infty$) nor it is a dual space (as is the case for $p = \infty$). So, it is in general not possible to extract weakly convergent subsequences from bounded sequences, and the minimization problem could turn out to be ill-posed if we consider $u \in W^{1,1}$. The natural framework in order to obtain the existence of a minimizer is to extend the problem to $u \in BV$ and replace the L^1 norm of the gradient with the mass of the measure ∇u (see Sects 1.5 and 1.6

for the first appearance of BV functions, and Box 6.1 for the multi-dimensional case). Exactly as in the 1D case (in particular, as it happened in Sect. 1.6), boundary data are not always well-defined for BV functions, as they could jump exactly on the boundary. Hence, it is the same to solve

$$\min\{\|\nabla u\|_{\mathcal{M}(\overline{\Omega})} : u \in BV(\Omega), u = g \text{ on } \partial\Omega\},$$

where the condition $u = g$ on $\partial\Omega$ can be intended as "u is a BV function on a larger domain $\tilde{\Omega}$ which contains Ω in its interior, which coincides with an extension of g on $\tilde{\Omega} \setminus \Omega$, this extension being continuous in a suitable sense in $\tilde{\Omega} \setminus \Omega$", or to solve

$$\min\{\|\nabla u\|_{\mathcal{M}(\Omega)} + \int_{\partial\Omega} |\text{Tr}[u] - g| \, d\mathcal{H}^{d-1} : u \in BV(\Omega)\}, \qquad (2.10)$$

where the trace of a BV function should also be suitably defined (and corresponds in this case to the trace *coming from inside* Ω).

In particular, it is possible for certain choices of g and of Ω that the solution of this problem (which is in general not unique, because the functional which is minimized is no longer strictly convex) is a smooth function inside Ω but does not take the value g on $\partial\Omega$.

Formally, the equation solved by the minimizers is the 1-Laplacian equation, i.e.

$$\Delta_1 u := \nabla \cdot \left(\frac{\nabla u}{|\nabla u|} \right) = 0$$

and functions which solve it are called, besides 1-harmonic, *least gradient* functions.

In order to treat points where $\nabla u = 0$, a possible interpretation of the above equation is the following: there exists a vector field $z : \Omega \to \mathbb{R}^d$ with $|z| \leq 1$ and $z \cdot \nabla u = |\nabla u|$ (i.e. $z = \nabla u/|\nabla u|$ whenever $\nabla u \neq 0$) such that $\nabla \cdot z = 0$. Since the condition $|z| \leq 1$ already implies $z \cdot \nabla u \leq |\nabla u|$, the equality $z \cdot \nabla u = |\nabla u|$ is equivalent to $\int z \cdot \nabla u = \int |\nabla u|$, a condition which has a meaning also when ∇u is a measure.

Note that, whenever u is a smooth function with non-vanishing gradient, the quantity $\Delta_1 u(x_0)$ coincides with the curvature (in the sense of the sum of the principal curvatures) of the codimension 1 surface $\{u = u(x_0)\}$ at the point x_0. In particular, imposing that it vanishes means imposing that these surfaces all have zero mean curvature; in dimension $d = 2$ they should be locally segments. Finally, also note that the level sets are invariant if one composes u with a monotone increasing function, and this is absolutely consistent with the fact that we have $\Delta_1 u = \Delta_1(f(u))$ for every monotone increasing function f, just because the quantity $\nabla u/|\nabla u|$ is 0-homogeneous and it is not affected by the multiplication by $f'(u)$ which appears in the composition.

We do not want to give general results on least gradient functions here, but we want to underline a nice transformation which is only possible in dimension $d = 2$. Indeed, in the two-dimensional case, if we set $\mathbf{v} := R\nabla u$, where R is a 90° rotation, we have $\nabla \cdot \mathbf{v} = 0$ in Ω. Moreover, the Dirichlet condition on u, which can be seen (up to additive constants) as a condition on its tangential derivative, becomes after this rotation a condition on the normal component of $\mathbf{v}$. Imposing the divergence in the interior of Ω and $\mathbf{v} \cdot \mathbf{n}$ on $\partial \Omega$ is the same as imposing the distributional divergence (which can have a singular part on $\partial \Omega$) on the whole space of the vector field obtained by extending $\mathbf{v}$ to 0 outside Ω. Since the norms of $\mathbf{v}$ and of ∇u are the same, the optimization problem defining least gradient functions can be reformulated as the minimization of the L^1 norm (or of the mass of a vector measure) under divergence constraints. This minimal flow problem will be discusses in Chap. 4 and in particular in the discussion Sect. 4.6, and it is equivalent to an optimal transport problem from the positive to the negative part of the prescribed divergence. In this case, it consists in an optimal transport problem between measures supported on $\partial \Omega$. This reformulation, suggested in [104], made it possible in [79] to prove the following results in the 2D case:

Theorem 2.21 *Assume that Ω is a 2D strictly convex domain and assume that g is a BV function on the 1D curve $\partial \Omega$. Then the solutions of Problem (2.10) satisfy $\mathrm{Tr}[u] = g$ on Ω. If moreover g is continuous on $\partial \Omega$, then the solution is unique.*

If furthermore Ω is uniformly convex and $g \in W^{1,p}(\partial \Omega)$ for $p \leq 2$, then the solution u satisfies $u \in W^{1,p}(\Omega)$. If $g \in C^{1,\alpha}(\partial \Omega)$, then u satisfies $u \in W^{1,p}(\Omega)$ for $p = \frac{2}{1-\alpha}$. If $g \in C^{1,1}$, then u is Lipschitz continuous.

The very last part of the above statement (Lipschitz regularity if the boundary data are $C^{1,1}$) is a particular case of what can be proven under the so-called *bounded slope condition*, which we do not detail here and for which we refer, for instance, to [64, 186].

2.5 Exercises

Exercise 2.1 Prove the existence and uniqueness of the solution of

$$\min \left\{ \int_\Omega \left(f(x)|u(x)| + |\nabla u(x)|^2 \right) \, dx \; : \; u \in H^1(\Omega), \int_\Omega u = 1 \right\}$$

when Ω is an open, connected and bounded subset of $\mathbb{R}^d$ and $f \in L^2(\Omega)$. Where do we use connectedness? Also prove that, if Ω is not connected but has a finite number of connected components and we add the assumption $f \geq 0$, then we have existence but maybe not uniqueness, and that if we withdraw both connectedness and positivity of f, then we might not even have existence.

Exercise 2.2 Consider $\Omega = (0, 1)^d \subset \mathbb{R}^d$ and $A = \{0\} \times (0, 1)^{d-1} \subset \partial \Omega$. Prove that the set of functions $u \in W^{1,p}(\Omega)$ such that $\mathrm{Tr}[u] = 0$ a.e. on A coincides with the closure in $W^{1,p}(\Omega)$ of the set $\{\varphi \in C^1(\overline{\Omega}) \; : \; \mathrm{spt}(\varphi) \cap A = \emptyset\}$.

Exercise 2.3 Given a smooth and connected open domain $\Omega \subset \mathbb{R}^d$ and two exponents $\alpha, \beta > 0$, consider the following problem:

$$\min \left\{ \int_\Omega \left(-|u|^\alpha + |\nabla u|^p \right) dx + \int_{\partial\Omega} |\mathrm{Tr}[u]|^\beta \, d\mathcal{H}^{d-1} : u \in W^{1,p}(\Omega) \right\}.$$

Prove that the above problem has a solution if $\alpha < \min\{\beta, p\}$ and that it has no solution if $\alpha > \min\{\beta, p\}$.

Exercise 2.4 Given a measurable function $c : \Omega \to \mathbb{R}$ such that $c_0 \leq c \leq c_1$ a.e., for two strictly positive constants c_0, c_1, a function $g \in W^{1,p}(\Omega)$, and a continuous function $F : \Omega \times \mathbb{R} \to \mathbb{R}_+$, prove that the following problem has a solution:

$$\min \left\{ \int_\Omega \left(c(x)|\nabla u|^p + F(x, u(x)) \right) dx : u - g \in W_0^{1,p}(\Omega) \right\}.$$

Exercise 2.5 Fully solve

$$\min \left\{ \int_Q \left(|\nabla u(x, y)|^2 + u(x, y)^2 \right) dx\, dy : u \in C^1(Q), u = \phi \text{ on } \partial Q \right\},$$

where $Q = [-1, 1]^2 \subset \mathbb{R}^2$ and $\phi : \partial Q \to \mathbb{R}$ is given by

$$\phi(x, y) = \begin{cases} 0 & \text{if } x = -1, \ y \in [-1, 1] \\ 2(e^y + e^{-y}) & \text{if } x = 1, \ y \in [-1, 1] \\ (x + 1)(e + e^{-1}) & \text{if } x \in [-1, 1], \ y = \pm 1. \end{cases}$$

Find the minimizer and the value of the minimum.

Exercise 2.6 Which among the functions u which can be written in polar coordinates as $u(\rho, \theta) = \rho^\alpha \sin(k\theta)$ are harmonic on the unit ball? (i.e. for which values of $\alpha, k \in \mathbb{R}$).

Exercise 2.7 Let $u \in \mathcal{D}'(\mathbb{R}^d)$ be a distributional solution of $\Delta u = b(x) \cdot \nabla u + f(x)u$, where $f : \mathbb{R}^d \to \mathbb{R}$ and $b : \mathbb{R}^d \to \mathbb{R}^d$ are given C^∞ functions. Prove $u \in C^\infty(\mathbb{R}^d)$.

Exercise 2.8 Prove, using the mean property, that any bounded harmonic function u on the whole space $\mathbb{R}^d$ is constant. Also prove that any harmonic function u on the whole space $\mathbb{R}^d$ satisfying a growth bound of the form $|u(x)| \leq C(1 + |x|)^p$ for some exponent p is actually a polynomial function.

Exercise 2.9 Prove that any smooth solution of $\Delta u = f$ in $\Omega \subset \mathbb{R}^d$, $d \geq 3$, satisfies

$$\fint_{\partial B(x_0,R)} u \, dx = u(x_0) + \frac{1}{d(d-1)\omega_d} \int_{B(x_0,R)} f(x) \left(|x - x_0|^{1-d} - R^{1-d} \right) dx.$$

Find an analogous formula for the case $d = 2$ and for the average on the ball instead of the sphere.

Exercise 2.10 Let $M \subset \mathbb{R}^N$ be the zero set of a smooth function $h : \mathbb{R}^N \to \mathbb{R}$ with $\nabla h \neq 0$ on M. Prove that if a function $u : \Omega \to M$ (with $\Omega \subset \mathbb{R}^d$) satisfies $\Delta u(x) = c(x)\nabla h(u(x))$ for a scalar function c, then we have

$$c = -\frac{\sum_{i=1}^{d} D^2 h(u(x))u_i(x) \cdot u_i(x)}{|\nabla h(u(x))|^2}.$$

Also prove that any smooth function $u : \Omega \to M$ satisfying $\Delta u(x) = c(x)\nabla h(u(x))$ for the above expression of c is a minimizer of $v \mapsto \int_\Omega |\nabla v|^2$ with prescribed boundary datum as soon as Ω is contained in a small enough ball.

Exercise 2.11 Determine for which values of $p > 1$ the function $u : \mathbb{R}^2 \to \mathbb{R}$ given by $u(x_1, x_2) := |x_1|^p - |x_2|^p$ is ∞-harmonic, and use this information to give bounds on the optimal $C^{1,\alpha}$ regularity of ∞-harmonic functions. Compare to Formula 2.9.

Exercise 2.12 For $p \in (1, \infty)$, let u be a solution of $\Delta_p u = 0$ on Ω. Let M be a constant such that $\mathrm{Tr}[u] \leq M$ a.e. on $\partial\Omega$. Prove $u \leq M$ a.e. in Ω.

Hints

Hint to Exercise 2.1 Minimizing sequences are bounded in H^1 because of the Poincaré–Wirtinger inequality. This cannot be applied when Ω is non-connected. In this case we use $f \geq 0$ and the term with $\int f|u|$ on each component to bound a suitable L^1 norm of u (distinguishing the components where f is identically 0, where we can replace u with a constant).

Hint to Exercise 2.2 We approximate u with $u(1 - \eta_n)$ where η_n is a cutoff function with $\eta_n = 1$ on A and $\mathrm{spt}\,\eta_n \subset [0, 1/n] \times [0, 1]^{d-1}$. We need to prove $u\eta_n \to 0$ at least weakly in $W^{1,p}$. The key point is bounding the norm $||u\nabla\eta_n||_{L^p}$ using the trace of u and suitable Poincaré inequalities.

Hint to Exercise 2.3 If α is large the infimum is $-\infty$ using either large constants if $\alpha > \beta$ or nu_0 with $u_0 \in W_0^{1,p}$ if $\alpha > p$. If α is small we can prove compactness.

Hint to Exercise 2.4 For a minimizing sequence, prove that ∇u_n is bounded and hence weakly convergent, both in $L^p(\,\mathrm{d}x)$ and in $L^p(c(x)\,\mathrm{d}x)$.

Hint to Exercise 2.5 Write and solve the Euler–Lagrange equation (by guessing the form of the solution).

Hint to Exercise 2.6 For smoothness, we need $\alpha \geq 0$ and $k \in \mathbb{Z}$. When $\alpha^2 = k^2$ we have the real part of a holomorphic function. In general, compute the gradient and its divergence in polar coordinates.

Hint to Exercise 2.7 Start from a distribution in a certain H^{-k} and prove that it actually belongs to H^{1-k}.

Hint to Exercise 2.8 Use Proposition 2.8 to prove $\nabla u = 0$. For the polynomial growth, prove $|\nabla u| \leq C(1 + |x|)^{p-1}$ and iterate.

Hint to Exercise 2.9 Re-perform the proof of the mean property keeping into account the non-vanishing Laplacian.

Hint to Exercise 2.10 Differentiate $h \circ u$ twice. For the optimality, act as we did for geodesics in Sect. 1.1.2.

Hint to Exercise 2.11 Check that only $p = 4/3$ works. This also coincides with the limit regularity for p-harmonic functions when $p \to \infty$, which gives $8/6$.

Hint to Exercise 2.12 Use the variational interpretation of p-harmonic functions as minimizers of the L^p norm of the gradient, and truncate them at M obtaining a better competitor.

Chapter 3
Lower Semicontinuity

As we saw in the two previous chapters, the standard way to prove the existence of a solution to a variational problem consists in what is called the *direct method* of the calculus of variations (several books have been devoted to this topic, see for instance [67] and [101]), which includes as a key ingredient the lower semicontinuity of the functional to be minimized w.r.t. a convergence for which we have compactness of a minimizing sequence. Lower semicontinuity results are then a crucial aspect in the calculus of variations, and in particular the lower semicontinuity w.r.t. weak and weak-* convergences, since it is usually for these kinds of convergences that we are able to obtain compactness. This chapter is devoted to this kind of result, the main goal being to be able to treat more general functionals than those we considered in Chap. 2, where as far as weak convergence was concerned only the semicontinuity of the norm was used. The first section is devoted to more abstract results, the second to general results on the semicontinuity of integral functionals for strong and weak convergence in L^p spaces (essentially based on Chap. 4 of [101]), and the third (presented for completeness—since this theory will appear less often in the rest of the book—and mainly based on the paper [35]) to the semicontinuity of some class of functionals for the weak-* convergence of measures. A short section on the application of these results to the existence of minimizers follows. The chapter closes with a discussion section on the case of vector-valued functional spaces, where the assumptions could be relaxed, but a much harder theory would be needed, followed by an exercise list.

F. Santambrogio, *A Course in the Calculus of Variations*, Universitext,
https://doi.org/10.1007/978-3-031-45036-5_3

3.1   Convexity and Lower Semicontinuity

We recall the sequential definition of lower semicontinuity, which will be the one we will use throughout this book:[1] a function $f : X \to \mathbb{R} \cup \{+\infty\}$ is said to be lower semicontinuous (l.s.c. for short) if for every sequence $x_n \to x$ in X we have $\liminf_n f(x_n) \geq f(x)$. The definition can also be re-written by assuming that $f(x_n)$ converges, since we can always extract a subsequence which realizes the liminf as a limit. In this case we could say that f is lower semicontinuous if for every sequence $x_n \to x$ such that $f(x_n) \to \ell \in \overline{\mathbb{R}}$ we have $f(x) \leq \ell$.

We start with a characterization of lower semicontinuity in terms of level sets.

Proposition 3.1 *A function $f : X \to \mathbb{R} \cup \{+\infty\}$ is l.s.c. if and only if for every ℓ the level set $A(\ell) := \{x \in X : f(x) \leq \ell\}$ is closed.*

Proof First we prove that the sequential lower semicontinuity implies the closedness of the level sets. Take a sequence $x_n \in A(\ell)$ and assume $x_n \to x$. From $f(x_n) \leq \ell$ and $f(x) \leq \liminf_n f(x_n)$ we deduce $f(x) \leq \ell$, hence $x \in A(\ell)$, so that $A(\ell)$ is closed.

Assume now that the sets $A(\ell)$ are all closed, and take a sequence $x_n \to x$. Setting $\ell_0 := \liminf_n f(x_n)$, for every $\ell > \ell_0$ we have $x_n \in A(\ell)$ infinitely often (i.e. there is a subsequence x_{n_k} such that $x_{n_k} \in A(\ell)$). From the closedness of $A(\ell)$ we deduce $x \in A(\ell)$, i.e. $f(x) \leq \ell$ for every $\ell > \ell_0$. This implies $f(x) \leq \ell_0 = \liminf_n f(x_n)$, and f is l.s.c.   □

A similar characterization can be given in terms of epigraphs. We define

$$\text{Epi}(f) := \{(x, t) \in X \times \mathbb{R} : t \geq f(x)\}$$

and we have:

Proposition 3.2 *A function $f : X \to \mathbb{R} \cup \{+\infty\}$ is l.s.c. if and only if $\text{Epi}(f)$ is closed.*

Proof First let us start from sequential lower semicontinuity and prove the closedness of the epigraph. Take a sequence $(x_n, t_n) \in \text{Epi}(f)$ and assume $x_n \to x$, $t_n \to t$. From $f(x_n) \leq t_n$ and $f(x) \leq \liminf_n f(x_n)$ we deduce $f(x) \leq t$, hence $(x, t) \in \text{Epi}(f)$ so $\text{Epi}(f)$ is closed.

Assume now that $\text{Epi}(f)$ is closed, and take a sequence $x_n \to x$. Setting $\ell_0 := \liminf_n f(x_n)$, for every $\ell > \ell_0$ we have $(x_n, \ell) \in \text{Epi}(f)$ infinitely often. From the closedness of the epigraph we deduce $(x, \ell) \in \text{Epi}(f)$, i.e. $f(x) \leq \ell$ for every $\ell > \ell_0$. Again, this implies $f(x) \leq \ell_0 = \liminf_n f(x_n)$, and f is l.s.c.   □

---

[1] Analogously, in the sequel the notion of closed set will be identified with that of sequentially closed; for the sake of the topologies that we shall consider, which are either metrizable or metrizable on bounded sets, this will be enough.

Another important property of lower semicontinuous functionals is that they are stable under sup. More precisely:

Proposition 3.3 *Given a family of functions* $f_\alpha : X \to \mathbb{R} \cup \{+\infty\}$ *which are all l.s.c., define* $f(x) := \sup_\alpha f_\alpha(x)$. *Then* f *is also l.s.c.*

Proof Two easy proofs can be given of this fact. One is based on the previous characterization: for every ℓ we have $\{f \le \ell\} = \{x : f_\alpha(x) \le \ell \text{ for every } \alpha\} = \bigcap_\alpha \{f_\alpha \le \ell\}$. If all functions f_α are l.s.c. all the sets $\{f_\alpha \le \ell\}$ are closed, and so is their intersection, so that $\{f \le \ell\}$ is closed and f is l.s.c.

Another proof is based on the definition via the liminf. Take $x_n \to x$ and write

$$f_\alpha(x) \le \liminf_n f_\alpha(x_n) \le \liminf_n f(x_n).$$

It is then enough to take the sup over α in the left-hand side in order to obtain

$$f(x) \le \liminf_n f(x_n),$$

which is the desired result. □

All these properties are highly reminiscent of those of convex functions. In order to define the notion of convexity we need the space X to be a vector space, which we will rather write as $\mathcal{X}$. We say that $f : \mathcal{X} \to \mathbb{R} \cup \{+\infty\}$ is convex if for every $x, y \in \mathcal{X}$ and $t \in [0, 1]$ we have

$$f((1 - t)x + ty) \le (1 - t)f(x) + tf(y).$$

We have the following properties

Proposition 3.4 *A function* $f : \mathcal{X} \to \mathbb{R} \cup \{+\infty\}$ *is convex if and only if* $\mathrm{Epi}(f)$ *is convex.*

Proof First let us start from the convexity of f and let us prove the convexity of the epigraph. Take two points $(x_0, t_0), (x_1, t_1) \in \mathrm{Epi}(f)$ and consider the convex combination $((1 - t)x_0 + tx_1, (1 - t)t_0 + tt_1)$. The convexity of f implies

$$f((1 - t)x_0 + tx_1) \le (1 - t)f(x_0) + tf(x_1) \le (1 - t)t_0 + tt_1,$$

so that $((1 - t)x_0 + tx_1, (1 - t)t_0 + tt_1) \in \mathrm{Epi}(f)$ and the epigraph is then convex.

Assume now that $\mathrm{Epi}(f)$ is convex, and take two points x_0 and x_1. If $f(x_0)$ or $f(x_1) = +\infty$ there is nothing to prove, otherwise we have $(x_i, f(x_i)) \in \mathrm{Epi}(f)$ and, by convexity $((1 - t)x_0 + tx_1, (1 - t)f(x_0) + tf(x_1)) \in \mathrm{Epi}(f)$, i.e. $f((1 - t)x_0 + tx_1) \le (1 - t)f(x_0) + tf(x_1)$, so that f is convex. □

Note that it is not possible to characterize convex functions in terms of their level sets only. It is true that the level sets $A(\ell) := \{x : f(x) \le \ell\}$ are convex if f is convex (simple exercise, see Exercise 3.1) but the converse is false, as any increasing function of a convex function also has convex level sets (also see Exercise 3.1). The functions which have convex level sets are sometimes called *level-convex*.[2]

Proposition 3.5 *Given a family of functions $f_\alpha : X \to \mathbb{R} \cup \{+\infty\}$ which are all convex, define $f(x) := \sup_\alpha f_\alpha(x)$. Then f is also convex.*

Proof Again, two easy proofs can be given of this fact. One is based on the fact that the epigraph of the sup is the intersection of the epigraphs (since the condition $t \geq \sup_\alpha f_\alpha(x)$ is equivalent to $t \geq f_\alpha(x)$ for every α), and on the convexity of the intersection of an arbitrary family of convex sets.

Another proof is based on the definition via the inequality: take x, y and write

$$f_\alpha((1-t)x + ty) \leq (1-t)f_\alpha(x) + tf_\alpha(y) \leq (1-t)f(x) + tf(y).$$

Again, taking the sup over α in the left-hand side provides the desired inequality.
□

Besides these analogies, we stress now a first connection between the two notions of lower semicontinuity and convexity. If the notion of convexity is independent of the topological or metric structure that we choose for the vector space x, the lower semicontinuity of course depends on the choice of the topology. For normed vector spaces, we will be mainly concerned with two natural choices: the strong topology and the weak (or weak-*) one. We will see that convexity is a crucial ingredient in the lower semicontinuity for the weak convergence (also called weak lower semicontinuity).

Note that having a strong notion of convergence makes it easier to be lower semicontinuous (since there are fewer convergent sequences of points), so that weak lower semicontinuity is actually a stronger notion than strong lower semicontinuity.

The first result involving convexity and lower semicontinuity is the following.

Theorem 3.6 *Assume that $f : X \to \mathbb{R} \cup \{+\infty\}$ is strongly lower semicontinuous and convex. Then f is also weakly lower semicontinuous.*

Proof In view of the characterization of lower semicontinuity in terms of level sets we just need to prove that each set $A(\ell) := \{x : f(x) \leq \ell\}$ is weakly closed. We know that it is convex and strongly closed. This implies that it is weakly closed because of a simple consequence of the Hahn–Banach theorem that we state in Box 3.1.
□

The above theorem required a functional analysis fact about convex sets, which we will state together with a well-known consequence.

Proposition 3.7 *Every convex subset C of a normed vector space X is strongly closed if and only if it is weakly closed.*

---

[2] Unfortunately sometimes, and in particular, in the economics literature, they are also called *quasi-convex*, a notion which has a completely different meaning in the calculus of variations, as we will see in Sect. 3.5.

Proof It is always true that weak closedness implies strong closedness (since any strongly convergent sequence is also weakly convergent). We need to prove the converse implication. Take a sequence $x_n \in C$ and assume $x_n \rightharpoonup x$. If $x \notin C$, we then consider the two closed convex sets C and $\{x\}$. They are disjoint and one is compact. They can be separated thanks to the Hahn–Banach theorem (see Box 3.1 below) by an element $\xi \in \mathcal{X}'$. Yet, this is not compatible with $\langle \xi, x_n \rangle \to \langle \xi, x \rangle$, which should be satisfied because of $x_n \rightharpoonup x$. Then $x \in C$ and C is weakly closed. $\qquad\square$

Another consequence of the last fact is the following, known as Mazur's Lemma.

Lemma 3.8 *For every weakly convergent sequence $x_n \rightharpoonup x$ there exists a sequence y_n of convex combinations of the points x_n (more precisely, y_n is a finite convex combination of points of the form x_k, with $k \geq n$) such that $y_n \to x$.*

Proof Given n, consider the set C_n defined as the set of finite convex combinations of $\{x_k : k \geq n\}$, which is a convex set. Then we consider its (strong) closure $\overline{C_n}$. Since $\overline{C_n}$ is convex and strongly closed, it is also weakly closed, and it contains the sequence x_n from a certain point on, so it also contains the weak limit x. Hence, by definition of the closure, there exists a point $y_n \in C_n$ such that $||x - y_n|| < 1/n$. This sequence proves the theorem. $\qquad\square$

We note that the above results concern the weak convergence in a space $\mathcal{X}$ (in duality with $\mathcal{X}'$) and not the weak-* convergence in a space $\mathcal{X}'$ (in duality with $\mathcal{X}$, instead of $\mathcal{X}''$). As an example we can consider $C \subset \mathcal{M}([-1, 1])$ defined as the set of all absolutely continuous measures. It is a convex set (since it is a vector space) and it is closed under the convergence of measures in the norm of $\mathcal{M}([-1, 1])$ (indeed, whenever $\mu_n \to \mu$, for every $A \subset [-1, 1]$ we have $\mu_n(A) \to \mu(A)$, so that the negligibility of Lebesgue-negligible sets passes to the limit). On the other hand, the sequence μ_n defined as the uniform measure on $[-1/n, 1/n]$ weakly-* converges to $\delta_0 \notin C$, which shows that C is not weakly-* closed.

Box 3.1 Memo—*Hahn–Banach Theorem*

A well-known theorem in functional analysis is the Hahn–Banach theorem, which has different equivalent statements, and which we will present in its "geometric" form dealing with the separation of suitable convex sets by means of hyperplanes.

Hahn–Banach Separation Theorem *Let A and B be two convex non-empty subsets of a normed vector space $\mathcal{X}$, with $A \cap B = \emptyset$.*

If A is open, there exists a linear functional $\xi \in \mathcal{X}'$ which separates A and B, i.e. there is a number $c \in \mathbb{R}$ such that $\langle \xi, a \rangle < c \leq \langle \xi, b \rangle$ for all $a \in A, b \in B$.

(continued)

Box 3.1 (continued)

If A is compact and B is closed, there exists a linear functional $\xi \in X'$ which strictly separates A and B, i.e. there are two numbers $c_1 < c_2$ such that $\langle \xi, a \rangle \leq c_1 < c_2 \leq \langle \xi, b \rangle$ for all $a \in A, b \in B$.

When the space X itself is a dual ($X = Y'$) a natural question is whether it is possible to perform these separations by using elements of Y instead of $X' = Y''$. Unfortunately this is in general not true and the last statement (strict separation of a closed and a compact sets) requires a slightly different assumption. It is indeed true that if $A, B \subset X = Y'$ are convex, if we assume A to be compact and B to be weakly-* closed (which is stronger than strongly closed), then there exists a $y \in Y$ which strictly separates A and B, i.e. there are two numbers $c_1 < c_2$ such that $\langle a, y \rangle \leq c_1 < c_2 \leq \langle b, y \rangle$ for all $a \in A, b \in B$.

The distinction between strong and weak-* closedness is necessary since it is not true that convex strongly closed subsets of Y' are weakly-* closed, differently from what happens with weak convergence.

We finish this section with two considerations on concrete integral functionals over functional spaces. First we provide an example of a semicontinuity result obtained from Theorem 3.6.

Proposition 3.9 *Let* $L : \Omega \times \mathbb{R} \to \mathbb{R}_+$ *be a measurable function, such that* $s \mapsto L(x, s)$ *is convex for a.e.* $x \in \Omega$. *Then the functional*

$$L^p(\Omega) \ni u \mapsto J(u) := \int_\Omega L(x, u(x)) \, \mathrm{d}x$$

is lower semicontinuous for the weak L^p *convergence, for every* $p \in [1, +\infty)$.

Proof From the convexity of L in the variable s we deduce the convexity of the functional J. Then, the result is proven if we prove that J is l.s.c. for the strong convergence of L^p. If we take $u_n \to u$ in L^p and $J(u_n) \to \ell$, we can extract a subsequence such that $u_n(x) \to u(x)$ a.e. The continuity of L in the second variable[3] and Fatou's lemma, together with the lower bound $L \geq 0$, imply $J(u) \leq \liminf_n J(u_n)$, hence J is strongly l.s.c.   □

We then see that the convexity in the variable s is necessary for weak lower semicontinuity, and hence is equivalent to it, but we will state the result, for simplicity, in the case where L does not depend explicitly on x.

---

[3] Note that we assumed $L(x, \cdot)$ to be real-valued for every x, and convex functions are known to be locally Lipschitz on $\mathbb{R}$, hence continuous.

Proposition 3.10 *Consider the functional* $F : L^p(\Omega; \mathbb{R}^d) \to \mathbb{R} \cup \{+\infty\}$ *given by* $F(v) = \int L(v(x)) \, dx$ *for a continuous function* $L : \mathbb{R}^d \to \mathbb{R}$, $L \geq 0$. *Then,* F *is l.s.c. for the weak* L^p *convergence if and only if* L *is convex.*

Proof In order to show that convexity of L is necessary, we will build a sequence $v_n \rightharpoonup v$ on which we will evaluate F. If we take a cube inside Ω and choose $v_n = 0$ outside this cube for every n, we can reduce to the case where Ω is a cube Q. We then fix two vectors $a, b \in \mathbb{R}^d$ and a number $t \in [0, 1]$ and define a sequence v_n in the following way: we divide the cube into n parallel stripes of equal width ℓ/n (ℓ being the side of the cube), and each stripe is further divided into two parallel sub-stripes, one of width $(1 - t)\ell/n$ where v_n takes the value a, and one of width $t\ell/n$ where v_n takes the value b. We easily see that in this case we have $v_n \rightharpoonup v$, where v is the constant vector function equal to $(1 - t)a + tb$, and that we have $F(v_n) = (1 - t)|Q|L(a) + t|Q|L(b)$, while $F(v) = |Q|L((1 - t)a + tb)$. The semicontinuity of F then implies the convexity of L.

The converse implication is a particular case of Proposition 3.9.   □

3.2 Integral Functionals

We consider in this section the lower semicontinuity of integral functionals with respect to the strong and weak convergence of two different variables. Later, in Sect. 3.4, this will be applied to the case where the variable which converges weakly is the gradient of the one which converges strongly.

More precisely, we consider

$$F(u, v) := \int_\Omega L(x, u(x), v(x)) \, dx \quad \text{for } u, v \in L^1(\Omega).$$

We assume $L \geq 0$ so that we avoid any integrability issues in order to define F: as the integral of a non-negative function, it can take the value $+\infty$, but it is always well-defined. The variables u and v can be vector-valued, and in many concrete applications, at least in what concerns v, this will be the case. We will analyze the lower semicontinuity of F w.r.t. the strong convergence of u and the weak convergence of v. In order to deal with the weak convergence in L^1, we first recall an important fact from functional analysis.

Box 3.2 Memo—*Weak Convergence in* L^1

Definition A sequence of functions $v_n \in L^1(\Omega)$ converges weakly in L^1 to a function $v \in L^1(\Omega)$ if for every L^∞ function φ we have $\int v_n \cdot \varphi \to \int v \cdot \varphi$.

(continued)

Box 3.2 (continued)
Definition A family of functions $(v_\alpha)_\alpha \in L^1(\Omega)$ is said to be equi-integrable if for every $\varepsilon > 0$ there exists a $\delta > 0$ such that on every set A with $|A| \le \delta$ we have $\int_A |v_\alpha| \le \varepsilon$ for any α.

The following equivalence characterizes compactness for the weak L^1 convergence (this result is known as the Dunford–Pettis theorem).
Theorem *Given a family of functions $(v_\alpha)_\alpha \in L^1(\Omega)$, the following are equivalent*

- *the family is precompact for the weak L^1 convergence (i.e. from any sequence v_{α_n} one can extract a weakly convergent subsequence $v_{\alpha_{n_k}}$);*
- *the family is equi-integrable;*
- *there exists a convex and superlinear function $\psi : \mathbb{R} \to \mathbb{R}_+$ and a constant $C < +\infty$ such that $\int_\Omega \psi(|v_\alpha|) \le C$ for every α.*

The goal of this section is to prove that the functional F is lower semicontinuous for the strong-weak convergence as above as soon as L is a Carathéodory function (measurable in x, continuous in (u, v)), convex in the last variable.

We start from a particular case.

Lemma 3.11 *Assume that L is continuous in all variables, nonnegative, and convex in v. Assume that Ω is compact. Take a sequence (u_n, v_n) with $u_n \to u$ in L^1 and $v_n \rightharpoonup v$ in L^1. Assume that u_n and v_n are also L^∞ functions, uniformly bounded. Then $\liminf_n F(u_n, v_n) \ge F(u, v)$.*

Proof First of all, we write $F(u_n, v_n) = F(u_n, v_n) - F(u, v_n) + F(u, v_n)$. We first consider $F(u_n, v_n) - F(u, v_n) = \int_\Omega L(x, u_n, v_n) - L(x, u, v_n)$ and we note that the continuity of L is uniform on $\Omega \times B_R \times B_R$, where R is an upper bound of $||u_n||_{L^\infty}, ||v_n||_{L^\infty}$. Hence, there exists a modulus of continuity ω (a continuous and bounded function with $\omega(0) = 0$) such that we can write $|L(x, u_n, v_n) - L(x, u, v_n)| \le \omega(|u_n - u|)$. The strong convergence $u_n \to u$ implies a.e. convergence up to subsequences, and hence we have $\omega(|u_n - u|) \to 0$ a.e. This convergence is dominated by a constant, so that we obtain $F(u_n, v_n) - F(u, v_n) \to 0$.

We then consider the term $F(u, v_n)$. We define $\tilde{L}_u(x, v) := L(x, u(x), v)$ and $\tilde{F}_u(v) := \int \tilde{L}(x, v(x)) \, dx = F(u, v)$, which is an integral functional, lower semicontinuous thanks to Proposition 3.9. We then obtain $\liminf_n F(u, v_n) \ge F(u, v)$, which concludes the proof. $\square$

We now want to remove the assumption on the boundedness of v_n, thanks to the following lemma.

Lemma 3.12 *For any constant $M > 0$, let $f_M : \mathbb{R}^k \to \mathbb{R}^k$ be defined via*

$$f_M(v) := v \mathbb{1}_{\overline{B(0,M)}}(v) = \begin{cases} v & \text{if } |v| \leq M, \\ 0 & \text{if } |v| > M. \end{cases}$$

Given a sequence $v_n \to v$ in $L^1(\Omega)$, consider for any $M > 0$ the sequence $v_{n,M} := f_M(v_n)$. It is possible to extract a subsequence and to find, for every $M \in \mathbb{N}$, a function $v_{(M)}$ such that, for every M, the sequence $v_{n,M}$ weakly converges in L^1 (actually, even weakly-* in L^∞) to $v_{(M)}$.

Then we have $\lim_{M \to \infty} ||v_{(M)} - v||_{L^1(\Omega)} = 0$.

Proof First, let us fix a convex and superlinear function ψ such that $\int \psi(|v_n|) \leq C$, which exists thanks to the weak convergence of v_n. We observe that we have, using the fact that $s \mapsto \psi(s)/s$ is nondecreasing,

$$\int_{|v_n|>M} |v_n| \, dx = \int_{|v_n|>M} \frac{|v_n|}{\psi(|v_n|)} \psi(|v_n|) \, dx \leq \frac{M}{\psi(M)} \int_\Omega \psi(|v_n|) \, dx \leq C \frac{M}{\psi(M)}.$$

We now take a test function $\phi \in L^\infty$ and write the equality

$$\left| \int_\Omega \phi \cdot (v_n - v_{n,M}) dx \right| = \left| \int_{|v_n|>M} \phi \cdot v_n dx \right| \leq ||\phi||_{L^\infty} \int_{|v_n|>M} |v_n| dx \leq C \frac{M||\phi||_{L^\infty}}{\psi(M)}.$$

We then take the limit $n \to \infty$ for fixed M and obtain, thanks to weak convergence,

$$\left| \int_\Omega \phi \cdot (v - v_{(M)}) \, dx \right| \leq C ||\phi||_{L^\infty} \frac{M}{\psi(M)}.$$

This estimate, valid for arbitrary ϕ, due to the duality between L^1 and L^∞, means precisely

$$||v_{(M)} - v||_{L^1(\Omega)} \leq C \frac{M}{\psi(M)}.$$

The superlinearity of ψ provides $\lim_{M \to \infty} \frac{M}{\psi(M)} = 0$ and hence the claim. $\qquad \square$

We can now extend the results of Lemma 3.11.

Proposition 3.13 *The conclusions of Lemma 3.11 also hold if removing the assumption on the uniform bound on $||v_n||_{L^\infty}$.*

Proof We first fix $M > 0$ and write $F(u_n, v_n) = (F(u_n, v_n) - F(u, v_{n,M})) + F(u, v_{n,M})$. By applying Lemma 3.11 we obtain $\liminf_n F(u, v_{n,M}) \geq F(u, v_{(M)})$.

We consider now the other term, for which we have

$$F(u_n, v_n) - F(u, v_{n,M}) = \int_{|v_n|>M} (L(x, u_n, v_n) - L(x, u_n, 0))\, dx$$

$$\geq - \int_{|v_n|>M} L(x, u_n, 0)\, dx,$$

where we used the positivity of L. We can then use

$$\int_{|v_n|>M} L(x, u_n, 0)\, dx \leq C_1 |\{|v_n| > M\}| \leq \frac{C}{M},$$

and C_1 denotes the maximum of L on $\Omega \times \overline{B_R} \times \{0\}$ (remember that u_n is equibounded, Ω is compact, and L is continuous), and $C = C_1 C_2$, where $C_2 := \sup_n \|v_n\|_{L^1}$. Indeed, since the sequence v_n is bounded in L^1, we have $|\{|v_n| > M\}| \leq C_2/M$.

Finally, we obtain

$$\liminf_n F(u_n, v_n) \geq \liminf_n \left(F(u_n, v_n) - F(u, v_{n,M}) \right) + \liminf_n F(u, v_{n,M})$$

$$\geq -\frac{C}{M} + F(u, v_{(M)}).$$

This estimate is valid for arbitrary M, and we can now send $M \to \infty$. Lemma 3.12 provides $v_{(M)} \to v$ in L^1 (in the strong sense, but weak convergence would have been enough) so that we can apply, exactly as at the end of Lemma 3.11, the same idea of considering the functional where we freeze u, and Proposition 3.9 provides the semicontinuity $\liminf_M F(u, v_{(M)}) \geq F(u, v)$. Since the term $\frac{C}{M}$ tends to 0, we obtain the claim.   □

We now need to get rid of the continuity assumption on L, and of the boundedness of u_n. This will require some classical results from measure theory.

Box 3.3 Memo—*Lusin's Theorem*

A well-known theorem in measure theory states that every measurable function f on a reasonable measure space (X, μ) is actually continuous on a set K with $\mu(X \setminus K)$ small. This set K can be taken to be compact. Actually, there are at least two statements: either we want f to be merely continuous on K, or we want f to coincide on K with a continuous function defined on X. This theorem is usually stated for real-valued functions, but we happen to need it for functions valued in more general spaces. Let us be more precise: take a

(continued)

Box 3.3 (continued)

topological space X endowed with a finite regular measure μ (i.e. any Borel set $A \subset X$ satisfies $\mu(A) = \sup\{\mu(K) : K \subset A, K \text{ compact}\} = \inf\{\mu(B) : B \supset A, B \text{ open}\}$). The target space Y will be assumed to be second-countable (i.e. it admits a countable family $(B_i)_i$ of open sets such that any other open set $B \subset Y$ may be expressed as a union of B_i; for instance, separable metric spaces are second-countable).

Theorem (Weak Lusin) *Under the above assumptions on X, Y, μ, if $f : X \to Y$ is measurable, then for every $\varepsilon > 0$ there exists a compact set $K \subset X$ such that $\mu(X \setminus K) < \varepsilon$ and the restriction of f to K is continuous.*

Proof For every $i \in \mathbb{N}$, set $A_i^+ = f^{-1}(B_i)$ and $A_i^- = f^{-1}(B_i^c)$. Consider compact sets $K_i^\pm \subset A_i^\pm$ such that $\mu(A_i^\pm \setminus K_i^\pm) < \varepsilon 2^{-i}$. Set $K_i = K_i^+ \cup K_i^-$ and $K = \bigcap_i K_i$. For each i we have $\mu(X \setminus K_i) < \varepsilon 2^{1-i}$. By construction, K is compact and $\mu(X \setminus K) < 4\varepsilon$. To prove that f is continuous on K it is sufficient to check that $f^{-1}(B) \cap K$ is relatively open in K for each open set B, and it is enough to check this for $B = B_i$. Equivalently, it is enough to prove that $f^{-1}(B_i^c) \cap K$ is closed, and this is true since it coincides with $K_i^- \cap K$.

Theorem (Strong Lusin) *Under the same assumptions on X, if $f : X \to \mathbb{R}$ is measurable, then for every $\varepsilon > 0$ there exists a compact set $K \subset X$ and a continuous function $g : X \to \mathbb{R}$ such that $\mu(X \setminus K) < \varepsilon$ and $f = g$ on K.*

Proof First apply the weak Lusin theorem, since $\mathbb{R}$ is second countable. Then we just need to extend $f_{|K}$ to a continuous function g on the whole X. This is possible since $f_{|K}$ is uniformly continuous (as a continuous function on a compact set) and hence has a modulus of continuity ω such that $|f(x) - f(x')| \leq \omega(d(x, x'))$ (the function ω can be taken to be sub-additive and continuous). Then define $g(x) = \inf\{f(x') + \omega(d(x, x')) : x' \in K\}$. It can be easily checked that g is continuous and coincides with f on K.

Note that this last proof strongly uses the fact that the target space is $\mathbb{R}$. It could be adapted to the case of $\mathbb{R}^d$ just by extending component-wise. On the other hand, it is clear that the strong version of Lusin's theorem cannot hold for any space Y: just take X connected and Y disconnected. A measurable function $f : X \to Y$ taking values in two different connected components on two sets of positive measure cannot be approximated by continuous functions in the sense of the strong Lusin theorem.

Another classical result, often presented together with Lusin's theorem, is the following.

Box 3.4 Memo—*Egoroff's Theorem*

Theorem (Egoroff) *Let X be a measure space with finite measure, and $u_n : X \to Y$ a sequence of functions, valued in a metric space Y, which is a.e. converging to a function $u : X \to Y$. Then, for every $\varepsilon > 0$, there exists a compact set $K \subset X$, with $|X \setminus K| < \varepsilon$, such that the convergence $u_n \to u$ is uniform on K.*

Note that the theorem is usually stated for functions valued in $\mathbb{R}$, but it is enough to apply it to $d(u_n(x), u(x))$ in order to generalize to arbitrary metric spaces.

An interesting observation is the fact that Egoroff's theorem implies the weak Lusin theorem in the Euclidean case (i.e. when X is an open subset of the Euclidean space and $Y = \mathbb{R}^k$). Indeed, it is enough to take $u \in L^1$ and approximate it with a sequence of smooth functions u_n, which exists because C_c^∞ is dense in L^1. Up to subsequences we can assume a.e. pointwise convergence and applying Egoroff's theorem transforms this into uniform convergence on K. Hence u is continuous on K as a uniform limit of continuous functions. The case where u is not L^1 can be treated by first truncating it.

A statement which we need in this framework is the following, known as the Scorza-Dragoni theorem, which we will present as a consequence of Lusin's theorem. We first provide an intermediate lemma.

Lemma 3.14 *Let S be a compact metric space, and $f : \Omega \times S \to \mathbb{R}$ be a measurable function, such that $f(x, \cdot)$ is continuous for a.e. x. Then for every $\varepsilon > 0$ there exists a compact subset $K \subset \Omega$ with $|\Omega \setminus K| < \varepsilon$ and f continuous on $K \times S$.*

Proof The continuity of a function of two variables, when the second lives in a compact space S, is equivalent to the continuity of the map $x \mapsto f(x, \cdot)$ seen as a map from Ω to the Banach space $C(S)$. We can apply the weak version of Lusin's theorem to this map and easily obtain the claim.   □

Theorem 3.15 (Scorza-Dragoni) *Let $f : \Omega \times \mathbb{R}^d \to \mathbb{R}$ be a measurable function, such that $f(x, \cdot)$ is continuous for a.e. x. Then for every $\varepsilon > 0$ there exists a compact subset $K \subset \Omega$ with $|\Omega \setminus K| < \varepsilon$ and f continuous on $K \times \mathbb{R}^d$.*

Proof The statement is more difficult to prove because of the non-compactness of the second space. We then fix a number $R \in \mathbb{N}$ and consider the restriction of f to $\Omega \times \overline{B_R}$, where $\overline{B_R}$ is the closed ball of radius R. We consider $\varepsilon_R = \varepsilon 2^{-R-1}$ and apply the previous lemma, thus obtaining a set K_R such that f is continuous on $K_R \times \overline{B_R}$ and $|\Omega \setminus K_R| < \varepsilon_R$. Set $K = \bigcap_R K_R$. We then have $|\Omega \setminus K| < \varepsilon$ and f is continuous on $K \times \overline{B_R}$ for every R. Since continuity is a local property, we deduce that f is continuous on $K \times \mathbb{R}^d$.   □

We are now in position to prove the main theorem of this section:

Theorem 3.16 *Assume that L is a Carathéodory function, continuous in (s, v) and measurable in x, nonnegative, and convex in v. For any sequence (u_n, v_n) with $u_n \to u$ in L^1 and $v_n \rightharpoonup v$ in L^1 we have $\liminf_n F(u_n, v_n) \geq F(u, v)$.*

Proof We start by supposing that Ω is of finite measure. We also assume, which is possible up to extracting a subsequence, that we have $u_n \to u$ a.e. Using together Lusin's (in its weak form), Egoroff's and Scorza-Dragoni's theorems, for any $\varepsilon > 0$ we can find a compact set $K \subset \Omega$ with $|\Omega \setminus K| < \varepsilon$ such that

- u is continuous (and hence bounded) on K;
- u_n converges uniformly to u on K, and hence they are equibounded;
- L is continuous on $K \times \mathbb{R}^m \times \mathbb{R}^k$.

If we consider $F_K(u, v) := \int_K L(x, u(x), v(x)) \, dx$ we can then apply Proposition 3.13 and obtain

$$\liminf_n F(u_n, v_n) \geq \liminf_n F_K(u_n, v_n) \geq F_K(u, v) = \int_K L(x, u(x), v(x)) \, dx.$$

We now choose a sequence $\varepsilon_j \to 0$ and for each j we consider the corresponding compact set K_j. By replacing each K_j with $\bigcup_{i \leq j} K_i$ it is not restrictive to assume that K_j is increasing in j, and all the properties satisfied on K_j are still valid, as they are stable when considering finite unions of sets. We then obtain

$$\liminf_n F(u_n, v_n) \geq \int_{K_j} L(x, u(x), v(x)) \, dx = \int_\Omega L(x, u(x), v(x)) \mathbb{1}_{K_j}(x) \, dx.$$

Since $\mathbb{1}_{K_j}$ converges increasingly a.e. to $\mathbb{1}_\Omega$ (as a consequence of the inclusion $K_j \subset K_{j+1}$ and of $|\Omega \setminus K_j| < \varepsilon_j$), sending $j \to \infty$ provides the desired semicontinuity

$$\liminf_n F(u_n, v_n) \geq \int_\Omega L(x, u(x), v(x)) \, dx.$$

The case where Ω is not of finite measure can be easily dealt with by writing

$$F = \sup_R F_R, \quad \text{where } F_R(u, v) := \int_{\Omega \cap B(0,R)} L(x, u(x), v(x)) \, dx.$$

Each functional in the sup has been proven to be lower semicontinuous, and this property is stable under taking the sup. $\qquad \square$

Note that in the above result the function L is assumed to be continuous in s. It would be possible to replace this assumption with its lower semicontinuity, but the proof is much harder, and we refer to [52] for these more refined semicontinuity results.

3.3 Functionals on Measures

In this section we prove the lower semicontinuity of a particular family of functionals defined on measures, with respect to the weak-* convergence of measures. We consider here vector-valued measures, which are the dual of vector-valued continuous functions.

Box 3.5 Memo—*Vector Measures*

Definition A finite vector measure λ on a metric space Ω is a map associating with every Borel subset $A \subset \Omega$ a value $\lambda(A) \in \mathbb{R}^N$ such that, for every countable disjoint union $A = \bigcup_i A_i$ (with $A_i \cap A_j = \emptyset$ for $i \neq j$), we have

$$\sum_i |\lambda(A_i)| < +\infty \quad \text{and} \quad \lambda(A) = \sum_i \lambda(A_i).$$

We denote by $\mathcal{M}^N(\Omega)$ the set of finite vector measures on Ω. With such measures we can associate a positive scalar measure $|\lambda| \in \mathcal{M}_+(\Omega)$ by

$$|\lambda|(A) := \sup \left\{ \sum_i |\lambda(A_i)| \ : \ A = \bigcup_i A_i \text{ with } A_i \cap A_j = \emptyset \text{ for } i \neq j \right\}.$$

This scalar measure is called the *total variation measure* of λ. For simplicity we only consider the Euclidean norm on $\mathbb{R}^N$, and write $|\lambda|$ instead of $\|\lambda\|$ (a notation that we keep for the total mass of the total variation measure, see below), but the same could be defined for other norms as well. The integral of a measurable function $\xi : \Omega \to \mathbb{R}^N$ w.r.t. λ is well-defined if $|\xi| \in L^1(\Omega, |\lambda|)$, is denoted $\int_\Omega \xi \cdot d\lambda$, and can be computed as $\sum_{i=1}^N \int \xi_i \, d\lambda_i$, thus reducing to integrals of scalar functions according to scalar measures. It could also be defined as a limit of integrals of piecewise constant functions. *Functional Analysis Facts* The quantity $\|\lambda\| := |\lambda|(\Omega)$ is a norm on $\mathcal{M}^N(\Omega)$, and this normed space is the dual of $C_0(\Omega; \mathbb{R}^N)$, the space of continuous function on Ω vanishing at infinity

$$C_0(\Omega; \mathbb{R}^N)$$

$$= \{\varphi : \Omega \to \mathbb{R}^N \ : \ \forall \varepsilon > 0 \ \exists K \subset \Omega \text{ compact such that } |\varphi| \leq \varepsilon \text{ on } \Omega \setminus K\}$$

endowed with the norm $\|\varphi\| := \max\{|\varphi(x)| \ : \ x \in \Omega\}$. Of course, when Ω itself is compact, $C_0(\Omega; \mathbb{R}^N)$ and $C(\Omega; \mathbb{R}^N)$ coincide. We use here the duality product $\langle \xi, \lambda \rangle := \int \xi \cdot d\lambda$.

(continued)

> **Box 3.5** (continued)
>   A clarifying fact is the following.
> **Proposition** *For every* $\lambda \in \mathcal{M}^N(\Omega)$ *there exists a measurable function* $u :$ $\Omega \to \mathbb{R}^N$ *such that* $\lambda = u \cdot |\lambda|$ *and* $|u| = 1$ *a.e. (for the measure* $|\lambda|$*). In particular,* $\int \xi \cdot d\lambda = \int (\xi \cdot u) \, d|\lambda|$.
>   Note that L^1 vector functions can be identified with vector measures, absolutely continuous w.r.t. the Lebesgue measure. Their L^1 norm coincides with the norm in $\mathcal{M}^N(\Omega)$.

For simplicity, throughout this section Ω will be assumed to be a compact domain.

As a first example, we consider functionals of the form

$$\lambda = \mathbf{v} \cdot \mu \mapsto \int_\Omega f(\mathbf{v}(x)) \, d\mu(x),$$

where μ is a given positive measure over Ω (for instance the Lebesgue measure).

Let us first see which are the natural conditions on f so as to ensure lower semicontinuity. Then, we will give a precise result.

As a first point, an easy adaptation of the construction of Proposition 3.10 shows that one needs to require f to be convex.

Another requirement concerns the growth of f. For simplicity it is always possible to assume $f(0) = 0$ (up to adding a constant to f). Suppose that f satisfies $f(v) \leq C|v|$ for all $v \in \mathbb{R}^N$ and $C < +\infty$. Consider the functional defined by

$$\mathcal{F}(\lambda) := \begin{cases} \int f(\mathbf{v}(x)) \, d\mu(x) & \text{if } \lambda = \mathbf{v} \cdot \mu, \\ +\infty & \text{otherwise.} \end{cases}$$

Take a sequence λ_n of absolutely continuous measures weakly converging to a singular measure λ: we then get $\mathcal{F}(\lambda_n) \leq C\|\lambda_n\| \leq C$ while $\mathcal{F}(\lambda) = +\infty$, thus violating the semicontinuity. This suggests that one should have $C = +\infty$, i.e. f should in some sense be superlinear. This is clarified in the following proposition.

Proposition 3.17 *Assume that f is convex, l.s.c., and superlinear (i.e. f satisfies $\lim_{|v| \to \infty} \frac{f(v)}{|v|} = +\infty$). Then the functional $\mathcal{F}$ defined by*

$$\mathcal{F}(\lambda) := \begin{cases} \int f(\mathbf{v}(x)) \, d\mu(x) & \text{if } \lambda = \mathbf{v} \cdot \mu, \\ +\infty & \text{otherwise} \end{cases}$$

is l.s.c. for the weak-* convergence of measures.

Proof We take a sequence $\lambda_n \overset{*}{\rightharpoonup} \lambda$ and we assume, up to extracting a subsequence, that we have $\mathcal{F}(\lambda_n) \leq C$ (if no such subsequence exists, then there is nothing to prove, since $\liminf \mathcal{F}(\lambda_n) = +\infty$). This means that the sequence of functions $\mathbf{v}_n$ defined by $\lambda_n = \mathbf{v}_n \cdot \mu$ is such that $\int f(\mathbf{v}_n)$ is bounded, which means (see Box 3.2) that $\mathbf{v}_n$ is equi-integrable and, up to extracting another subsequence, we can assume $\mathbf{v}_n \rightharpoonup \mathbf{v}$ in L^1. We then have $\int \varphi \cdot \mathbf{v}_n \, d\mu \to \int \varphi \cdot d\lambda$ for every continuous test function φ (because of $\lambda_n \overset{*}{\rightharpoonup} \lambda$) but also $\int \varphi \cdot \mathbf{v}_n \, d\mu \to \int \varphi \cdot \mathbf{v} \, d\mu$ for every $\varphi \in L^\infty(\mu)$ (because of $\mathbf{v}_n \rightharpoonup \mathbf{v}$ in $L^1(\mu)$). Hence, $\lambda = \mathbf{v} \cdot \mu$, and $\mathcal{F}(\lambda) = \int f(\mathbf{v}) \, d\mu$. The semicontinuity results of the previous sections are then enough to obtain $\mathcal{F}(\lambda) \leq \liminf \mathcal{F}(\lambda_n)$.   □

Next, we consider a more general statement which includes the previous one as a particular case, but also considers non-superlinear integrand functions f. In this case we need to "compensate" for their non-superlinearity. In order to do so, we first need to introduce the notion of a recession function (see Sect. 4.2.1 as well).

Definition 3.18 Given a proper convex function $f : \mathbb{R}^N \to \mathbb{R} \cup \{+\infty\}$ its recession function f^∞ is defined by

$$f^\infty(v) := \sup_{x : f(x) < +\infty} f(x + v) - f(x)$$

$$= \sup_{x : f(x) < +\infty} \lim_{t \to +\infty} \frac{f(x + tv)}{t}$$

$$= \sup_{t > 0, x : f(x) < +\infty} \frac{f(x + tv) - f(x)}{t}.$$

The function f^∞ is 1-homogeneous, i.e. $f^\infty(cv) = cf^\infty(v)$ for every $c \geq 0$.

Given a 1-homogeneous function f and a vector measure λ on Ω, we define the integral $\int_\Omega f(\lambda)$ as follows: take an arbitrary reference measure μ, positive and scalar, such that $\lambda \ll \mu$, define $\int_\Omega f(\lambda) := \int_\Omega f(\mathbf{v}(x)) \, d\mu(x)$ where $\mathbf{v}$ is the density of λ w.r.t. μ, i.e. $\lambda = \mathbf{v} \cdot \mu$. The result does not depend on μ, since, if we take μ_1 and μ_2 such that $\lambda = \mathbf{v}_i \cdot \mu_i$, setting $\mu_3 := \mu_1 + \mu_2$, then λ, μ_1 and μ_2 are absolutely continuous w.r.t. μ_3. If we call α_i the density of μ_i w.r.t. μ_3 and $\mathbf{v}_3$ that of λ w.r.t. μ_3, we have $\alpha_i \mathbf{v}_i = \mathbf{v}_3$. Thus, we get

$$\int_\Omega f(\mathbf{v}_1) \, d\mu_1 = \int_\Omega f(\mathbf{v}_1) \alpha_1 \, d\mu_3 = \int_\Omega f(\mathbf{v}_3) \, d\mu_3 = \int_\Omega f(\mathbf{v}_2) \alpha_2 \, d\mu_3 = \int_\Omega f(\mathbf{v}_2) \, d\mu_2.$$

In particular, it is always possible to choose $\mu = |\lambda|$ in order to guarantee $\lambda \ll \mu$.

We then have the following result

Proposition 3.19 *Let* $f : \mathbb{R}^N \to \mathbb{R}$ *be a proper, convex, and l.s.c. function, and* f^∞ *its recession function. Let* μ *be a fixed finite positive measure on* Ω. *For every vector measure* λ *write* $\lambda = \mathbf{v} \cdot \mu + \lambda^s$, *where* $\mathbf{v} \cdot \mu$ *is the absolutely continuous part of* λ *and* λ^s *is the singular part (w.r.t.* μ*). Then, the functional defined by*

$$\mathcal{F}(\lambda) = \int_\Omega f(\mathbf{v}(x)) \, \mathrm{d}\mu(x) + \int_\Omega f^\infty(\lambda^s) \tag{3.1}$$

is l.s.c. for the weak-* convergence of measures.

Proof We will see in Sects. 4.2.1 and 4.2.3 that there exists a convex function f^* (called Fenchel–Legendre transform of f) satisfying the two following equalities

$$f(v) = \sup \left\{ v \cdot w - f^*(w) \; : \; w \in \mathbb{R}^N \right\}; \quad f^\infty(v) = \sup \left\{ v \cdot w : f^*(w) < \infty \right\}.$$

Let us consider the following functional

$$\widetilde{\mathcal{F}}(\lambda) := \sup_{K, a \in C(\Omega; K)} \int_\Omega a(x) \cdot \mathrm{d}\lambda(x) - \int_\Omega f^*(a(x)) \, \mathrm{d}\mu(x),$$

where K ranges among compact and convex subsets of $\{f^* < +\infty\}$ (on each of these sets f^* is bounded, so that the integral $\int_\Omega f^*(a(x)) \, \mathrm{d}\mu(x)$ is well-defined). The functional $\widetilde{\mathcal{F}}$ is obviously l.s.c. since it is the supremum of a family of affine functionals, each continuous w.r.t. the weak-* convergence. We want to prove $\widetilde{\mathcal{F}} = \mathcal{F}$.

In order to do so, first note that, thanks to Lusin's theorem (see Box 3.3), applied here to the measure $|\lambda| + \mu$, it is not difficult to replace continuous functions valued in K with measurable functions valued in the same set. Lusin's theorem does not guarantee that the values can be taken inside K, but since K is convex it is possible to compose with the projection onto K in order to ensure that it is so. We then get

$$\widetilde{\mathcal{F}}(\lambda) := \sup_{K, a : \Omega \to K} \int_\Omega a(x) \cdot \mathrm{d}\lambda(x) - \int_\Omega f^*(a(x)) \, \mathrm{d}\mu(x).$$

Then take a set A such that $|\lambda^s|(\Omega \setminus A) = \mu(A) = 0$: this allows us to write

$$\widetilde{\mathcal{F}}(\lambda) := \sup_{K, a : \Omega \to K} \int_{\Omega \setminus A} \left[a(x) \cdot \mathbf{v}(x) - f^*(a(x)) \right] \, \mathrm{d}\mu(x) + \int_A a(x) \cdot \mathrm{d}\lambda^s(x).$$

The values of $a(x)$ may be chosen independently on A and $\Omega \setminus A$ and we can check that we have

$$\sup_{K,a:\Omega \to K} \int_{\Omega \setminus A} \left[a(x) \cdot \mathbf{v}(x) - f^*(a(x)) \right] \, d\mu(x)$$

$$= \int_{\Omega \setminus A} \left(\sup_{w \,:\, f^*(w) < +\infty} \mathbf{v}(x) \cdot w - f^*(w) \right) d\mu(x)$$

$$= \int_{\Omega \setminus A} f(\mathbf{v}(x)) \, d\mu(x) = \int_{\Omega} f(\mathbf{v}(x)) \, d\mu(x),$$

as well as, writing $\lambda^s = \mathbf{v}_0 \cdot |\lambda^s|$,

$$\sup_{K,a:\Omega \to K} \int_A a(x) \cdot d\lambda^s(x) = \sup_{K,a:\Omega \to K} \int_A a(x) \cdot \mathbf{v}_0 \, d|\lambda^s|(x)$$

$$= \int_A \left(\sup_{w \,:\, f^*(w) < +\infty} \mathbf{v}_0(x) \cdot w \right) d|\lambda^s|(x)$$

$$= \int_A f^\infty(\mathbf{v}_0(x)) \, d|\lambda^s|(x) = \int f^\infty(\lambda^s).$$

This allows us to conclude $\widetilde{\mathcal{F}} = \mathcal{F}$.   □

As we saw in the previous sections of this chapter, the convexity of the integrand plays a crucial role in the lower semicontinuity of integral functionals. Yet, in the case of functionals defined on measures a small exception is possible, which concerns the presence of atoms. Given a measure λ we will define the atoms of λ as the points x such that $\lambda(\{x\}) \neq 0$ (equivalently, $|\lambda|(\{x\}) \neq 0$), and their set is denoted by $A(\lambda) := \{x \,:\, \lambda(\{x\}) \neq 0\}$. Note that this set is always at most countable, since $\sum_{x \in A(\lambda)} |\lambda|(\{x\}) \leq |\lambda|(\Omega)$. The atomic part of λ is then defined as

$$\lambda^{\#} := \sum_{x \in A(\lambda)} \lambda(\{x\})\delta_x.$$

Measures λ such that $\lambda = \lambda^{\#}$ are said to be purely atomic.

We now consider a functional depending on the masses of the atoms of λ. As we did before, let us first consider a simple example,

$$\mathcal{G}(\lambda) := \begin{cases} \sum_{x \in A(\lambda)} g(\lambda(\{x\})) & \text{if } \lambda = \lambda^{\#}, \\ +\infty & \text{if } \lambda \text{ is not purely atomic.} \end{cases}$$

As before, let us first understand which are the basic properties of g so as to guarantee semicontinuity. Let us assume $g(0) = 0$ (which is by the way necessary if we want to avoid ambiguities due to zero-mass atoms). First, take $\lambda_n = a\delta_{x_n} + b\delta_x$. Assume $x_n \to x$. Then $\lambda_n \overset{*}{\rightharpoonup} \lambda = (a+b)\delta_x$. In this case semicontinuity requires

$$g(a+b) \leq g(a) + g(b),$$

i.e. we need g to be subadditive.

Assume that g satisfies $g(v) \leq C|v|$ for all v and take, similarly to what we did before, a sequence $\lambda_n \overset{*}{\rightharpoonup} \lambda$ where λ_n is purely atomic but λ is not. Then we have $G(\lambda_n) \leq C\|\lambda_n\| \leq C$ but $G(\lambda) = +\infty$. This is a contradiction with the semicontinuity and suggests that we should impose $C = +\infty$, i.e. g should not be bounded by something with linear growth. Note that subadditive functions, due to $g(nv) \leq ng(v)$, are indeed sublinear at infinity, so the superlinear growth should occur near 0. This is consistent with the most common examples of subadditive functions (think of $g(v) = |v|^\alpha$ for $\alpha \in (0,1)$) and with what occurs when approximating a non-atomic measure by atomic measures: in this case, we need to have more and more atoms, and their masses to tend to 0.

It is useful to establish the following.

Lemma 3.20 *Assume that g is an l.s.c. and subadditive function with $g(0) = 0$. Assume that $\lambda_n = \sum_{i=1}^M a_{i,n}\delta_{x_{i,n}}$ is a sequence of atomic vector measures with a bounded number of atoms and bounded mass. Then, up to a subsequence, we have $\lambda_n \overset{*}{\rightharpoonup} \lambda$ for a certain measure λ which is also atomic, has at most M atoms, and satisfies $G(\lambda) \leq \liminf G(\lambda_n)$.*

Proof Let us note that we can assume that the points $x_{i,n}$ are distinct. If a measure λ_n has fewer than M atoms then we can choose arbitrary points $x_{i,n}$ to complete the list of atoms and set $a_{i,n} = 0$ for those extra indices.

Now, extract a subsequence (not relabeled) such that for each $i = 1, \ldots, M$ one has $a_{i,n} \to a_i$ and $x_{i,n} \to x_i$ (this is possible because we assumed Ω to be compact). For this subsequence one has $\lambda_n \overset{*}{\rightharpoonup} \lambda := \sum_{i=1}^M a_i\delta_{x_i}$. It is possible that the points x_i are not distinct. If they are distinct we have $G(\lambda) = \sum_{i=1}^M g(a_i)$, otherwise we have (thanks to subadditivity) $G(\lambda) \leq \sum_{i=1}^M g(a_i)$. Anyway we have

$$G(\lambda) \leq \sum_{i=1}^M g(a_i) = \lim_n \sum_{i=1}^M g(a_{i,n}) = \lim_n G(\lambda_n),$$

which proves the desired result (one can choose the subsequence so that it realizes the liminf of the whole sequence). □

It is now possible to prove the following.

Lemma 3.21 *Assume that g is a non-negative subadditive and l.s.c. function such that $g(0) = 0$ and $\lim_{v \to 0} g(v)/|v| = +\infty$. Then G is l.s.c.*

Proof Assume without loss of generality that all the λ_n are purely atomic. Fix a number $M > 0$ and use $\lim_{v \to 0} g(v)/|v| = +\infty$: this implies that there exists an ε_0 such that for all v with $|v| < \varepsilon_0$ we have $g(v) > M|v|$. Consider a sequence $\lambda_n \overset{*}{\rightharpoonup} \lambda$, assume $G(\lambda_n) \leq C < +\infty$ and decompose it into $\lambda_n = \lambda_n^s + \lambda_n^b$, where $\lambda_n^b = \sum_{i=1}^N a_{i_n} \delta_{x_{i,n}}$ is the sum of the atoms of λ_n with mass at least ε_0. In particular, there are no more than $N := C\varepsilon_0^{-1}$ of these atoms, where C is a bound for $||\lambda_n||$. The other part λ_n^s (the "small" part, λ_n^b being the "big" one) is just defined as the remaining atoms (every λ_n is purely atomic since $G(\lambda_n) < +\infty$).

If we write

$$C \geq G(\lambda_n) \geq G(\lambda_n^s) = \sum_i g(a_{i,n}) \geq M \sum_i |a_{i,n}| = M|\lambda_n^s|(\Omega)$$

we get an estimate on the mass of the "small" part. Hence it is possible to get, up to subsequences,

$$\lambda_n^b \rightharpoonup \lambda^b \quad \text{and} \quad \lambda_n^s \rightharpoonup \lambda^s, \ |\lambda^s|(\Omega) \leq \frac{C}{M}.$$

We can now apply Lemma 3.20 to prove that λ^b is purely atomic and that $G(\lambda^b) \leq \liminf_n G(\lambda_n^b) \leq \liminf_n G(\lambda_n)$.

This proves that λ must be purely atomic, since the possible atomless part of λ must be contained in λ^s, but $|\lambda^s|(\Omega) \leq C/M$. Hence the mass of the non-atomic part of λ must be smaller than C/M for every $M > 0$, i.e. it must be zero.

We have now proven that λ is purely atomic and we have an estimate on $G(\lambda^b)$, where λ^b is a part of λ that depends on M. If we write $(a_i)_i$ for the masses of λ and $(a_i^M)_i$ for those of λ^b we have

$$\sum_i g(a_i^M) \leq \liminf_n G(\lambda_n) := \ell.$$

We want to prove $\sum_i g(a_i) \leq \ell$ and, with this aim, it is enough to let $M \to \infty$. Actually, $|\lambda^s|(\Omega) = \sum_i |a_i - a_i^M| \leq C/M \to 0$ implies that for each i we have $a_i - a_i^M \to 0$ and thus $a_i^M \to a_i$. Using the semicontinuity of g we have $g(a_i) \leq \liminf_{M \to \infty} g(a_i^M)$. If we fix an arbitrary number N we get

$$\sum_{i=1}^N g(a_i) \leq \liminf_{M \to \infty} \sum_{i=1}^N g(a_i^M) \leq \ell.$$

By passing to the supremum over N we finally get

$$\sum_{i=1}^\infty g(a_i) \leq \ell,$$

which is the claim.   □

Also for this class of functionals we are able to provide a more general semicontinuity result which allows g to be non-superlinear at 0. We first need to discuss the slope at 0 of subadditive l.s.c. functions.

Lemma 3.22 *Let $g : \mathbb{R}^N \to \mathbb{R}$ be an l.s.c. subadditive function with $g(0) = 0$. Then for every v there exists the limit $\lim_{t \to 0^+} \frac{g(tv)}{t}$ which also equals $\sup_{t>0} \frac{g(tv)}{t}$. This limit is denoted by $g^0(v)$.*

Proof It is not restrictive to consider the case $n = 1$, as the statement only concerns the behavior of g on the half-line $\{tv, t \in \mathbb{R}_+\}$. Set $\ell := \liminf_{t \to 0^+} \frac{g(tv)}{t}$. If $\ell = +\infty$ then it is clear that we also have $\limsup_{t \to 0^+} \frac{g(tv)}{t} = \sup_{t>0} \frac{g(tv)}{t} = \ell$ and there is nothing else to prove. If $\ell < +\infty$, then for every ε there are infinitely many small values of t such that $g(tv) \le (\ell + \varepsilon)t$. Because of sub-additivity, if a certain t belongs to $\{t : g(t) \le Ct\}$ for some C, then all its multiples, of the form nt for positive integer n, also belong to the same set. Hence, if such a set contains arbitrary small numbers t it also contains a dense set. Because of the lower semicontinuity of g, if it contains a dense set it actually contains every point, so that we can deduce $g(tv) \le (\ell + \varepsilon)t$ for every t. The value of $\varepsilon > 0$ being arbitrary, we also find $g(tv) \le \ell t$, so that $\ell \ge \sup_{t>0} \frac{g(tv)}{t} = \ell$ and the result is proven.   □

Note that the function $t \mapsto \frac{g(tv)}{t}$ is not always monotone when g is subadditive but not concave. For instance, one can use as an example the function $g(s) = |\sin s|$ in dimension one, and, if we want an example where there is not even local monotonicity in a neighborhood of the origin, $g(s) = \sum_k 2^{-k}|\sin(ks)|$. Moreover, we also observe that the lower semicontinuity of g was crucial in the above result, which would be false for non-l.s.c. subadditive functions g such as $g = \mathbb{1}_{\mathbb{R} \setminus \mathbb{Q}}$.

We now look at the function g^0:

Proposition 3.23 *The function g^0 defined via $g^0(v) = \lim_{t \to 0^+} \frac{g(tv)}{t}$ is convex, 1-homogeneous, and lower semicontinuous.*

Proof The subadditivity of g implies

$$\frac{g(t(v + w))}{t} \le \frac{g(tv)}{t} + \frac{g(tw)}{t}$$

and, taking the limit $t \to 0$, we obtain the subadditivity of g^0.

In order to prove that g^0 is 1-homogeneous we take v and $c > 0$ and write

$$\frac{g(t(cv))}{t} = c \frac{g((tc)v)}{tc}.$$

and, taking the limit $t \to 0$ (hence $ct \to 0$), we obtain $g^0(cv) = cg^0(v)$.

The convexity of g^0 then follows from the next Lemma 3.24.

Its semicontinuity is due, instead, to its representation as a sup: for each t the function $v \mapsto \frac{g(tv)}{t}$ is l.s.c. and hence g^0 is also l.s.c. as a sup of l.s.c. functions.   □

Lemma 3.24 *Given a function* $f : X \to \mathbb{R}$ *on a vector space* X *with* $f(0) = 0$, *any two of the following conditions imply the third one:*

1. *f is convex;*
2. *f is 1-homogeneous;*
3. *f is subadditive.*

Proof In order to prove 1+2⇒3 we write

$$f(x + y) = f\left(\frac{1}{2}(2x + 2y)\right) \leq \frac{1}{2}f(2x) + \frac{1}{2}f(2y) = f(x) + f(y),$$

using first the convexity and then the 1-homogeneity of f.

In order to prove 2+3⇒1 we write

$$f((1 - t)x + ty) \leq f((1 - t)x) + f(ty) = (1 - t)f(x) + tf(y),$$

using first the subadditivity and then the 1-homogeneity of f.

In order to prove 3+1⇒2 we consider the ratio $h(t) := \frac{f(tv)}{t}$. This ratio is nondecreasing, as a consequence of the convexity of f and of $f(0) = 0$. Moreover, f is continuous on the line $\{tv : t \in \mathbb{R}\}$ as a consequence of convexity in finite dimensions, so that we can apply Lemma 3.22 since we have semicontinuity and subadditivity. Hence, we have $h(t_0) \geq \lim_{t \to 0^+} h(t) = \sup_{t > 0} h(t) \geq h(t_0)$ for every $t_0 > 0$, which proves that h is constant, and hence f is 1-homogeneous.   □

For the proof of the next semicontinuity result it is useful to approximate g^0 from below by convex 1-homogeneous functions g^ε such that $g^\varepsilon \leq g$ on the ball $B(0, \varepsilon)$. We then prove the following.

Lemma 3.25 *Let* $g : \mathbb{R}^N \to \mathbb{R}$ *be an l.s.c. subadditive function such that* $g(v) \geq c|v|$ *for some constant* $c > 0$. *Set* $C_\varepsilon := \{w : w \cdot v \leq g(v) \text{ for every } v \text{ s.t. } |v| \leq \varepsilon\}$ *and* $g^\varepsilon(v) := \sup\{w \cdot v : w \in C_\varepsilon\}$. *We then have* $g^\varepsilon \leq g \leq g^0$ *on* $B(0, \varepsilon)$ *and* $g^\varepsilon(v) \to g^0(v)$ *for every* $v \in \mathbb{R}^N$.

Proof Note that by definition we have $g^\varepsilon(v) \leq g(v)$ for every $v \in B(0, \varepsilon)$. We then use the subadditivity of g to write $g(v) \leq 2^k g(2^{-k}v)$ and, taking the limit $k \to \infty$, we obtain $g \leq g^0$.

In order to find the limit of g^ε let us first consider the sets C_ε. It is clear that these sets increase as ε decreases. Let us define $C_0 := \{w : w \cdot v \leq g^0(v) \text{ for every } v\}$. The assumption $g(v) \geq c|v|$ implies $g^0(v) \geq c|v|$, so that $B(0, c) \subset C_0$. Hence, C_0 is a convex set with non-empty interior, and it is then the closure of its interior, which we denote by C_0°. We now prove $\bigcup_\varepsilon C_\varepsilon \supset C_0^\circ$. Suppose by contradiction that a certain point w belongs to C_0° but not to $\bigcup_\varepsilon C_\varepsilon$. The condition $w \in C_0^\circ$ means that there exists a $\delta > 0$ such that $w + \delta e \in C_0$ for all vectors e with $|e| = 1$, i.e.

$w \cdot v + \delta|v| \leq g^0(v)$ for all v. On the other hand, the condition $w \notin \bigcup_\varepsilon C_\varepsilon$ means that for any j, we can find a v_j such that $|v| \leq 2^{-j}$ but $w \cdot v_j > g(v_j)$. Because of the subadditivity of g this also implies $w \cdot (nv_j) > g(nv_j)$ for every n. Hence, we can find for every j a vector $\hat{v}_j$ such that $1 \leq |\hat{v}_j| \leq 2$ and such that

$$w \cdot (2^{-h}\hat{v}_j) > g(2^{-h}\hat{v}_j) \tag{3.2}$$

for every $h = 1, \ldots, j$. Up to subsequences, we can assume that the vectors $\hat{v}_j$ converge as $j \to \infty$ to a certain vector v with $1 \leq |v| \leq 2$. We now fix an integer h and take the limit $j \to \infty$ in (3.2) so that we obtain

$$w \cdot (2^{-h}v) \geq g(2^{-h}v).$$

Multiplying by 2^h and taking now the limit $h \to \infty$ we obtain $w \cdot v \geq g_0(v)$, which contradicts $w \cdot v + \delta|v| \leq g^0(v)$.

Hence, we obtain

$$\lim_{\varepsilon \to 0} g_\varepsilon(v) \geq \sup\{w \cdot v : v \in C_0^\circ\} = \sup\{w \cdot v : v \in C_0\}.$$

The claim will be proven if we can show that the last expression above coincides with $I_{C_0}^*$, but $I_{C_0} = g_0^*$ and $I_{C_0}^* = g_0^{**} = g_0$, and for this we refer again to Sect. 4.2.1. □

We can now state the following semicontinuity result.

Theorem 3.26 *Given a subadditive and l.s.c. nonnegative function* $g : \mathbb{R}^N \to \mathbb{R}$, *the functional* $\mathcal{G}$ *defined as follows:*

$$\mathcal{G}(\lambda) = \sum_{x \in A(\lambda)} g(\lambda(\{x\})) + \int_\Omega g^0(\lambda - \lambda^\#) \tag{3.3}$$

is lower semicontinuous on $\mathcal{M}^d(\Omega)$ *for the weak* convergence of measures.*

Proof First, we assume for now that g satisfies $g(v) \geq c|v|$ for some constant $c > 0$.

Let us take a sequence $\lambda_n \overset{*}{\rightharpoonup} \lambda$. Fix a number $\varepsilon > 0$ and decompose λ_n into $\lambda_n = \lambda_n^s + \lambda_n^b$, where $\lambda_n^b = \sum_{i=1}^N a_{in}\delta_{x_{i,n}}$ is the sum of the atoms of λ_n with mass at least ε. In particular, there are no more than $N := C\varepsilon^{-1}$ of these atoms. The other part λ_n^s (again, the "small" part) is just defined as the remaining atoms together with the non-atomic part $\lambda_n - \lambda_n^\#$. The definition of $\mathcal{G}$, which is additive on measures concentrated on disjoint sets, allows us to write $\mathcal{G}(\lambda_n) = \mathcal{G}(\lambda_n^b) + \mathcal{G}(\lambda_n^s)$.

Up to subsequences, we can assume $\lambda_n^b \overset{*}{\rightharpoonup} \lambda^b$, $\lambda_n^s \overset{*}{\rightharpoonup} \lambda^s$, and $\lambda = \lambda^b + \lambda^s$. Lemma 3.20 tells us that λ^b is finitely atomic, and we have $\mathcal{G}(\lambda^b) \leq \liminf_n \mathcal{G}(\lambda_n^b)$.

In what concerns the other part, we use the inequalities $g \geq g^\varepsilon$ on $B(0, \varepsilon)$ and $g^0 \geq g^\varepsilon$ everywhere, in order to obtain

$$\liminf_n \mathcal{G}(\lambda_n^s) \geq \liminf_n \int_\Omega g^\varepsilon(\lambda_n^s) \geq \int_\Omega g^\varepsilon(\lambda^s),$$

where we used the semicontinuity of the functional $\lambda \mapsto \int g^\varepsilon(\lambda)$ as a particular case of Proposition 3.19.

We then observe that the atoms of λ are split into the parts λ^b and λ^s, while the non-atomic part is completely contained in λ^s. Hence we have obtained

$$\liminf_n \mathcal{G}(\lambda_n) \geq \sum_{x \in A(\lambda)} g(\lambda^b(\{x\})) + g^\varepsilon(\lambda^b(\{x\})) + \int_\Omega g^\varepsilon(\lambda - \lambda^\#).$$

We now take the limit $\varepsilon \to 0$. By monotone convergence, using Lemma 3.25, we have $\int g^\varepsilon(\lambda - \lambda^\#) \to \int g^0(\lambda - \lambda^\#)$. For each atom $x \in A(\lambda)$, we have

$$g(\lambda^b(\{x\})) + g^\varepsilon(\lambda^b(\{x\})) \geq h_\varepsilon(\lambda(\{x\})),$$

where the function $h_\varepsilon : \mathbb{R}^N \to \mathbb{R}$ is defined via

$$h(v) := \inf\{g(v_1) + g_\varepsilon(v_2) : v_1 + v_2 = v\}.$$

We will see that, pointwisely, we have $\lim_{\varepsilon \to 0} h_\varepsilon(v) \geq g(v)$, from which the result follows, again by monotone convergence in the sum.

In order to prove $\lim_{\varepsilon \to 0} h_\varepsilon(v) \geq g(v)$ we fix v and take $v_1^\varepsilon, v_2^\varepsilon$ realizing the inf in the definition of h_ε. Such two values exist because the functions which are minimized are l.s.c. and the lower bound $g(v), g^\varepsilon(v) \geq c|v|$ allows us to bound any minimizing sequences. Such a bound is uniform in ε, so we can assume, up to subsequences, $v_1^\varepsilon \to v_1, v_2^\varepsilon \to v_2$ as $\varepsilon \to 0$, with $v_1 + v_2 = v$. We fix $\varepsilon_0 > 0$ and use $g^\varepsilon \geq g^{\varepsilon_0}$ for $\varepsilon < \varepsilon_0$, so that we have $h_\varepsilon(v) \geq g(v_1^\varepsilon) + g^{\varepsilon_0}(v_2^\varepsilon)$. Taking the limit as $\varepsilon \to 0$, we get $\lim_{\varepsilon \to 0} h_\varepsilon(v) \geq g(v_1) + g^{\varepsilon_0}(v_2)$. We then take the limit $\varepsilon_0 \to 0$, thus obtaining

$$\lim_{\varepsilon \to 0} h_\varepsilon(v) \geq g(v_1) + g^0(v_2) \geq g(v_1) + g(v_2) \geq g(v),$$

where we also used $g^0 \geq g$ and the subadditivity of g. This concludes the proof in the case $g(v) \geq c|v|$. The general case can be obtained using Lemma 3.27. $\qquad \square$

Lemma 3.27 *A functional $F : X \to \mathbb{R} \cup \{+\infty\}$ defined on a Banach space X is l.s.c. for the weak (or weak-*, if X is a dual space) convergence if and only if for every $c > 0$ the functional F_c defined through $F_c(x) = F(x) + c\|x\|$ is l.s.c. for the same convergence.*

Proof Obviously, since the norm is always l.s.c., if F is l.s.c. so is F_c for any c. Assume now that F_c is l.s.c. for every $c > 0$ and take a sequence $x_n \overset{*}{\rightharpoonup} x$ (or $x_n \overset{*}{\rightharpoonup} x$). Since any weakly (or weakly-*) convergent sequence is bounded (see Box 4.2), we have $\|x_n\| \leq M$ for a suitable constant M. Then, we observe that we have

$$F(x) \leq F_c(x) \leq \liminf_n F_c(x_n) \leq \liminf_n F(x_n) + cM.$$

This shows $\liminf_n F(x_n) \geq F(x) - cM$ and the result follows by taking the limit $c \to 0$. $\square$

Finally, we can combine the two results, that for a convex function f and that for a sub-additive function g, into the following lower semicontinuity result.

Theorem 3.28 *Given a positive non-atomic finite measure μ, a subadditive and l.s.c. nonnegative function $g : \mathbb{R}^N \to \mathbb{R}$ and a convex function $f : \mathbb{R}^N \to \mathbb{R}$, assume $f^\infty = g^0 = \varphi$. Then the functional F defined as follows is weakly-* lower semicontinuous on $\mathcal{M}^n(\Omega)$:*

$$F(\lambda) = \sum_{x \in A(\lambda)} g(\lambda(\{x\})) + \int_\Omega \varphi(\lambda^s - \lambda^\#) + \int_\Omega f(\mathbf{v}(x)) \, d\mu(x),$$

where we write $\lambda = \mathbf{v} \cdot \mu + \lambda^s$ and λ^s is the singular part of λ, which includes a possible atomic part $\lambda^\#$.

Proof Let us take a sequence $\lambda_n \overset{*}{\rightharpoonup} \lambda$ and decompose it into $\lambda_n = \mathbf{v}_n \cdot \mu + \lambda_n^s$. We then have

$$F(\lambda_n) = \mathcal{G}(\lambda_n^s) + \mathcal{F}(\mathbf{v}_n \cdot \mu),$$

where $\mathcal{G}$ and $\mathcal{F}$ are the functionals associated with the functions g and f, respectively, according to Eqs. (3.3) and (3.1). Up to subsequences we have $\lambda_n^s \overset{*}{\rightharpoonup} \lambda_g$ and $\mathbf{v}_n \cdot \mu \overset{*}{\rightharpoonup} \lambda_f$, with $\lambda = \lambda_g + \lambda_f$. Note that nothing guarantees that λ_g and λ_f are the singular and absolutely continuous parts of λ. Using the semicontinuity of $\mathcal{F}$ and $\mathcal{G}$ we obtain

$$\liminf_n F(\lambda_n) \geq \mathcal{G}(\lambda_g) + \mathcal{F}(\lambda_f)$$

$$= \sum_{x \in A(\lambda_g)} g(\lambda_g(\{x\})) + \int_\Omega \varphi(\lambda_g^s - \lambda_g^\# + \lambda_f^s) + \int_\Omega f(\mathbf{w}(x)) \, d\mu(x),$$

where $\lambda_f = \mathbf{w}(x) \cdot \mu + \lambda_f^s$, and we used $f^\infty = g^0 = \varphi$. We observe that the measure $\lambda_g^s - \lambda_g^\# + \lambda_f^s$ can contain atoms (in the part λ_f^s) and absolutely continuous parts (in $\lambda_g^s - \lambda_g^\#$). On the other hand, all the part of λ which is neither atomic nor absolutely continuous is there. We then write

$$\lambda_g^s - \lambda_g^\# + \lambda_f^s = (\lambda^\# - \lambda_g^\#) + (\lambda^s - \lambda^\#) + (\mathbf{v} - \mathbf{w}) \cdot \mu.$$

Then we use $g(a) + \varphi(b) = g(a) + g^0(b) \geq g(a) + g(b) \geq g(a+b)$, so that we obtain

$$\sum_{x \in A(\lambda_g)} g(\lambda_g(\{x\})) + \int_\Omega \varphi(\lambda^\# - \lambda_g^\#) \geq \sum_{x \in A(\lambda)} g(\lambda(\{x\})).$$

We also use $f(v) \leq f(w) + f^\infty(v - w)$ to obtain

$$\int_\Omega \varphi((\mathbf{v} - \mathbf{w}) \cdot \mu) + \int_\Omega f(\mathbf{w}(x)) \, d\mu(x) \geq \int_\Omega f(\mathbf{v}(x)) \, d\mu(x)$$

and this concludes the proof of the semicontinuity of F. □

3.4 Existence of Minimizers

This short section is devoted to some existence results which extend those of Sect. 2.1 and can be obtained by combining the techniques already used there with the semicontinuity results of this chapter.

A general statement that we can obtain is, for instance, the following.

Theorem 3.29 *Let Ω be an open domain and $g \in W^{1,p}(\Omega)$ and an integrand $L : \Omega \times \mathbb{R} \times \mathbb{R}^d \to \mathbb{R}$ such that $L(x, s, v)$ is continuous in (s, v) and convex in v for a.e. x and such that we have*

$$c|v|^p - a(x) - C|s|^{q_1} \leq L(x, s, v) \leq C|v|^p + a(x) + C|s|^{q_2}$$

for some constants $c, C > 0$, a function $a \in L^1(\Omega)$, and some exponents p, q_1, q_2 with $1 \leq q_1 < p \leq q_2 \leq p^*$. Then the optimization problem

$$\min \left\{ J(u) := \int_\Omega L(x, u(x), \nabla u(x)) \, dx \ : \ u - g \in W_0^{1,p}(\Omega) \right\} \quad (3.4)$$

admits a solution.

Proof The assumption on the upper bound on L has been introduced to guarantee $J(u) < +\infty$ for every $u \in W^{1,p}(\Omega)$. The function $u = g$ satisfies the boundary constraint, and provides a finite value to J, which shows that the inf in the optimization problem is not $+\infty$. We now take a minimizing sequence u_n with

$J(u_n) \leq C < +\infty$. Using the lower bound on L we find

$$c||\nabla u_n||_{L^p}^p \leq C(1 + ||u||_{L^{q_1}}^{q_1}).$$

Using the same estimates as in Theorem 2.1 we have the inequality $||u||_{L^p} \leq C||\nabla u||_{L^p} + C(g)$ for every admissible u. We can moreover bound the L^{q_1} norm with the L^p norm and obtain

$$c||\nabla u_n||_{L^p}^p \leq C(1 + ||\nabla u_n||_{L^p}^{q_1}),$$

which proves that ∇u_n is bounded in L^p since $q_1 < p$. This implies that u_n is bounded in $W^{1,p}$, which is a reflexive space since we assumed $p > 1$. We can then extract a subsequence weakly converging to a limit u. This limit is also admissible (i.e. $u - g \in W_0^{1,p}(\Omega)$) and the compact injection of $W^{1,p}$ into L^{q_1} implies pointwise a.e. convergence of u_n to u.

We now want to apply Theorem 3.16 in order to show the semicontinuity of J. The only lacking assumption is the positivity of L. Hence, we write $\tilde{L}(x, s, v) = L(x, s, v) + a(x) + C|s|^{q_1}$ and we have $\tilde{L} \geq 0$; moreover, $\tilde{L}$ satisfies the same assumptions as L in terms of continuity or convexity. Hence, the functional

$$(u, v) \mapsto \int_\Omega \tilde{L}(x, u(v), v(x))\,dx$$

is l.s.c. for the strong L^1 convergence of u and the weak L^1 convergence of v, so that we have

$$\liminf_n \int_\Omega \tilde{L}(x, u_n(v), \nabla u_n(x))\,dx \geq \int_\Omega \tilde{L}(x, u(v), \nabla u(x))\,dx.$$

Moreover, we have

$$\lim_n \int_\Omega \left(a(x) + C|u_n(x)|^{q_1}\right)\,dx = \int_\Omega \left(a(x) + C|u(x)|^{q_1}\right)\,dx$$

because of the strong L^{q_1} convergence of u_n to u, from which we conclude the semicontinuity of J and hence the optimality of u.   □

The main difference between this result and those of Sect. 2.1 is that we do not require that the dependence of the functional on the gradient is exactly through its L^p norm. Hence, to obtain its semicontinuity we need to use the tools for more general functionals developed in this chapter, and to obtain the compactness of minimizing sequences we need to require a lower bound on the growth of the integrand which mimics this L^p norm.

We do not develop the details here, but it would of course be possible to obtain an analogue of Theorem 2.4 with boundary penalization, as well as results which

use the functionals whose semicontinuity was proven in Sect. 3.3. Some examples
in this direction are available in the exercises (see Exercises 3.8 and 3.9).

3.5   Discussion: Lower Semicontinuity for Vector-Valued Maps

The semicontinuity results that we introduced in this chapter to study the existence
of minimizers of integral functionals of the form $u \mapsto \int L(x, u, \nabla u)$ actually
concern more general functionals of the form $(u, v) \mapsto \int L(x, u, v)$ and they do
not exploit the fact that we only look at $v = \nabla u$. Actually, they do not even
exploit the fact that the variable v will be a gradient. Due to this generality, the
assumptions on L for the functional to be l.s.c. impose convexity in the variable v.
For many applications it would be important to relax this assumption, in particular
in compressible mechanics, where one would like to study $u : \mathbb{R}^d \to \mathbb{R}^d$
as a deformation and consider functionals penalizing the compression of such a
deformation, hence functionals depending on $\det(Du)$. The determinant being a d-
degree polynomial in the entries of Du, it is in general not a convex function of
Du. This motivates the study of more general functionals. In particular, we look at
functionals of the form $u \mapsto F(u) := \int f(Du) \, dx$ for $u : \Omega \to \mathbb{R}^k$, $\Omega \subset \mathbb{R}^d$
and we want to understand the natural conditions on $f : \mathbb{R}^{d \times k} \to \mathbb{R}$ for F to be
lower semicontinuous for the weak $W^{1,p}$ convergence. For simplicity and in order
to avoid growth conditions, one can consider $p = \infty$.

Charles Morrey found in [152] the sharp condition, as he observed that the
assumption that f is *quasi-convex* is necessary, and also sufficient.

Definition 3.30 A function $f : \mathbb{R}^{d \times k} \to \mathbb{R}$ is said to be quasi-convex if for every
matrix A and every function $\varphi \in C^1_c(Q)$ we have $\fint_Q f(A + \nabla \varphi(x)) \, dx \geq f(A)$,
where $Q = (0, 1)^d$ is the unit cube in $\mathbb{R}^d$.

It is clear that convex functions are quasi-convex, since $\int \nabla \varphi = 0$ and then
Jensen's inequality gives $\fint_Q f(A + \nabla \varphi(x)) \, dx \geq f \left(\fint_Q (A + \nabla \varphi(x)) \, dx \right) = f(A)$.

This definition could be somehow disappointing as it cannot be given by only
looking at the behavior of f as a function on matrices, but one needs to look
at matrix fields and it involves the notion of gradient... we could even say that
it is a way of cheating as the final goal is to produce necessary and sufficient
conditions for the l.s.c. behavior of F and the assumption is given in terms of
F instead of f. Yet, this is exactly the sharp condition. One can easily see that
quasiconvexity is necessary for the semicontinuity by considering $u(x) = Ax$ and
$u_n(x) = Ax + \frac{1}{n} \varphi(nx)$. It is clear that Du_n is bounded in L^∞ and weakly-*
converges to $Du = A$, and that u_n uniformly converges to u, and when $\Omega = Q$ from
$F(u_n) = \fint_Q f(A + \nabla \varphi(x)) \, dx$ (which can be obtained via a change-of-variables)
we see that the lower semicontinuity of F on this sequence is equivalent to the
inequality defining quasiconvexity.

We do not prove here the fact that this condition is also sufficient but it is the case, and it is also possible to prove the semicontinuity of F when we add dependence on x and u. In particular, $u \mapsto \int L(x, u(x), Du(x)) \, dx$ is lower semicontinuous for the Lipschitz weak convergence ($u_n \to u$ uniformly together with $\|Du_n\|_{L^\infty}$ bounded) if $f(x, u, v)$ is continuous in (x, u) and quasiconvex in v for every (x, u) (see [152]). Once we know this result the choice of the definition of quasi-convexity is not disappointing anymore, as we are characterizing the lower semicontinuity of the functional in terms of pointwise (for every (x, u) properties of $f(x, u, \cdot)$) and not only in terms of the integrals of given test functions.

The notion of quasiconvexity can be compared to other notions, in particular to that of polyconvexity and rank-1-convexity.

Definition 3.31 A function $f : \mathbb{R}^{d \times k} \to \mathbb{R}$ is said to be polyconvex if there exists a convex function $g : \mathbb{R}^N \to \mathbb{R}$ such that $f(A) = g((\det(A_\alpha))_\alpha)$ where the matrices A_α stand for all the square submatrices that one can extract from a $d \times k$ matrix and N is the number of those submatrices (we have $N = \sum_{j=1}^{\min\{k,d\}} \binom{d}{j}\binom{k}{j}$). When $d = k = 2$ this means $f(A) = g(A, \det A)$ and when $d = k = 3$ this means $f(A) = g(A, \mathrm{Cof}(A), \det A)$ where $\mathrm{Cof}(A)$ is the cofactor matrix of A, characterized by the condition $A\mathrm{Cof}(A) = \det(A)\mathbf{I}$.

Definition 3.32 A function $f : \mathbb{R}^{d \times k} \to \mathbb{R}$ is said to be rank-1-convex if $t \mapsto f(A + tB)$ is convex in t whenever B is a matrix with rank one.

We have the following result.

Theorem 3.33 *If f is convex, then it is polyconvex. If f is polyconvex, then it is quasi-convex. If f is quasi-convex, then it is rank-1-convex. If $\min\{k, d\} = 1$, then rank-1-convex functions are also convex.*

The only non-trivial statements hereabove are the quasi-convexity of polyconvex functions, and the rank-1-convexity of quasi-convex functions.

For the latter, we just observe that rank-1 matrices B can be written as $a \otimes b$ for $a \in \mathbb{R}^k$ and $b \in \mathbb{R}^d$ and we can construct a function φ whose gradient $D\varphi$ is only composed of multiples of $a \otimes b$. Given a function $\phi : \mathbb{R} \to \mathbb{R}$ which is 1-periodic and satisfies $\phi' = 1 - t$ on $(0, t)$ and $\phi' = -t$ on $(t, 1)$, we can define $\varphi(x) = a\phi(b \cdot x)$ and see (up to correcting boundary effects due to the fact that φ is not compactly supported) that the quasi-convexity of f implies $f(A) \leq tf(A + (1-t)a \otimes b) + (-t)f(A + ta \otimes b)$, which gives the rank-1-convexity of f.

The fact that polyconvex functions are quasi-convex relies again on Jensen's inequality, once we find that $\varphi \mapsto \int \det(D\varphi) \, dx$ (when $\varphi : \mathbb{R}^d \to \mathbb{R}^d$, so that $D\varphi$ is a square matrix) is a null-Lagrangian, i.e. it only depends on the boundary value of ϕ. This can be seen by writing the determinant as a divergence. Remember that we have

$$\det(A) = \sum_j A^{1j} z^j(A), \quad \text{where } z^j(A) = \sum_{\sigma : \sigma_1 = j} \varepsilon(\sigma) A^{2\sigma_2} A^{3\sigma_3} \cdots A^{d\sigma_d},$$

$$(3.5)$$

where the sum is over permutations σ and $\varepsilon(\sigma)$ stands for the signature of the permutation. The terms $z^j(A)$ are the entries of the cofactor matrix $\mathrm{Cof}(A)$ (here those on the first line of this matrix). An important point is the fact that cofactors of gradients have zero-divergence, as a consequence of Schwartz's theorem:

Proposition 3.34 *If φ is a vector-valued C^2 function, then*

$$\sum_j \frac{\partial}{\partial x_j} z^j(D\varphi) = 0.$$

Proof Computing the above sum of derivatives we obtain

$$\sum_j \frac{\partial}{\partial x_j} z^j(D\varphi) = \sum_\sigma \varepsilon(\sigma) \frac{\partial}{\partial x_{\sigma_1}} \left(\frac{\partial \varphi^2}{\partial x_{\sigma_2}} \frac{\partial \varphi^3}{\partial x_{\sigma_3}} \cdots \frac{\partial \varphi^d}{\partial x_{\sigma_d}} \right).$$

Expanding the derivatives further, we find the sum of all terms which are products of $(d-2)$ first derivatives of components of φ (the component i differentiated with respect to σ_i for $i \geq 2$) times a unique factor with two derivatives (one w.r.t. to σ_i, one w.r.t. to σ_1). Since we are considering all permutations, these terms can be paired and each term appears exactly twice, using the commutativity of the second derivatives. Indeed, the same term where φ^i is differentiated w.r.t. x_j and to x_k appears when $(\sigma_i, \sigma_1) = (j, k)$ but also when $(\sigma_i, \sigma_1) = (k, j)$. These two permutations differ by a transposition and so their signatures are opposite, so that they cancel in the sum, thus proving the result.   □

We then obtain the following result.

Proposition 3.35 *Assume $d = k$. Then, the functional F defined via $F(u) := \int_\Omega \det(Du)\, dx$ is such that $F(u) = F(\tilde{u})$ whenever $u - \tilde{u} \in W_0^{1,d}(\Omega)$.*

Proof First, we observe that, for smooth φ, we have

$$\det(D\varphi) = \nabla \cdot \mathbf{v}[\varphi], \quad \text{where } \mathbf{v}[\varphi]^j := \varphi^1 z^j(D\varphi).$$

Indeed, we have

$$\nabla \cdot \mathbf{v}[\varphi] = \sum_j \frac{\partial}{\partial x_j} (\varphi^1 z^j(D\varphi))$$

$$= \sum_j \frac{\partial \varphi^1}{\partial x_j} z^j(D\varphi) + \varphi^1 \sum_j \frac{\partial}{\partial x_j} z^j(D\varphi) = \det(D\varphi) + 0,$$

where in the last inequality we used the characterization (3.5) of the determinant and the result of Proposition 3.34. This shows that $\int_\Omega \det(D\varphi)$ only depends on φ and its derivatives on $\partial\Omega$, so that we have $F(u) = F(\tilde{u})$ whenever $u, \tilde{u} \in C^\infty$ with

$u - \tilde{u} \in C_c^\infty(\Omega)$. Then we observe that F is continuous for the strong convergence of $W^{1,d}$ so that this stays true whenever $u, \tilde{u} \in W^{1,d}$ with $u - \tilde{u} \in W_0^{1,d}(\Omega)$, by approximating u with C^∞ functions and $u - \tilde{u}$ with functions in C_c^∞. Note that the dependence on higher-order derivatives on $\partial\Omega$ disappears, in the end. $\qquad\square$

It is not difficult to adapt the above result to the case of determinants of smaller minors $h \times h$, possibly in the case of rectangular matrices $k \times d$. This allows us to prove the quasi-convexity of polyconvex functionals, since we have in this case

$$\fint_Q f(A + D\varphi) \, dx = \fint_Q g((\det((A + D\varphi)_\alpha))_\alpha) \, dx$$

$$\geq g\left(\left(\fint_Q \det((A + D\varphi)_\alpha) \, dx\right)_\alpha\right)$$

$$= g\left((\det(A_\alpha))_\alpha\right) = f(A),$$

using the convexity of g and the result of Proposition 3.35 adapted to all minors.

The question of whether or not rank one convexity implies quasiconvexity was left open by Morrey. The question is natural, as it can be proven that rank one affine functions (i.e. functions f such that both f and $-f$ are rank one convex) are indeed quasi-affine functions (both f and $-f$ are quasi-convex). However, Morrey conjectured that quasi-convexity and rank one convexity should be different notions. This has been proven thanks to a family of counterexamples introduced by Dacorogna and Marcellini and later improved by Alibert and Dacorogna ([6, 68]). Consider for $A \in \mathbb{R}^{2\times 2}$

$$f(A) = ||A||^2(||A||^2 + c\det(A)),$$

where the norm is the Euclidean norm in $\mathbb{R}^4$ (i.e. $||A||^2 = \text{Tr}(A^T A)$). This function is a 4-homogeneous polynomial and it can be proven that

- f is convex if and only if $|c| \leq \frac{4}{3}\sqrt{2}$;
- f is polyconvex if and only if $|c| \leq 2$;
- f is rank-one convex if and only if $|c| \leq \frac{4}{\sqrt{3}}$;
- f is quasi-convex if and only if $|c| \leq M$ for a constant M whose precise value is unknown but satisfies $M \in (2, \frac{4}{\sqrt{3}})$.

3.6 Exercises

Exercise 3.1 Prove that the level sets $\{x \in X : f(x) \leq \ell\}$ are all convex if f is convex, but find an example of a function f which is not convex, despite having convex level sets.

Exercise 3.2 Assume that u_n is a sequence weakly converging to u in $L^1(\Omega)$ and assume $\int_\Omega \sqrt{1 + |u_n|^2}\, dx \to \int_\Omega \sqrt{1 + |u|^2}\, dx$. Prove $u_n \to u$ a.e. and in L^1.

Exercise 3.3 Assume that the functional $u \mapsto F(u) := \int_\Omega L(x, u(x))\, dx$ is l.s.c. for the weak convergence in L^p, and that $L : \Omega \times \mathbb{R} \to \mathbb{R}$ is continuous. Prove that $L(x, \cdot)$ is convex for every x.

Exercise 3.4 Assume that u_n is a sequence weakly converging to u in $L^p(X)$ and set $v_n := H(u_n)$ where $H : \mathbb{R} \to \mathbb{R}$ is a given convex function. Assume that v_n weakly converges to v in $L^1(X)$. Prove $v \geq H(u)$, and find examples where the inequality is strict.

Exercise 3.5 Consider the functional $J(u) := \int_\Omega \arctan(u + |\nabla u|^2)\, dx$ and the minimization problem

$$\min\{J(u) \ : \ \|u\|_{L^\infty}, \|\nabla u\|_{L^\infty} \leq c_0\}.$$

Prove that this problem admits a solution if $c_0 > 0$ is small enough.

Exercise 3.6 Show that for every l.s.c. function $f : \mathbb{R} \to \mathbb{R}_+$ there exists a sequence of functions $f_k : \mathbb{R} \to \mathbb{R}_+$, each k−Lipschitz, such that for every $x \in \mathbb{R}$ the sequence $(f_k(x))_k$ increasingly converges to $f(x)$.

Use this fact and the theorems of this chapter to prove the semicontinuity, w.r.t. to the weak convergence in $H^1(\Omega)$, of the functional

$$J(u) = \int_\Omega f(u(x))|\nabla u(x)|^p\, dx,$$

where $p \geq 1$ and $f : \mathbb{R} \to \mathbb{R}_+$ is l.s.c.

Exercise 3.7 Let Ω be an open, connected and smooth subset of $\mathbb{R}^d$. Prove that the following minimization problem admits a solution

$$\min\left\{\int_\Omega \left(\frac{1}{2}|\nabla u|^2 - (1 + v^2)\sin\left(\frac{u}{1 + v^2}\right) + (2 + \arctan(u))|\nabla v|^2\right) dx \ : \right.$$

$$\left. u, v \in H_0^1(\Omega)\right\}.$$

Exercise 3.8 Prove that the following problem admits a minimizer

$$\min\left\{\int_0^1 L(t, u(t), u'(t))\, dt + J(u) \ : \ \begin{array}{l} u : [0, 1] \to \mathbb{R}^d \text{ piecewise } H^1, \\ |u| \leq M \text{ a.e.} \end{array}\right\},$$

where the set of piecewise H^1 functions is defined as those functions $u \in L^2([0, 1])$ for which there exists a decomposition of $[0, 1]$ into finitely many intervals $[t_i, t_{i+1}]$ with $0 = t_0 < t_1 < \cdots < t_N = 1$ such that the restriction of u to each of these intervals is H^1; u then admits a piecewise continuous representative, admitting right and left limit at each point t_i, and the functional J is defined as the number of jumps of u, i.e. the number of indices i such that these two limits are different. We assume here that $L(t, s, v)$ is continuous in (s, v) and C^2 in v, with $D^2_{vv}L \geq c_0\mathbf{I}$ for $c_0 > 0$ and M is any given constant.

Exercise 3.9 Prove that the following problem admits a minimizer

$$\min \left\{ \int_0^1 L(t, u(t), u'(t))\, dt + H(u) \ : \ u : [0, 1] \to \mathbb{R}^d \text{ piecewise } H^1 \right\},$$

where the set of piecewise H^1 functions is defined as in Exercise 3.8 and the functional H is defined as $\sum_i \left(1 + \sqrt{|u(t_i)^+ - u(t_i^-)|}\right)$, where the points t_i are, again, the jump points of u, i.e. those points where the right and left limits are different. We assume here as well that $L(t, s, v)$ is continuous in (s, v) and C^2 in v, with $D^2_{vv}L \geq c_0\mathbf{I}$ for $c_0 > 0$.

Hints

Hint to Exercise 3.1 Use the definition of convexity of a function to prove the convexity of level sets. As a counter-example to the converse implication, any monotone function on $\mathbb{R}$ can be used.

Hint to Exercise 3.2 Use the inequality

$$\sqrt{1 + a^2} \geq \sqrt{1 + b^2} + \frac{b}{\sqrt{1 + b^2}}(a - b) + c(b)\min\{(a - b)^2, 1\}$$

for a number $c(b) > 0$, applied to $a = u_n$ and $b = u$, to deduce $u_n \to u$ a.e. In order to get L^1 strong convergence, prove that equi-integrability and a.e. convergence imply L^1 convergence.

Hint to Exercise 3.3 Localize the example of Proposition 3.10 on a small cube of side ε around a point x and take the limit $\varepsilon \to 0$.

Hint to Exercise 3.4 Fix a non-negative function φ and consider the functional defined by $u \mapsto \int \varphi(x)H(u(x))\, dx$.

Hint to Exercise 3.5 Prove that the function $v \mapsto \arctan(u + v^2)$ is convex on $(-\varepsilon, \varepsilon)$ if u and ε are small.

Hint to Exercise 3.6 Use $f_k(x) := \inf_y f(y) + kd(x, y)$. Then write the functional as a sup.

Hint to Exercise 3.7 Prove boundedness in H^1 and use the semicontinuity results of Sect. 3.2.

Hint to Exercise 3.8 Subtract $\frac{c_0}{2}v^2$ from $L(x, u, v)$ and prove separately the semi-continuity of two parts of the functional using Sects. 3.2 and 3.3. Obtain a bound of a minimizing sequence u_n in BV, thus proving weak-* convergence of the derivatives u'_n.

Hint to Exercise 3.9 In the previous exercise the intensity $u(t_i^+) - u(t_i^-)$ of the jump was not penalized but the L^∞ norm was bounded. Here we can only have finitely many jumps, with a bound on their intensity, and this allows us to obtain compactness. The conclusion follows again by Sects. 3.2 and 3.3.

Chapter 4
Convexity and its Applications

This chapter is devoted to several applications of convexity other than the semicontinuity of integral functionals that we already saw in Chap. 3. After a first section about the identification of the minimizers (discussing their uniqueness and the sufficient conditions for optimality, both in a general setting and in the calculus of variations, i.e. when we face integral functionals on functional spaces) the attention turns to many notions coming from abstract convex analysis, in particular related to Legendre transforms and duality. Section 4.2 will be quite general, while Sects. 4.3 and 4.4 will concentrate on some precise problems in the calculus of variations which can be described through the language of convex duality and where tools from convex duality will help in establishing regularity results. Section 4.5 will come back to an abstract setting, exploiting the ideas presented in concrete cases in order to discuss the well-known Fenchel–Rockafellar duality theorem. As usual, the chapter contains a discussion section (devoted here to questions from the theory of optimal transport and of some congestion games) and closes with a (quite long) exercise section.

4.1 Uniqueness and Sufficient Conditions

We discuss in this section two important roles that convexity plays in optimization: the fact that for convex functions necessary conditions for optimality are also sufficient, and the fact that strict convexity implies uniqueness of the minimizers.

We start with the relation between necessary and sufficient optimality conditions and recall what happens in the easiest case, i.e. the one-dimensional case: if $f : \mathbb{R} \to \mathbb{R}$ is convex and $x_0 \in \mathbb{R}$ is such that $f'(x_0) = 0$, then x_0 is a minimizer for f (which would not necessarily be the case if f was not convex).

This can be seen as a consequence of the *above-the-tangent inequality*: if $f \in C^1$ is convex we have $f(x) \geq f(x_0) + f'(x_0)(x - x_0)$ so that $f'(x_0) = 0$ implies $f(x) \geq$

© The Author(s), under exclusive license to Springer Nature Switzerland AG 2023
F. Santambrogio, *A Course in the Calculus of Variations*, Universitext,
https://doi.org/10.1007/978-3-031-45036-5_4

$f(x_0)$, for every x, x_0. In higher (but finite) dimensions this can be expressed in terms of ∇f (of course, only if f is differentiable: the case of non-smooth convex functions will be discussed in the next section), and we have the inequality

$$f(x) \geq f(x_0) + \nabla f(x_0) \cdot (x - x_0),$$

so that a point x_0 minimizes f if and only if $\nabla f(x_0) = 0$.

In the abstract (and possibly infinite-dimensional) setting, the situation is essentially the same, but we need to be precise about the notion of necessary optimality conditions, as we never defined a proper notion of gradient or differential in abstract spaces. The way we obtained, for instance, the necessary optimality conditions in Chaps. 1 and 2 (the Euler–Lagrange equation in the calculus of variations) was the following: given $f : X \rightarrow \mathbb{R}$ and $x_0 \in X$, we choose a perturbation v and compute $j(\varepsilon) := f(x_0 + \varepsilon v)$. We take as a necessary optimality condition the fact that $j'(0) = 0$ for every v. The 1D above-the-tangent inequality can be written as $j(\varepsilon) \geq j(0) + \varepsilon j'(0)$, so that $j'(0) = 0$ implies $j(\varepsilon) \geq j(0)$. Taking $\varepsilon = 1$ we obtain $f(x_0 + v) \geq f(x_0)$ for arbitrary v, i.e. the minimality of x_0.

In order to apply this argument to the functional case, where we consider an integral functional $F(u) := \int L(x, u(x), \nabla u(x)) \, dx$ (with possible boundary penalizations or constraints), we need a minor clarification: when writing necessary conditions, we often consider a class of smooth perturbations, for instance $\varphi \in C_c^\infty(\Omega)$, and impose $j'(0) = 0$ only for those φ. This means that we have

$$\int_\Omega (\partial_s L(x, u, \nabla u)\varphi + \nabla_v L(x, u, \nabla u) \cdot \nabla\varphi) \, dx = 0 \qquad (4.1)$$

for every $\varphi \in C_c^\infty(\Omega)$. This is, by definition, what we need to write the Euler–Lagrange equation as a PDE in the sense of distributions: $\nabla \cdot (\nabla_v L) = \partial_s L$. Yet, we need for the necessary conditions to be sufficient that the same integral version of the PDE is satisfied for every φ in the same vector space as the optimizer, so that $u + \varphi$ covers all possible competitors. As we saw in Sect. 2.2, if we know information on the summability of $\partial_s L(x, u, \nabla u)$ and $\nabla_v L(x, u, \nabla u)$ we can usually obtain, by density, that if (4.1) holds for $\varphi \in C_c^\infty$ then it also holds for any φ. More precisely, if $X = W^{1,p}(\Omega)$ and $\partial_s L(x, u, \nabla u), \nabla_v L(x, u, \nabla u) \in L^{p'}$ (p' being the dual exponent of p, i.e. $p' = p/(p-1)$) then (4.1) for every $\varphi \in C_c^\infty$ implies the validity of the same equality for every $\varphi \in W_0^{1,p}$, which is enough to consider the minimization of F in X with prescribed boundary conditions.

A similar argument can be extended to the case of boundary penalizations. Consider the optimization problem

$$\min \left\{ J(u) := \int_\Omega L(x, u(x), \nabla u(x)) dx + \int_{\partial\Omega} \psi(x, \mathrm{Tr}[u]) d\mathcal{H}^{d-1} : \begin{array}{l} u \in W^{1,p}(\Omega), \\ \mathrm{Tr}[u] = g \text{ on } A \end{array} \right\},$$

where $A \subset \partial\Omega$ is a fixed part of the boundary. Then, if J is a convex functional, the condition

$$\int_\Omega (\partial_s L(x, u, \nabla u)\varphi + \nabla_v L(x, u, \nabla u) \cdot \nabla\varphi) \, dx + \int_{\partial\Omega} \partial_s \psi(x, \text{Tr}[u])\varphi d\mathcal{H}^{d-1} = 0$$

for all $\varphi \in W^{1,p}(\Omega)$ such that $\text{Tr}[\varphi] = 0$ on A is sufficient for the optimality of u. Note that this condition is a weak version of (2.7), but is slightly stronger than (2.6).

It now becomes important to understand when J is convex. Of course, the convexity of the function $(x, s, v) \mapsto L(x, s, v)$ in the variables (s, v) for a.e. x is sufficient for the convexity of J, but it is not necessary.

Example 4.1 Consider the quadratic form $J(u) = \int_0^T (|u'(t)|^2 - |u(t)|^2) \, dt$ for $u \in H_0^1([0, T])$. We can easily compute

$$J(u + h) = J(u) + 2 \int_0^T (u'(t) \cdot h'(t) - u(t) \cdot h(t)) \, dt + J(h).$$

We will use Lemma 4.2 below. The second term in the last expression is of the form $A(h)$, thus we just have to see whether the last one is non-negative. This is the case, thanks to the Poincaré inequality, for small T, and more precisely for $T \leq \pi$. Hence, this functional J is convex despite the integrand $L(x, s, v)$ being clearly non-convex in s.

Lemma 4.2 *If a function $f : X \to \mathbb{R} \cup \{+\infty\}$ is such that for every $x \in X$ there exists a linear function $A : X \to \mathbb{R}$ satisfying $f(x + h) \geq f(x) + A(h)$ for every $h \in X$, then f is convex.*

Proof To see this, we apply the assumption to $x = (1 - t)x_0 + tx_1$ and $h = (1 - t)(x_1 - x_0)$, thus obtaining $f(x_1) = f(x + h) \geq f(x) + (1 - t)A(x_1 - x_0)$. We then apply the same condition to the same point x but $h = -t(x_1 - x_0)$, thus obtaining $f(x_0) \geq f(x) - tA(x_1 - x_0)$. We multiply the first inequality by t, the second by $(1 - t)$, and we sum up, and obtain the desired convexity inequality. □

Unfortunately, the above discussion has to be modified in the case where there are constraints which cannot be written in the form of belonging to an affine subspace. To stick to the finite-dimensional case, we can say that the minimization of a smooth convex function f on $\mathbb{R}^N$ is equivalent to $\nabla f = 0$; for the case where the problem is $\min\{f(x) : x \in H\}$ where $H \subset \mathbb{R}^N$ is a given affine subspace, of the form $H = x_0 + V$, V being a vector subspace of $\mathbb{R}^N$, the optimality is equivalent to ∇f being orthogonal to V, but a simpler approach is just to re-write the problem as an optimization problem in $\mathbb{R}^k$, $k = \dim(V) < N$, and impose that the gradient vanishes. This corresponds to choosing a basis of V and just considering perturbations in the direction of the vector of the basis. In the particular case where $V = \text{span}(e_1, e_2, \ldots, e_k)$, the vectors e_i being vectors of the canonical basis of $\mathbb{R}^N$, this is essentially implicit: given a function $f : \mathbb{R}^N \to \mathbb{R}$ and some numbers $a_{k+1}, \ldots, a_N$ the optimality conditions for the minimization problem $\min_{x_1,\dots,x_k} f(x_1, \dots, x_k, a_{k+1}, \dots, a_N)$ are of course given by the fact that the partial derivatives $\partial_{x_i} f$ vanish for $i = 1, \dots, k$, as this can be considered as the minimization of a function g defined over $\mathbb{R}^k$ with $g(x) = f(x, a)$. This is what is implicitly done in the calculus of variations when the values of a competitor u are fixed on some subset A (typically, $A \subset \partial\Omega$): we just change the vector space we are minimizing on, and consider perturbations which vanish on the set where values are prescribed.

The situation is much more complicated when the constraint is given by a convex set which is not an affine space (for instance in the case of inequality constraints). In this case, the finite-dimensional procedure is simple, and based as usual on the reduction to the 1D case. Assume we are minimizing a convex function f on a convex set $K \subset X$. Then, a point $x_0 \in K$ is optimal if and only if it satisfies

$$\lim_{\varepsilon \to 0^+} \frac{f((1-\varepsilon)x_0 + \varepsilon x) - f(x_0)}{\varepsilon} \geq 0 \quad \text{for every } x \in K.$$

This condition is of course necessary since $(1 - \varepsilon)x_0 + \varepsilon x \in K$ (because of the convexity of K) and then $f((1 - \varepsilon)x_0 + \varepsilon x) - f(x_0) \geq 0$, and it is sufficient since the one-variable function $[0, 1] \ni \varepsilon \mapsto f((1 - \varepsilon)x_0 + \varepsilon x)$ is convex and then it is minimal at $\varepsilon = 0$ if its right-derivative at $\varepsilon = 0$ is non-negative. Note that the limit defining the right derivative always exists when facing a 1D convex functional, since the incremental ratio is a monotone function (but could be equal to $-\infty$).

When f is differentiable at x_0 the above condition can be written using the gradient as $\nabla f(x_0) \cdot (x - x_0) \geq 0$ and characterizes the minimizers. In other words, a point optimizes a convex function on a convex set if and only if it minimizes the linearization of the same function around such a point on the same set.

For the calculus of variations, a possible (quite technical) statement (which also includes the non-constrained case) is the following.

Theorem 4.3 *Consider the minimization problem*

$$\min \left\{ J(u) := \int_\Omega L(x, u(x), \nabla u(x)) \, dx + \int_{\partial\Omega} \psi(x, \mathrm{Tr}[u]) d\mathcal{H}^{d-1} : u \in K \right\},$$
(4.2)

where $K \subset W^{1,p}(\Omega)$ is a convex set of Sobolev functions and L and ψ are C^1 in the variables s and v. Assume that the functional J is convex on K. Consider a function $u_0 \in K$ which satisfies the conditions (2.3) and (2.5) and such that we have

$$\partial_s L(x, u_0, \nabla u_0) \in L^{(p*)'}(\Omega),$$

$$\nabla_v L(x, u_0, \nabla u_0) \in L^{p'}(\Omega),$$

$$\partial_s \psi(x, \mathrm{Tr}[u_0]) \in L^{p'}(\partial\Omega).$$

Also assume the following properties:

- $(u_0 + \mathrm{Lip}(\Omega)) \cap K$ *is dense (for the strong $W^{1,p}$ convergence)*[1] *in K;*
- *either $L(x, \cdot, \cdot)$ and $\psi(x, \cdot)$ are convex, or J is continuous on K for the strong $W^{1,p}$ convergence.*

Then u_0 is a solution of (4.2) if and only if

$$\int_\Omega (\partial_s L(x, u_0, \nabla u_0)(u - u_0) + \nabla_v L(x, u_0, \nabla u_0) \cdot \nabla(u - u_0)) \, \mathrm{d}x$$

$$+ \int_{\partial\Omega} \partial_s \psi(x, \mathrm{Tr}[u_0]) \mathrm{Tr}[u - u_0] \, \mathrm{d}\mathcal{H}^{d-1} \geq 0$$

$$(4.3)$$

for all $u \in K$.

Proof First let us prove that (4.3) is necessary for optimality. The conditions (2.3) and (2.5) allow us to differentiate w.r.t. ε the function $\varepsilon \mapsto J(u_0 + \varepsilon\varphi)$ for $\varphi \in \mathrm{Lip}(\Omega)$ (indeed, in order to apply dominated convergence to the differentiation under the integral sign, we need to guarantee that the perturbation $\varepsilon\varphi$ is bounded with bounded gradient). Hence, for every $u \in K$ of the form $u = u_0 + \varphi$, $\varphi \in \mathrm{Lip}(\Omega)$, we have (4.3) because of the optimality of u_0. Taking an arbitrary $u \in K$ and approximating it through a sequence $u_0 + \varphi_n$, $\varphi_n \in \mathrm{Lip}(\Omega)$, the summability that we assumed on $\partial_s L(x, u_0, \nabla u_0)$, $\nabla_v L(x, u_0, \nabla u_0)$ and $\partial_s \psi(x, \mathrm{Tr}[u_0])$ allows us to obtain the same result for u, because we have $\varphi_n \rightharpoonup u - u_0$ in L^{p^*}, $\nabla\varphi_n \rightharpoonup \nabla\varphi$ in L^p, and $\mathrm{Tr}[\varphi_n] \rightharpoonup \mathrm{Tr}[u - u_0]$ in $L^p(\partial\Omega)$.

For the sufficiency of (4.3) we distinguish two cases. If $L(x, \cdot, \cdot)$ and $\psi(x, \cdot)$ are convex, we just write

$$J(u) = \int_\Omega L(x, u(x), \nabla u(x)) \, \mathrm{d}x + \int_{\partial\Omega} \psi(x, \mathrm{Tr}[u]) \, \mathrm{d}\mathcal{H}^{d-1}$$

$$\geq J(u_0) + \int_\Omega (\partial_s L(x, u_0, \nabla u_0)(u - u_0) + \nabla_v L(x, u_0, \nabla u_0) \cdot \nabla(u - u_0)) \, \mathrm{d}x$$

$$+ \int_{\partial\Omega} \partial_s \psi(x, \mathrm{Tr}[u_0]) \mathrm{Tr}[u - u_0] \, \mathrm{d}\mathcal{H}^{d-1}$$

and we use (4.3) to obtain $J(u) \geq J(u_0)$.

If, instead, we assume that J is continuous on K, it is enough to prove $J(u_0 + \varphi) \geq J(u_0)$ for every $\varphi \in \mathrm{Lip}(\Omega)$ such that $u_0 + \varphi \in K$. In this case, (4.3) shows that $\frac{\mathrm{d}}{\mathrm{d}\varepsilon}(J(u_0 + \varepsilon\varphi))$ is nonnegative for $\varepsilon = 0$, hence $[0, 1] \ni \varepsilon \mapsto J(u_0 + \varepsilon\varphi))$

---

[1] Or for the weak $W^{1,p}$ convergence, the two notions being equivalent thanks to Mazur's lemma 3.8.

is minimal at $\varepsilon = 0$, which proves the claim. By continuity and density of $(u_0 + \mathrm{Lip}(\Omega)) \cap K$ the inequality extends to $J \geq J(u_0)$ on K.   □

Note that in the above statement we preferred to re-use the conditions (2.3) and (2.5) in order to guarantee differentiability of J but in some practical cases this could be avoided if we are able to differentiate J along perturbations φ which are not necessarily Lipschitz continuous. For instance, in Example 4.1, it is possible to explicitly compute and differentiate $J(u + \varepsilon\varphi)$ for every u and every φ. This would allow us to simplify the argument without using the density of Lipschitz perturbations; by the way, this is also a case where L is not convex in (s, v), but J is continuous for the strong $W^{1,2}$ convergence.

Remark 4.4 Another possible assumption on L and ψ could be

$$|\partial_s L(x, s, v))| \leq C(1 + |s|^{p^*-1}),$$

$$|\nabla_v L(x, s, v)| \leq C(1 + |v|^{p-1}),$$

$$|\partial_s \psi(x, s))| \leq C(1 + |s|^{p-1}).$$

In this case it would be easy to prove that $\varepsilon \mapsto J(u + \varepsilon\varphi)$ is differentiable for any $\varphi \in W^{1,p}$ with no need for the L^∞ bounds on φ and $\nabla\varphi$ that we used so far. This would simplify the analysis of the optimality conditions under constraints. On the other hand, it is easy to see that these bounds imply $|L(x, s, v)| \leq a_0(x) + C|s|^{p^*} + C|v|^p$ and $|\psi(x, v)| \leq a_1(x) + C|s|^p$ (with $a_0 \in L^1(\Omega)$, $a_1 \in L^1(\partial\Omega)$), which automatically makes the functional J continuous for the strong $W^{1,p}$ convergence, and allows us to apply the strategy that we discussed.

We then move to the second topic of the section, also related to convexity, which is the uniqueness of the optimizer.

An easy and classical statement is the following.

Proposition 4.5 *If $K \subset X$ is a convex set in a vector space X, and $f : X \to \mathbb{R} \cup \{+\infty\}$ is strictly convex and not identically $+\infty$ on K, then the problem $\min\{f(x) : x \in K\}$ admits at most one minimizer.*

Proof If two minimizers $x_0 \neq x_1$ existed, then $x = (x_0 + x_1)/2$ would provide a strictly better value for f.   □

In the above result, of course the strict convexity of f on K (and not on the whole space X) would be enough.

We list some sufficient conditions for strict convexity, in the calculus of variations case $J(u) = \int_\Omega L(x, u(x), \nabla u(x))\,dx + \int_{\partial\Omega} \psi(x, \mathrm{Tr}[u](x))$. In all the cases below we assume anyway that for a.e. $x \in \Omega$ the function L is convex in (s, v) and for a.e. $x \in \partial\Omega$ the function ψ is convex in s, which guarantees at least convexity of J (though not being necessary for this convexity). Then

- if for a.e. x the function L is strictly convex in (s, v), then J is strictly convex;
- if for a.e. x the function L is strictly convex in s, then J is strictly convex;

- if for a.e. x the function L is strictly convex in v, then the only possibility for having equality in $J((1 - t)u_0 + tu_1) \leq (1 - t)J(u_0) + tJ(u_1)$ is $\nabla u_0 = \nabla u_1$. If we assume Ω to be connected, then we get $u_0 - u_1 = const$. Then, if the constraints are such that the minimization is performed over a set which does not contain two functions whose difference is a constant, we have strict convexity of J on the set K on which we are minimizing. This happens for instance if the boundary values are fixed, or if the average is fixed. As another option, if there exists a subset $A \subset \partial\Omega$ with $\mathcal{H}^{d-1}(A) > 0$ and $\psi(x, \cdot)$ strictly convex for all $x \in A$, then we should also have $\mathrm{Tr}[u_0] = \mathrm{Tr}[u_1]$ on A, which finally implies $u_0 = u_1$.

Of course, looking again at the Example 4.1, it is possible to build cases where J is strictly convex but L is not even convex in s.

4.2 Some Elements of Convex Analysis

This section is devoted to the main notions in convex analysis which are useful for our purposes, and in particular to the notion of the Fenchel–Legendre transform, which is crucial for duality theorems.

4.2.1 The Fenchel–Legendre Transform

In this section we consider a pair (in the sense of a non-ordered couple) of normed vector spaces which are in duality, i.e. one of the spaces is the dual of the other. Given a space X in this duality pair, we will call X^* the other one, so that we have either $X^* = X'$ or $X = (X^*)'$. By definition, we have $X^{**} = X$ for any X, and this should not be confused with the fact that X is reflexive or not (see Box 2.3).

We denote by $\langle \xi, x \rangle$ the duality between an element $\xi \in X^*$ and $x \in X$. This gives rise to a notion of weak convergence on each of the two spaces, which is the weak convergence on one of them and the weak-* convergence on the other. We call this form of weak convergence (X, X^*)-weak convergence. The (X, X')-weak convergence is the standard weak convergence on X, while the (X', X)-weak convergence is the weak-* convergence on X'. Moreover, on each of these two spaces we have a notion of strong convergence, induced by their norms.

By affine functions we will mean all functions of the form $X \ni x \mapsto \langle \xi, x \rangle + c$ for $\xi \in X^*$. In this way affine functions are also continuous, both for the strong convergence and for the (X, X^*)-weak convergence. Note that the set of affine functions on a space depends on the pair we choose. If we take a non-reflexive space X with its dual X' and its bi-dual $X'' := (X')'$, we can consider the duality pair $\{X', X''\}$ or $\{X', X\}$ and the set of affine functions on the space X' will not be the same (in the second case, we only consider functions of the form

$X' \ni \xi \mapsto \langle \xi, x \rangle + c$ for $x \in X$, while in the first one we can replace $x \in X$ with an arbitrary element of $X'' \supset X$.

Definition 4.6 We say that a function valued in $\mathbb{R} \cup \{+\infty\}$ is *proper* if it is not identically equal to $+\infty$. The set $\{f < +\infty\}$ is called the *domain* of f.

Definition 4.7 Given a duality pair of spaces and a proper function $f : X \to \mathbb{R} \cup \{+\infty\}$ we define its Fenchel–Legendre transform $f^* : X^* \to \mathbb{R} \cup \{+\infty\}$ via

$$f^*(\xi) := \sup_x \langle \xi, x \rangle - f(x).$$

Remark 4.8 We observe that we trivially have $f^*(0) = -\inf_X f$.

We note that f^*, as a sup of affine continuous functions, is both convex and l.s.c., as these two notions are stable under sup.

We prove the following results.

Proposition 4.9 *We consider a duality pair and a function $f : X \to \mathbb{R} \cup \{+\infty\}$ which is proper, convex, and (X, X^*)-weakly l.s.c. Then*

1. *there exists an affine function ℓ such that $f \geq \ell$;*
2. *f is a sup of affine functions;*
3. *there exists a $g : X^* \to \mathbb{R} \cup \{+\infty\}$ such that $f = g^*$;*
4. *we have $f^{**} = f$.*

Proof We consider the epigraph $\mathrm{Epi}(f) := \{(x, t) \in X \times \mathbb{R} : t \geq f(x)\}$ of f, which is a convex and closed set in $X \times \mathbb{R}$. More precisely, this set is also weakly closed if $X^* = X'$ (just because in this case strongly closed sets are weakly closed when they are convex), and weakly-* closed if $X = (X^*)'$ because we assumed lower semicontinuity for the (X, X^*)-weak convergence, which is in this case the weak-* convergence. We take a point x_0 such that $f(x_0) < +\infty$ and consider the singleton $\{(x_0, f(x_0) - 1)\}$ which is a convex and compact set in $X \times \mathbb{R}$. The Hahn–Banach separation theorem (see Box 3.1) allows us to strictly separate this set from the epigraph. It is important for us to separate them through an affine function in the sense of our definition, i.e. using an element of X^* and not of X'. The only case where this is different is when $X^* \neq X'$, i.e. when $X = (X^*)'$. Yet, with our assumption the epigraph would be weakly-* closed, and this allows us to use elements of the predual instead of the dual of X.

Hence, we deduce the existence of a pair $(\xi, a) \in X^* \times \mathbb{R}$ and a constant c such that $\langle \xi, x_0 \rangle + a(f(x_0) - 1) < c$ and $\langle \xi, x \rangle + at > c$ for every $(x, t) \in \mathrm{Epi}(f)$. Note that this last condition implies $a \geq 0$ since we can take $t \to \infty$. Moreover, we should also have $a > 0$, otherwise taking any point $(x, t) \in \mathrm{Epi}(f)$ with $x = x_0$ we have a contradiction. If we then take $t = f(x)$ for all x such that $f(x) < +\infty$ we obtain $af(x) \geq -\langle \xi, x \rangle + \langle \xi, x_0 \rangle + a(f(x_0) - 1)$ and, dividing by $a > 0$, we obtain the first claim.

We now take an arbitrary $x_0 \in X$ and $t_0 < f(x_0)$ and separate again the singleton $\{(x_0, t_0)\}$ from $\mathrm{Epi}(f)$, thus getting a pair $(\xi, a) \in X^* \times \mathbb{R}$ and a constant c such that $\langle \xi, x_0 \rangle + at_0 < c$ and $\langle \xi, x \rangle + at > c$ for every $(x, t) \in \mathrm{Epi}(f)$. Again, we have $a \geq 0$. If $f(x_0) < +\infty$ we obtain as before $a > 0$ and the inequality $f(x) > -\frac{\xi}{a} \cdot (x - x_0) + t_0$. We then have an affine function ℓ with $f \geq \ell$ and $\ell(x_0) = t_0$. This shows that the sup of all affine functions smaller than f is, at the point x_0, at least t_0. Hence this sup equals f on $\{f < +\infty\}$. In the case $f(x_0) = +\infty$ we can perform the very same argument as long as we can guarantee that for t_0 arbitrarily large the corresponding coefficient a is strictly positive. Suppose now that we have a point x_0 such that $f(x_0) = +\infty$ and $a = 0$ in the construction above. Then we find $\langle \xi, x_0 \rangle < c$ and $\langle \xi, x \rangle \geq c$ for every x such that $(x, t) \in \mathrm{Epi}(f)$ for at least one $t \in \mathbb{R}$. Hence, $\langle \xi, x \rangle \geq c$ for every $x \in \{f < +\infty\}$. Consider now $\ell_n(x) = \ell(x) - n(\langle \xi, x \rangle - c)$, where ℓ is the affine function smaller than f previously found. We have $f \geq \ell \geq \ell_n$ since on $\{f < +\infty\}$ we have $\langle \xi, x \rangle - c \geq 0$. Moreover $\lim_n \ell_n(x_0) = +\infty$ since $\langle \xi, x_0 \rangle - c < 0$. This shows that at such a point x_0 the sup of the affine functions smaller than f equals $+\infty = f(x_0)$.

Once we know that f is a sup of affine functions we can write

$$f(x) = \sup_\alpha \langle \xi_\alpha, x \rangle + c_\alpha$$

for a family of indices α. We then set $c(\xi) := \sup\{c_\alpha : \xi_\alpha = \xi\}$. The set in the sup can be empty, which would mean $c(\xi) = -\infty$. Anyway, the sup is always finite: fix a point x_0 with $f(x_0) < +\infty$ and use $c_\alpha \leq f(x_0) - \langle \xi, x_0 \rangle$. We then define $g = -c$ and we see $f = g^*$.

Finally, before proving $f = f^{**}$ we prove that for any function f we have $f \geq f^{**}$ even if f is not convex or l.s.c. Indeed, we have $f^*(\xi) + f(x) \geq \langle \xi, x \rangle$, which allows us to write $f(x) \geq \langle \xi, x \rangle - f^*(\xi)$, an inequality true for every ξ. Taking the sup over ξ we obtain $f \geq f^{**}$. We now want to prove that this inequality is an equality if f is convex and l.s.c. We write $f = g^*$ and transform this into $f^* = g^{**}$. We then have $f^* \leq g$ and, transforming this inequality (which changes its sign), $f^{**} \geq g^* = f$, which proves $f^{**} = f$. $\qquad \square$

Remark 4.10 Note that the above result allows us to define a convex function f on a dual space X' and use as a duality pair (X', X) instead of (X', X''). Under the assumption that f is weakly-* lower semicontinuous we proved $f^{**} = f$, i.e. $f(\xi) = \sup\{\langle \xi, x \rangle - f^*(x) : x \in X\}$. If we had used the pair (X', X'') we would have obtained $f(\xi) = \sup\{\langle \xi, x \rangle - f^*(x) : x \in X''\}$, which is not the same result (on the other hand, this result only requires f to be weakly lower semicontinuous on X', which is equivalent to the strong lower semicontinuity on X').

We now know that, on the class of convex and (X, X^*)-weakly l.s.c. functions, the Fenchel–Legendre transform is an involution. Two convex and l.s.c. functions which are the Fenchel–Legendre transform of each other will be said to be two *conjugate* convex functions. We can also obtain a characterization of the double transform when we are not in this setting (i.e. f is not convex and l.s.c.).

Corollary 4.11 *Given an arbitrary proper function $f : X \to \mathbb{R} \cup \{+\infty\}$ we have $f^{**} = \sup\{g : g \leq f, g$ is convex and (X, X^*)-weakly l.s.c. $\}$.*

Proof Let us call h the function obtained as a sup on the right-hand side. Since f^{**} is convex, (X, X^*)-weakly l.s.c., and smaller than f, we have $f^{**} \leq h$. Note that h, as a sup of convex and l.s.c. functions, is also convex and l.s.c. for the same convergence, and it is of course smaller than f. We write $f \geq h$ and double transform this inequality, which preserves the sign. We then have $f^{**} \geq h^{**} = h$, and the claim is proven.   □

We finally discuss the relations between the behavior at infinity of a function f and its Legendre transform. We give two definitions.

Definition 4.12 A function $f : X \to \mathbb{R} \cup \{+\infty\}$ defined on a normed vector space X is said to be *coercive* if $\lim_{\|x\| \to \infty} f(x) = +\infty$; it is said to be *superlinear* if $\lim_{\|x\| \to \infty} \frac{f(x)}{\|x\|} = +\infty$.

We note that the definition of coercive does not include any speed of convergence to ∞, but that for convex functions this should be at least linear:

Proposition 4.13 *A proper, convex, and l.s.c. function $f : X \to \mathbb{R} \cup \{+\infty\}$ is coercive if and only there exist two constants $c_0, c_1 > 0$ such that $f(x) \geq c_0\|x\| - c_1$.*

Proof We just need to prove that c_0, c_1 exist if f is coercive, the converse being trivial. Take a point x_0 such that $f(x_0) < +\infty$. Using $\lim_{\|x\| \to \infty} f(x) = +\infty$ we know that there exists a radius R such that $f(x) \geq f(x_0) + 1$ as soon as $\|x - x_0\| \geq R$. By convexity, we have, for each x with $\|x - x_0\| > R$, the inequality $f(x) \geq f(x_0) + \|x - x_0\|/R$ (it is enough to use the definition of convexity on the three points x_0, x and $x_t = (1 - t)x_0 + tx \in \partial B(x_0, R)$). Since f is bounded from below by an affine function, it is bounded from below by a constant on $B(x_0, R)$, so that we can write $f(x) \geq c_2 + \|x - x_0\|/R$ for some $c_2 \in \mathbb{R}$ and all $x \in X$. We then use the triangle inequality and obtain the claim with $c_0 = 1/R$ and $c_1 = c_2 - \|x_0\|/R$.   □

Note that in the above proposition the lower semicontinuity is only used to prove that f is bounded from below on every ball.

Proposition 4.14 *A proper, convex, and (X, X^*)-weakly l.s.c. function $f : X \to \mathbb{R} \cup \{+\infty\}$ is coercive if and only if f^* is bounded in a neighborhood of 0; it is superlinear if and only if f^* is bounded on each bounded ball of X^*.*

Proof We know that f is coercive if and only if there exist two constants $c_0, c_1 > 0$ such that $f \geq g_{c_0, c_1}$, where $g_{c_0, c_1}(x) := c_0\|x\| - c_1$. Since both f and g_{c_0, c_1} are convex and (X, X^*)-weakly l.s.c., this inequality is equivalent to the opposite inequality for their transforms, i.e. $f^* \leq g^*_{c_0, c_1}$. We can compute the transform and obtain

$$ g^*_{c_0, c_1}(\xi) = \begin{cases} c_1 & \text{if } \|\xi\| \leq c_0, \\ +\infty & \text{if not.} \end{cases} $$

This shows that f is coercive if and only if there exist two constants R, C (with $R = c_0^{-1}, C = c_1$) such that $f^* \leq C$ on the ball of radius R of $\mathcal{X}'$, which is the claim.

We follow a similar procedure for the case of superlinear functions. We first note that a convex l.s.c. function f is superlinear if and only if for every c_0 there exists a c_1 such that $f \geq g_{c_0,c_1}$. Indeed, it is clear that, should this condition be satisfied, we would have $\liminf_{||x|| \to \infty} f(x)/||x|| \geq c_0$, and hence $\liminf_{||x|| \to \infty} f(x)/||x|| = +\infty$ because c_0 is arbitrary, so that f would be superlinear. On the other hand, if f is superlinear, for every c_0 we have $f(x) \geq c_0||x||$ for large $||x||$, say outside of $B(0, R)$. If we then choose $-c_1 := \min\{\inf_{B(0,R)} f - c_0 R, 0\}$ (a value which is finite since f is bounded from below by an affine function), the inequality $f(x) \geq c_0||x|| - c_1$ is true everywhere.

We then deduce that f is superlinear if and only if for every $R = c_0$ there is a constant c_1 such that $f^* \leq c_1$ on the ball of radius R of $\mathcal{X}'$, which is, again, the claim. $\qquad\square$

4.2.2 Subdifferentials

The above-the-tangent property of convex functions inspired the definition of an extension of the notion of differential, called the sub-differential, as a set-valued map:

Definition 4.15 Given a function $f : \mathcal{X} \to \mathbb{R} \cup \{+\infty\}$ we define its subdifferential at x as the set

$$\partial f(x) = \{\xi \in \mathcal{X}' : f(y) \geq f(x) + \langle \xi, y - x \rangle \ \forall y \in \mathcal{X}\}.$$

We observe that $\partial f(x)$ is always a closed and convex set, whatever f is. Moreover, if f is l.s.c. we easily see that the graph of the subdifferential multi-valued map is closed:

Proposition 4.16 Assume that f is l.s.c. and take a sequence $x_n \to x$. Assume $\xi_n \overset{*}{\rightharpoonup} \xi$ and $\xi_n \in \partial f(x_n)$. Then $\xi \in \partial f(x)$.

Proof For every y we have $f(y) \geq f(x_n) + \langle \xi_n, y - x_n \rangle$. We can then use the strong convergence of x_n and the weak convergence of ξ_n, together with the lower semicontinuity of f, to pass to the limit and deduce $f(y) \geq f(x) + \langle \xi, y - x \rangle$, i.e. $\xi \in \partial f(x)$. $\qquad\square$

Note that in the above proposition we could have exchanged strong convergence for x_n and weak-* for ξ_n for weak convergence for x_n (but f needed in this case to be weakly l.s.c., and f has not been assumed to be convex) and strong for ξ_n.

When dealing with arbitrary functions f, the subdifferential is in most cases empty, as there is no reason for the inequality defining $\xi \in \partial f(x)$ to be satisfied for

y very far from x. The situation is completely different when dealing with convex functions, which is the standard case where subdifferentials are defined and used. In this case we can prove that $\partial f(x)$ is never empty if x lies in the interior of the set $\{f < +\infty\}$ (note that outside $\{f < +\infty\}$ the subdifferential of a proper function is clearly empty).

We first provide an insight about the finite-dimensional case. In this case we simply write $\xi \cdot x$ for the duality product, which coincides with the Euclidean scalar product on $\mathbb{R}^N$.

We start from the following property.

Proposition 4.17 *Given $f : \mathbb{R}^N \to \mathbb{R} \cup \{+\infty\}$ assume that f is differentiable at a point x_0. Then $\partial f(x_0) \subset \{\nabla f(x_0)\}$. If moreover f is convex, then $\partial f(x_0) = \{\nabla f(x_0)\}$.*

Proof From the definition of sub-differential we see that $\xi \in \partial f(x_0)$ means that $y \mapsto f(y) - \xi \cdot y$ is minimal at $y = x_0$. Since we are supposing that f is differentiable at such a point, we obtain $\nabla f(x_0) - \xi = 0$, i.e. $\xi = \nabla f(x_0)$. This shows $\partial f(x_0) \subset \{\nabla f(x_0)\}$. The inclusion becomes an equality if f is convex since the above-the-tangent property of convex functions exactly provides $f(y) \geq f(x_0) + \nabla f(x_0) \cdot (y - x_0)$ for every y.   □

Proposition 4.18 *Assume that $f : \mathbb{R}^N \to \mathbb{R} \cup \{+\infty\}$ is convex and take a point x_0 in the interior of $\{f < +\infty\}$. Then $\partial f(x_0) \neq \emptyset$.*

Proof It is well-known that convex functions in finite dimensions are locally Lipschitz in the interior of their domain, and Lipschitz functions are differentiable Lebesgue-a.e. because of the Rademacher theorem (see Box 4.1). We can then take a sequence of points $x_n \to x_0$ such that f is differentiable at x_n. We then have $\nabla f(x_n) \in \partial f(x_n)$ and the Lipschitz behavior of f around x_0 implies $|\nabla f(x_n)| \leq C$. It is then possible to extract a subsequence such that $\nabla f(x_n) \to v$. Proposition 4.16 implies $v \in \partial f(x_0)$.   □

Box 4.1 Good to Know!—*Differentiability of Lipschitz and Convex Functions*

Theorem (Rademacher) *Any Lipschitz continuous function $f : \mathbb{R}^N \to \mathbb{R}$ is differentiable at Lebesgue-a.e. point.*

The proof of this non-trivial but crucial theorem, which is based on the a.e. differentiability of Lipschitz functions in 1D (a fact which can be proven independently of the Rademacher Theorem: absolutely continuous functions u in 1D coincide with the antiderivative of their distributional derivative $u' \in L^1$ and they are differentiable at Lebesgue points of u') can be found, for instance, in [15].

(continued)

Box 4.1 (continued)

Of course the same result holds for locally Lipschitz functions, and hence for convex functions, as they are always locally Lipschitz in the interior of their domain. On the other hand, convex functions are actually differentiable on a much larger set of points, and more precisely we have (see [2–4]):

Theorem *If $f : \mathbb{R}^N \to \mathbb{R}$ is convex then the set of non-differentiability points of f is of dimension at most $N - 1$. More precisely, it is contained—up to $\mathcal{H}^{N-1}$-negligible sets (see Box 6.4)—in a countable union of $(N - 1)$-dimensional Lipschitz hypersurfaces, i.e., in suitable coordinates, graphs of Lipschitz functions of $N - 1$ variables. Moreover, for every $k \geq 1$, the set of points x where the convex set $\partial f(x)$ has dimension at least k is at most of dimension $n - k$. Of course the same result is true for semi-convex functions, i.e. functions f such that $x \mapsto f(x) + \lambda |x|^2$ is convex for some $\lambda > 0$, and for locally semi-convex functions.*

We can easily see, even from the 1D case, that different situations can occur at the boundary of $\{f < +\infty\}$. If we take for instance the proper function f defined by

$$f(x) = \begin{cases} x^2 & \text{if } x \geq 0, \\ +\infty & \text{if } x < 0, \end{cases}$$

we see that we have $\partial f(0) = [-\infty, 0]$ so that the subdifferential can be "fat" on these boundary points. If we take, instead, the proper function f defined by

$$f(x) = \begin{cases} -\sqrt{x} & \text{if } x \geq 0, \\ +\infty & \text{if } x < 0, \end{cases}$$

we see that we have $\partial f(0) = \emptyset$, a fact related to the infinite slope of f at 0: of course, infinite slope can only appear at boundary points.

The proof of the fact that the sub-differential is non-empty[2] in the interior of the domain is more involved in the general (infinite-dimensional) case, and is based on the use of the Hahn–Banach Theorem. It will require the function to be convex and l.s.c., this second assumption being useless in the finite-dimensional case, since convex functions are locally Lipschitz in the interior of their domain. It also requires us to discuss whether the function is indeed locally bounded around some points. We state the following clarifying proposition.

---

[2] Note that, instead, we do not discuss the relations between subdifferential and gradients as we decided not to discuss the differentiability in the infinite-dimensional case.

Proposition 4.19 *If $f := X \to \mathbb{R} \cup \{+\infty\}$ is a proper convex function, bounded from below on each ball of X, then the following facts are equivalent*

1. *the interior of the domain of f is non-empty and f is locally Lipschitz continuous on it (i.e. for every point in the interior of the domain there exists a ball centered at such a point where f is finite-valued and Lipschitz continuous);*
2. *there exists a non-empty ball where f is finite-valued and Lipschitz continuous;*
3. *there exists a point where f is continuous and finite-valued;*
4. *there exists a point which has a neighborhood where f is bounded from above by a finite constant.*

Moreover, all the above facts hold if X is a Banach space, f is l.s.c., and the domain of f has non-empty interior.

Proof It is clear that 1 implies 2, which implies 3, which implies 4. Let us note that, if f is bounded from above on a ball $B(x_0, R)$, then necessarily f is Lipschitz continuous on $B(x_0, R/2)$. Indeed, we assumed that f is also bounded from below on $B(0, R)$. Then, suppose there are two points $x_1, x_2 \in B(x_0, R/2)$ with an incremental ratio equal to L and $f(x_1) > f(x_2)$, i.e.

$$L = \frac{f(x_1) - f(x_2)}{\|x_1 - x_2\|}.$$

Following the half-line going from x_2 to x_1 we find two points $x_3 \in \partial B(x_0, R/2)$ and $x_4 \in \partial B(x_0, 3R/4)$ with $|x_3 - x_4| \geq R/4$ and an incremental ratio which is also at least L. This implies $f(x_4) > f(x_3) + LR/4$ but the upper and lower bounds on f on the ball $B(0, R)$ imply that L cannot be too large, so that f is Lipschitz continuous on $B(x_0, R/2)$.

Hence, in order to prove that 4 implies 1 we just need to prove that every point in the interior of the domain admits a ball centered at such a point where f is bounded from above. We start from the existence of a ball $B(x_0, R)$ where f is bounded from above and we take another point x_1 in the interior of the domain of f. For small $\varepsilon > 0$, the point $x_2 := x_1 - \varepsilon(x_0 - x_1)$ also belongs to the domain of f and every point of the ball $B(x_1, r)$ with $r = \frac{\varepsilon R}{1+\varepsilon}$ can be written as a convex combination of x_2 and a point in $B(x_0, R)$: indeed we have

$$x_1 + v = \frac{1}{1+\varepsilon}x_2 + \frac{\varepsilon}{1+\varepsilon}\left(x_0 + \frac{1+\varepsilon}{\varepsilon}v\right)$$

so that $|v| < r$ implies $x_0 + \frac{1+\varepsilon}{\varepsilon}v \in B(x_0, R)$. Then, f is bounded from above on $B(x_1, r)$ by $\max\{f(x_2), \sup_{B(x_0, R)} f\}$, which shows the local bound around x_1.

Finally, we want to prove that f is necessarily locally bounded around every point of the interior of its domain if X is complete and f is l.s.c. Consider a closed ball $\overline{B}$ contained in $\{f < +\infty\}$ and write $\overline{B} = \bigcup_n \{f \leq n\} \cap \overline{B}$. Since f is l.s.c., each set $\{f \leq n\} \cap \overline{B}$ is closed. Since their countable union has non-empty interior, Baire's theorem (see Box 4.2) implies that at least one of these sets also has non-empty interior, so there exists a ball contained in a set $\{f \leq n\}$, hence a point where f is locally bounded from above. Then we satisfy condition 4., and consequently also conditions 1., 2. and 3. □

Remark 4.20 In the above proposition, the assumption that f is bounded from below on each ball of X is useless when f is l.s.c., as we know that convex l.s.c. functions are bounded from below by at least an affine function, which implies that they are bounded from below by a constant on each ball. Yet, we prefer not to add f l.s.c. in the assumptions, as this proposition could be used as a way to prove that some convex functions are indeed l.s.c., by proving that they are locally Lipschitz continuous.

Box 4.2 Memo—*Baire's Theorem and Applications*

Theorem (Baire) *In a complete metric space any countable union of closed sets with empty interior has empty interior.*

Among the many applications of this very important theorem in functional analysis we cite the Banach–Steinhaus theorem:

Theorem (Banach–Steinhaus) *If E is a Banach space and F a normed vector space then a family $(f_\alpha)_\alpha$ of continuous linear maps from E to F is uniformly bounded on the unit ball of E if and only if $(f_\alpha(x))_\alpha$ is bounded for every $x \in E$.*

This theorem has itself a well-known consequence: if a sequence of vectors weakly (or weakly-*) converges, then it is necessarily bounded.

We can now prove the following theorem.

Theorem 4.21 *Assume that $f : X \to \mathbb{R} \cup \{+\infty\}$ is a proper convex and l.s.c. function and take a point x_0 in the interior of $\{f < +\infty\}$. Also assume that there exists at least a point x_1 (possibly different from x_0) where f is finite and continuous. Then $\partial f(x_0) \neq \emptyset$.*

Proof Let us consider the set A given by the interior of Epi(f) in $X \times \mathbb{R}$ and $B = \{(x_0, f(x_0))\}$. They are two disjoint convex sets, and A is open. Hence, by the Hahn–Banach Theorem, there exists a pair $(\xi, a) \in X' \times \mathbb{R}$ and a constant such that $\langle \xi, x_0 \rangle + af(x_0) \leq c$ and $\langle \xi, x \rangle + at > c$ for every $(x, t) \in A$.

The assumption on the existence of a point x_1 where f is continuous implies that f is bounded (say by a constant M) on a ball around x_1 (say $B(x_1, R)$, and hence the set A is non-empty since it includes $B(x_1, R) \times (M, \infty)$). Then, the convex set Epi(f) has non-empty interior and it is thus the closure of its interior (to see this it is enough to connect every point of a closed convex set to the center of a ball contained in the set, and see that there is a cone composed of small balls contained in the convex set approximating such a point). In particular, since $(x_0, f(x_0))$ belongs to Epi(f), it is in the closure of A, so necessarily we have $\langle \xi, x_0 \rangle + af(x_0) \geq c$ and hence $\langle \xi, x_0 \rangle + af(x_0) = c$.

As we did in Proposition 4.9, we must have $a > 0$. Indeed, we use Proposition 4.19 to see that f is also locally bounded around x_0 so that points of the form $(x, t) = (x_0, t)$ for t large enough should belong to A and should satisfy $a(t - f(x_0)) > 0$, which implies $a > 0$.

Then, we can write the condition $\langle \xi, x \rangle + at > c = \langle \xi, x_0 \rangle + af(x_0)$ in a different form: dividing by a and using $\tilde{\xi} := \xi/a$ we get

$$\langle \tilde{\xi}, x \rangle + t > \langle \tilde{\xi}, x_0 \rangle + f(x_0) \text{ for every } (x, t) \in A.$$

The inequality becomes large on the closure of A but implies, when applied to $(x, t) = (x, f(x)) \in \text{Epi}(f) = \overline{A}$,

$$\langle \tilde{\xi}, x \rangle + f(x) \geq \langle \tilde{\xi}, x_0 \rangle + f(x_0),$$

which means precisely $-\tilde{\xi} \in \partial f(x_0)$.   □

Remark 4.22 Of course, thanks to the last claim in Proposition 4.19, when X is a Banach space we obtain $\partial f(x_0) \neq \emptyset$ for every x_0 in the interior of $\{f < +\infty\}$.

We list some other properties of subdifferentials.

Proposition 4.23

1. *A point x_0 solves* $\min\{f(x) : x \in X\}$ *if and only if we have* $0 \in \partial f(x_0)$.
2. *The subdifferential satisfies the monotonicity property*

$$\xi_i \in \partial f(x_i) \text{ for } i = 1, 2 \Rightarrow \langle \xi_1 - \xi_2, x_1 - x_2 \rangle \geq 0.$$

3. *If f is convex and l.s.c., the subdifferentials of f and f^* are related through*

$$\xi \in \partial f(x) \Leftrightarrow x \in \partial f^*(\xi) \Leftrightarrow f(x) + f^*(\xi) = \langle \xi, x \rangle.$$

Proof Point 1 is a straightforward consequence of the definition of subdifferential, since $0 \in \partial f(x_0)$ means that for every y we have $f(y) \geq f(x_0)$.

Point 2 is also straightforward if we sum up the inequalities

$$f(x_2) \geq f(x_1) + \langle \xi, x_2 - x_1 \rangle; \quad f(x_1) \geq f(x_2) + \langle \xi, x_1 - x_2 \rangle.$$

As for point 3, once we know that for convex and l.s.c. functions we have $f^{**} = f$, it is enough to prove $\xi \in \partial f(x) \Leftrightarrow f(x) + f^*(\xi) = \langle \xi, x \rangle$ since then, by symmetry, we can also obtain $x \in \partial f^*(\xi) \Leftrightarrow f(x) + f^*(\xi) = \langle \xi, x \rangle$. We now look at the definition of subdifferential, and we have

$$\xi \in \partial f(x) \Leftrightarrow \text{ for every } y \in X \text{ we have } f(y) \geq f(x) + \langle \xi, y - x \rangle$$
$$\Leftrightarrow \text{ for every } y \in X \text{ we have } \langle \xi, x \rangle - f(x) \geq \langle \xi, y \rangle - f(y)$$
$$\Leftrightarrow \langle \xi, x \rangle - f(x) \geq \sup_{y} \langle \xi, y \rangle - f(y)$$
$$\Leftrightarrow \langle \xi, x \rangle - f(x) \geq f^*(\xi).$$

This shows that $\xi \in \partial f(x)$ is equivalent to $\langle \xi, x \rangle \geq f(x) + f^*(\xi)$, which is in turn equivalent to $\langle \xi, x \rangle = f(x) + f^*(\xi)$, since the opposite inequality is always true by definition of f^*.   □

Remark 4.24 Point 2 in the above proposition includes a very particular case: if f and f^* are two conjugate smooth convex functions on $\mathbb{R}^N$, then $\nabla f^* = (\nabla f)^{-1}$. In particular, the inverse map of the gradient of a convex function is also a gradient of a convex function. This was the motivation for Legendre to introduce this notion of transform, and his intention was to apply it to the Euler–Lagrange equation: the condition $(\nabla_v L(t, \gamma(t), \gamma'(t)))' = \nabla_x L(t, \gamma(t), \gamma'(t))$ could indeed be written as $w'(t) = \nabla_x L(t, \gamma(t), \gamma'(t))$, where $w = \nabla_v L(t, \gamma(t), \gamma'(t))$ and, inverting, $\gamma'(t) = \nabla_w L^*(t, \gamma(t), w(t))$, which allows us in some cases to transform the second-order equation into a Hamiltonian system of order one.

We also state another property, but we prefer to stick to the finite-dimensional case for simplicity.

Proposition 4.25 *A function $f : \mathbb{R}^N \to \mathbb{R} \cup \{+\infty\}$ is strictly convex if and only if $\partial f(x_0) \cap \partial f(x_1) = \emptyset$ for all $x_0 \neq x_1$.*

Proof If we have $\partial f(x_0) \cap \partial f(x_1) \neq \emptyset$ for two points $x_0 \neq x_1$, take $\xi \in \partial f(x_0) \cap \partial f(x_1)$ and define $\tilde{f}(x) := f(x) - \xi \cdot x$. Then both x_0 and x_1 minimize $\tilde{f}$ and this implies that f is not strictly convex. If, instead, we assume that f is not strictly convex, we need to find two points with the same vector in the subdifferential. Consider two points x_0 and x_1 on which the strict convexity fails, i.e. f is affine on $[x_0, x_1]$. We then have a segment $S = \{(x, f(x)) : x \in [x_0, x_1]\}$ in the graph of f and we can separate it from the interior A of $\text{Epi}(f)$ exactly as we did in Theorem 4.21. Assume for now that the domain of f (and hence the epigraph) has non-empty interior.

Following the same arguments as in Theorem 4.21, we obtain the existence of a pair $(\xi, a) \in \mathbb{R}^N \times \mathbb{R}$ such that

$$\langle \xi, x \rangle + a f(x) < \xi \cdot x' + at \text{ for every } (x', t) \in A \text{ and every } x \in [x_0, x_1].$$

We then prove $a > 0$, divide by a, pass to a large inequality on the closure of A, and deduce $-\xi/a \in \partial f(x)$ for every $x \in [x_0, x_1]$.

If the domain of f has empty interior, it is enough to restrict the space where f is defined to the minimal affine space containing its domain. Then the domain, and the epigraph, would have (in this new lower-dimensional space) non-empty interior. This does not affect the conclusion, since if we want then to define subdifferentials in the original space it is enough to add arbitrary components orthogonal to the affine space containing the domain.   □

Limiting once more to the finite-dimensional case (also because we do not want to discuss differentiability in other settings) we can deduce from the previous proposition and from point 3. of Proposition 4.23 the following fact.

Proposition 4.26 *Take two proper, convex and l.s.c. conjugate functions f and f^* (with $f = f^{**}$); then f is a real-valued C^1 function on $\mathbb{R}^N$ if and only if f^* is strictly convex and superlinear.*

Proof We first prove that a convex function is C^1 if and only if $\partial f(x)$ is a singleton for every x. If it is C^1, and hence differentiable, we already saw that this implies $\partial f(x) = \{\nabla f(x)\}$. The converse implication can be proven as follows: first we observe that, if we have a map $v : \mathbb{R}^N \to \mathbb{R}^N$ with $\partial f(x) = \{v(x)\}$, then v is necessarily locally bounded and continuous. Indeed, the function f is necessarily finite everywhere (since the subdifferential is non-empty at every point) and hence locally Lipschitz; local boundedness of v comes from $v(x) \cdot e \leq f(x+e) - f(x) \leq C|e|$ (where we use the Local Lipschitz behavior of f): using e oriented as $v(x)$ we obtain a bound on $|v(x)|$. Continuity comes from local boundedness and from Proposition 4.16: when $x_n \to x$ then $v(x_n)$ admits a convergent subsequence, but the limit can only belong to $\partial f(x)$, i.e. it must equal $v(x)$. Then we prove $v(x) = \nabla f(x)$, and for this we use the definition of $\partial f(x)$ and $\partial f(y)$ so as to get

$$v(y) \cdot (y - x) \geq f(y) - f(x) \geq v(x) \cdot (y - x).$$

We then write $v(y) \cdot (y - x) = v(x) \cdot (y - x) + o(|y - x|)$ and we obtain the first-order expansion $f(y) - f(x) = v(x) \cdot (y - x) + o(|y - x|)$ which characterizes $v(x) = \nabla f(x)$.

This shows that f is C^1 as soon as subdifferentials are singletons. On the other hand, point 3. in Proposition 4.23 shows that this is equivalent to having, for each $x \in \mathbb{R}^N$, exactly one point ξ with $x \in \partial f^*(\xi)$. The fact that no more than one point ξ has the same vector in the subdifferential is equivalent (Proposition 4.25) to being strictly convex. The fact that each point is taken at least once as a subdifferential is, instead, equivalent to being superlinear (see Lemma 4.27 below)   □

Lemma 4.27 *A convex and l.s.c. function $f : \mathbb{R}^N \to \mathbb{R} \cup \{+\infty\}$ is superlinear if and only if ∂f is surjective.*

Proof Let us assume that f is convex, l.s.c., and superlinear. Then for every ξ the function $\tilde{f}$ given by $\tilde{f}(x) := f(x) - \xi \cdot x$ is also l.s.c., and superlinear, and it admits a minimizer x_0. Such a point satisfies $\xi \in \partial f(x_0)$, which proves that ξ is in the image of ∂f, which is then surjective.

Let us assume now that ∂f is surjective. Let us fix a number $L > 0$ and take the $2N$ vectors $\xi_i^{\pm} := \pm L e_i$, where the vectors e_i form the canonical basis of $\mathbb{R}^N$. Since each of these vectors belong to the image of the subdifferential, there exist points $x_i^{\pm}$ such that $\xi_i^{\pm} \in \partial f(x_i^{\pm})$. This implies that f satisfies $2N$ inequalities of the form $f(x) \geq \xi_i^{\pm} \cdot x - C_i^{\pm}$ for some constants $C_i^{\pm}$. We then obtain $f(x) \geq L\|x\|_{\infty} - C$, where $\|x\|_{\infty} = \max\{|x_i|\} = \max_i x \cdot (\pm e_i)$ is the norm on $\mathbb{R}^N$ given by the maximal modulus of the components, and $C = \max C_i^{\pm}$. From the equivalence of the norms in finite dimensions we get $f(x) \geq C(N)L\|x\| - C$, which shows $\liminf_{\|x\| \to \infty} \frac{f(x)}{\|x\|} \geq C(N)L$. The arbitrariness of L concludes the proof. $\qquad\square$

Remark 4.28 Note that this very last statement is false in infinite dimensions. Indeed, as one can see for instance in [34], the fact that all functions $f : X \to \mathbb{R} \cup \{+\infty\}$ such that f^* is finite everywhere (a condition which is equivalent to $\partial f^*(\xi) \neq \emptyset$ for every $\xi \in X'$, and then to the surjectivity of ∂f) are superlinear is equivalent to the fact that X is finite-dimensional! Exercises 4.11 and 4.12 discuss some subtle points about this fact.

4.2.3 Recession Functions

Recalling the use we made of it in Sect. 3.3, we discuss in this short subsection another notion in convex analysis, that of a *recession function*. Given a function $f : X \to \mathbb{R} \cup \{+\infty\}$ we define its recession function f^{∞} via

$$f^{\infty}(v) := \sup_{x \in X : f(x) < +\infty} f(x + v) - f(x).$$

We start with the following properties and characterizations.

Proposition 4.29 *If $f : X \to \mathbb{R} \cup \{+\infty\}$ is a proper, convex, and l.s.c. function, then the function f^{∞} is proper, convex, l.s.c. and 1-homogeneous. Moreover, it is also characterized by*

$$f^{\infty}(v) := \sup_{x \in X : f(x) < +\infty} \sup_{t > 0} \frac{f(x + tv) - f(x)}{t} = \sup_{x \in X : f(x) < +\infty} \lim_{t \to +\infty} \frac{f(x + tv)}{t}.$$

Proof We clearly have $f^{\infty}(0) = 0$ so that f^{∞} is proper. As a sup of the functions $v \mapsto f(x + v) - f(x)$, which are just translations of f, the function f^{∞} inherits convexity and lower semicontinuity from f. The fact that it is 1-homogeneous will be clear after proving the equality $f^{\infty}(v) = \sup_{x \in X : f(x) < +\infty} \lim_{t \to +\infty} \frac{f(x + tv)}{t}$.

Indeed, this would imply, for every $c > 0$:

$$f^\infty(cv) = \sup_{x \in \mathcal{X}\,:\, f(x) < +\infty} \lim_{t \to +\infty} \frac{f(x + tcv)}{t}$$

$$= \sup_{x \in \mathcal{X}\,:\, f(x) < +\infty} \lim_{s \to +\infty} \frac{f(x + sv)}{s/c} = c f^\infty(v),$$

where we used a change of variable $s = ct$.

In order to prove the equality in the last part of the claim we look at the behavior of f on each line directed as v: the function h_x defined via $h_x(t) := f(x + tv)$ is convex, and its incremental ratios are non-decreasing. This means that, studying the behavior of 1D convex functions, the sup defining $f^\infty(v)$ can be re-written as

$$\sup_{x \in \mathcal{X}\,:\, f(x) < +\infty} h_x(1) - h_x(0) = \sup_{x \in \mathcal{X}\,:\, f(x) < +\infty} \sup_{t > 0} h_x(t + 1) - h_x(t)$$

$$= \sup_{x \in \mathcal{X}\,:\, f(x) < +\infty} \sup_{t > 0} h'_x(t)$$

$$= \sup_{x \in \mathcal{X}\,:\, f(x) < +\infty} \lim_{t \to +\infty} h'_x(t)$$

$$= \sup_{x \in \mathcal{X}\,:\, f(x) < +\infty} \lim_{t \to +\infty} \frac{h_x(t) - h_x(0)}{t}$$

$$= \sup_{x \in \mathcal{X}\,:\, f(x) < +\infty} \sup_{t > 0} \frac{h_x(t) - h_x(0)}{t}.$$

This gives the first part of the required equality. For the last equality, we just need to observe that we have $f(x)/t \to 0$ as $t \to +\infty$. □

Proposition 4.30 *Given a duality pair and a function $f : \mathcal{X} \to \mathbb{R} \cup \{+\infty\}$, assume that f is a proper, convex, and $(\mathcal{X}, \mathcal{X}^*)$-weakly l.s.c. function. Then we have*

$$f^\infty(v) = \sup\{\langle \xi, v \rangle \,:\, f^*(\xi) < +\infty\}.$$

Proof We denote the function $v \mapsto \sup\{\langle \xi, v \rangle \,:\, f^*(\xi) < +\infty\}$ by $\widetilde{f^\infty}$. First, we prove that we always have $f^\infty \geq \widetilde{f^\infty}$. Let ξ be a point where $f^*(\xi) < +\infty$. We have $f(x + tv) + f^*(\xi) \geq \langle \xi, (x + tv) \rangle$, so that we have

$$f^\infty(v) = \sup_{x \in \mathcal{X}\,:\, f(x) < +\infty} \lim_{t \to +\infty} \frac{f(x + tv)}{t}$$

$$\geq \sup_{x \in \mathcal{X}\,:\, f(x) < +\infty} \lim_{t \to +\infty} \frac{\langle \xi, x \rangle - f^*(\xi)}{t} + \langle \xi, v \rangle = \langle \xi, v \rangle,$$

and the fact that ξ is arbitrary gives

$$f^\infty(v) \geq \sup\{\langle \xi, v \rangle \ : \ f^*(\xi) < +\infty\}.$$

We want to prove the opposite inequality. We start from the case where f is locally bounded on X, so that $\partial f(x) \neq \emptyset$ for every x (see Proposition 4.21). In this case we consider $\xi \in \partial f(x + v)$ and we have $f(x) \geq f(x + v) - \langle \xi, v \rangle$, hence $f(x + v) - f(x) \leq \langle \xi, v \rangle$. Since ξ is in the image of ∂f, it is in the domain of ∂f^* (due to Point 3. in Proposition 4.23), and hence $f^*(\xi) < +\infty$. We then obtain $f(x+v) - f(x) \leq \langle \xi, v \rangle \leq \widetilde{f^\infty}(v)$ and, passing to the sup in x, we get $f^\infty \leq \widetilde{f^\infty}$.

The general case is obtained via approximation. Given an arbitrary nonnegative, convex, X^*-weakly l.s.c. and superlinear function[3] $h : X^* \to \mathbb{R}$ (note that we take h finite everywhere), let f_ε be the function on X such that $(f_\varepsilon)^* = f^* + \varepsilon h$ (i.e. $f_\varepsilon = (f^* + \varepsilon h)^*$). Since $f^* + \varepsilon h$ is superlinear, Proposition 4.14 guarantees that f_ε is locally bounded on X. Moreover, we observe that the set where f_ε^* is finite is the same as the one where f^* is finite. We then have, for every x such that $f(x) < +\infty$,

$$\widetilde{f^\infty}(v) = (f_\varepsilon)^{\widetilde{\infty}}(v) = (f_\varepsilon)^\infty(v) \geq f_\varepsilon(x + v) - f_\varepsilon(x) \geq f_\varepsilon(x + v) - f(x).$$

We then take the limit $\varepsilon \to 0$ (using Exercise 4.2) in order to obtain $\widetilde{f^\infty}(v) \geq f(x + v) - f(x)$ and the sup over x in order to conclude $\widetilde{f^\infty}(v) \geq f^\infty(v)$. □

Remark 4.31 In Remark 4.28 we saw that $f^* < \infty$ is not enough for the superlinearity of f, while the previous proposition implies that when $f^* < \infty$ we have $f^\infty(v) = +\infty$ for every $v \neq 0$. It is important to note that this last condition is very different from superlinearity, as it corresponds to superlinearity on each line $x + tv$, but lacks uniformity in v.

4.2.4 Formal Duality for Constrained and Penalized Optimization

In this section we introduce the notion of the dual problem of a convex optimization problem through an inf-sup exchange procedure. This often requires us to write possible constraints as a sup penalization, and we will then see how to adapt to more general problems. No proof of duality results will be given, and we will analyze in detail in the next section the most relevant examples in the calculus of variations. The proofs presented in Sect. 4.3 will then inspire a more general theory in Sect. 4.5.

We start from the following problem:

$$\min \{f(x) \ : \ x \in X, \ Ax = b\},$$

---

[3] For instance $h(\xi) := \|\xi\|^2$.

where $A : X \to Y$ is a linear map between two normed vector spaces, $b \in Y$ is a fixed vector and $f : X \to \mathbb{R} \cup \{+\infty\}$ is a given convex and l.s.c. function. We will denote by A^t the transpose operator of A, a linear mapping defined on Y', the dual of Y, and valued in X', the dual of X, and characterized by

$$\langle A^t \xi, x \rangle := \langle \xi, Ax \rangle \quad \text{for all } \xi \in Y' \text{ and } x \in X.$$

We can see that the above problem is equivalent to

$$\min \left\{ f(x) + \sup_{\xi \in Y'} \langle \xi, Ax - b \rangle \ : \ x \in X \right\},$$

since we can compute the value of the expression $\sup_{\xi \in Y'} \langle \xi, Ax - b \rangle$ by distinguishing two cases: either $Ax = b$, in which case $\langle \xi, Ax - b \rangle = 0$ for every ξ and the sup equals 0, or $Ax \neq b$, in which case there exists an element $\xi \in Y'$ such that $\langle \xi, Ax - b \rangle \neq 0$ and, by multiplying ξ by arbitrarily large constants, positive or negative depending on the sign of $\langle \xi, Ax - b \rangle$, we can see that the sup is $+\infty$. Hence, adding this sup means adding 0 if the constraint is satisfied or adding $+\infty$ if not; since in a minimization problem adding the value $+\infty$ has the same effect as imposing a constraint, we can see the equivalence between the problem with the constraint $Ax = b$ and the problem with the sup over ξ.

We now arrive at a problem of the form

$$\inf_{x} \sup_{\xi} L(x, \xi), \quad \text{where } L(x, \xi) = f(x) + \langle \xi, Ax - b \rangle.$$

This is an inf-sup problem, and we can associate with it a second optimization problem, obtained by switching the order of the inf and the sup. We can consider

$$\sup_{\xi} \inf_{x} L(x, \xi),$$

which means maximizing over ξ the function obtained as the value of the inf over x. Remember that we have forgotten the constraint $Ax = b$, since it took part in the definition of L, and we now minimize over all x. We can then give a better expression of this new problem, that we will call the dual problem. Indeed we have

$$\sup_{\xi} \inf_{x} L(x, \xi) = \sup_{\xi} -\langle \xi, b \rangle + \inf_{x} f(x) + \langle \xi, Ax \rangle.$$

We then rewrite $\langle \xi, Ax \rangle$ as $\langle A^t \xi, x \rangle$ and change the sign in the inf so as to write it as a sup. We obtain

$$\sup_{\xi} \inf_{x} L(x, \xi) = \sup_{\xi} -\langle \xi, b \rangle - \sup_{x} -f(x) + \langle -A^t \xi, x \rangle.$$

We now recognize in the sup over x the form of a Fenchel–Legendre transform and we finally obtain

$$\sup_{\xi} \inf_{x} L(x, \xi) = \sup_{\xi} -\langle \xi, b \rangle - f^*(-A^t \xi).$$

This is a convex optimization problem in the variable ξ (the maximization of the sum of a linear functional and the opposite of a convex function, f^*, applied to a linear function of ξ), involving the Legendre transform of the original objective function f.

We would like the above two optimization problems ("inf sup" and "sup inf") to be related to each other, and for instance their values to be the same.

Given an arbitrary function L the values of inf sup and of sup inf are in general different, as we can see from this very simple example: take $L : A \times B \to \mathbb{R}$ with $A = B = \{\pm 1\}$ and $L(a, b) = ab$. In this case we have inf sup $= 1 >$ sup inf $= -1$. Here the two values are different, and the inf sup is larger than the sup inf. This is a general inequality, that we prove here.

Proposition 4.32 *Given an arbitrary function $L = A \times B \to \mathbb{R}$ we have*

$$\inf_{a} \sup_{b} L(a, b) \geq \sup_{b} \inf_{a} L(a, b).$$

Proof Take $(a_0, b_0) \in A \times B$ and write $L(a_0, b_0) \geq \inf_a L(a, b_0)$. We then take the sup over b_0 on both sides, thus obtaining $\sup_{b_0} L(a_0, b_0) \geq \sup_{b_0} \inf_a L(a, b_0)$. We have now a number on the right-hand side, and a function of a_0 on the left-hand side. We then take the inf over a_0 and get

$$\inf_{a_0} \sup_{b_0} L(a_0, b_0) \geq \sup_{b_0} \inf_{a} L(a, b_0),$$

which is exactly the same as the claim up to renaming the variables.   □

Although in general it is not possible to connect the two problems obtained as the inf-sup and sup-inf of the same function, this can be the case when some conditions are met. The main tool to achieve this is a theorem of Rockafellar (see [173, Section 37]) requiring concavity in the variable on which we maximize, convexity in the other one, and some compactness assumption. In our precise case concavity and convexity are met, since L is convex in x and linear in ξ, and hence concave. Yet, Rockafellar's statement concerns finite-dimensional spaces, and moreover we still have to deal with the compactness properties we would need. Hence, we will not provide any proof here that $\min\{f(x) : x \in X, Ax = b\}$ and $\max\{-\langle \xi, b \rangle - f^*(-A^t \xi) : \xi \in \mathcal{Y}'\}$ are equal, and we will wait until the next section for a proof in a very particular case.

We only discuss here some consequences and some variants of this duality approach.

A first consequence concerns sufficient optimality conditions. Assume we have two optimization problems issued by an inf-sup/sup-inf procedure, i.e. $\min\{f(a) : a \in A\}$ and $\max\{g(b) : b \in B\}$ with $f(a) := \sup_b L(a, b)$ and $g(b) := \inf_a L(a, b)$. Then, if $a_0 \in A$ and $b_0 \in B$ are such that $f(a_0) = g(b_0)$, automatically a_0 minimizes f and b_0 maximizes g, just because Proposition 4.32 guarantees $f(a_0) \geq \inf f \geq \sup g \geq g(b_0)$ and all inequalities must be equalities here. In our precise case the functional f should be replaced with f plus the constraint $Ax = b$, and g is given by $g(\xi) := -\langle \xi, b \rangle - f^*(-A^t \xi)$.

A second consequence concerns instead necessary optimality conditions. We need now to believe in $\inf f = \sup g$. Take a pair (a_0, b_0) where a_0 minimizes f and b_0 maximizes g. Then we deduce $f(a_0) = g(b_0)$, but this equality is a very strong piece of information in many cases. For instance, in our case this means that, if x_0 and ξ_0 are optimal, then we have

$$f(x_0) = -\langle \xi_0, b \rangle - f^*(-A^t \xi_0) \quad \text{and} \quad Ax_0 = b.$$

This can be re-written as

$$f(x_0) + f^*(-A^t \xi_0) = -\langle \xi_0, b \rangle = -\langle \xi_0, Ax_0 \rangle = -\langle A^t \xi_0, x_0 \rangle,$$

i.e. we have equality in the inequality $f(x) + f^*(y) \geq \langle x, y \rangle$. This is equivalent to

$$x_0 \in \partial f^*(-A^t \xi_0) \quad \text{and} \quad -A^t \xi_0 \in \partial f(x_0).$$

We can note the similarity with Lagrange multipliers, where optimizing a function f under a linear constraint of the form $Ax = b$ can be translated into the fact that ∇f should belong to a subspace, orthogonal to the affine space of the constraints, which is indeed the image of A^t.

Before moving on to variants of the previous pair of dual problems we want to insist that writing an equality constraint as a sup over test elements of a dual space is exactly what is always done in the weak formulation of PDEs. In the next section we will see as an example what happens when the constraint is of the form $\nabla \cdot \mathbf{v} = f$, which can be written as $\int \nabla \phi \cdot \mathbf{v} + \phi f = 0$ for every test function ϕ, and it is very natural to replace the constraint with a sup over ϕ. In this case, the dual problem turns out to be a maximization over scalar functions ϕ. Moreover, the transpose of the divergence operator $\nabla \cdot$ is the opposite of the gradient, since $\int \phi \nabla \cdot \mathbf{v} = -\int \nabla \phi \cdot \mathbf{v}$ by integration by parts, as soon as boundary conditions are taken care of. In this way, the functional $f^*(-A^t \phi)$ will be a very classical functional in the calculus of variations.

Among variants, we first want to discuss the case of inequality constraints instead of equalities. A constraint of the form $Ax \leq b$ only has a meaning if we give a notion of inequality among vectors, which in general is not canonically defined. In finite dimensions a general convention, mainly used in the Operational Research community, is that we can consider the inequality component-wise. In the calculus of variations, we can expect both Ax and b to be functions in a certain functional space, and we can require the inequality to be satisfied pointwise (or a.e.). What is important is that we should characterize the inequality in terms of test functions. For instance, the inequality $f \le g$ a.e. is equivalent to $\int \phi(f - g) \le 0$ for every $\phi \ge 0$ and a similar equivalence can be stated in the finite-dimensional componentwise case. We then write

$$\min\{f(x) \,:\, x \in X, \, Ax \le b\} = \min\left\{f(x) + \sup_{\xi \in \mathcal{Y}', \xi \ge 0} \langle \xi, Ax - b \rangle \,:\, x \in X\right\},$$

since

$$\sup_{\xi \in Y', \xi \ge 0} \langle \xi, y \rangle = \begin{cases} 0 & \text{if } y \le 0, \\ +\infty & \text{if not.} \end{cases}$$

We can then go on with the very same procedures and obtain the dual problem

$$\max\{-\langle \xi, b \rangle - f^*(-A^t\xi) \,:\, \xi \in \mathcal{Y}, \xi \ge 0\}.$$

Finally, once we know how to build dual problems out of constrained optimization problems, we could consider a more general case, such as

$$\min\{f(x) + g(Ax)\},$$

where $g = I_{\{b\}}$ corresponds to the previous example. In this case we do not have constraints to write as a sup, but we can decide to write one of the two functions f or g as a sup thanks to the double Legendre transform. We then set

$$L(x, \xi) := f(x) + \langle \xi, Ax \rangle - g^*(\xi)$$

and we easily see that we have

$$\min\{f(x) + g(Ax)\} = \inf_x \sup_\xi L(x, \xi).$$

We then interchange inf and sup, thus obtaining the dual problem

$$\sup_\xi \inf_x L(x, \xi) = \sup_\xi -g^*(\xi) + \inf_x f(x) + \langle \xi, Ax \rangle$$

$$= \sup_\xi -g^*(\xi) - \sup_x -f(x) + \langle -A^t\xi, x \rangle$$

$$= \sup_\xi -g^*(\xi) - f^*(-A^t\xi).$$

As we said, the equality constraint $Ax = b$ corresponds to $g = I_{\{b\}}$, so that we have $g^*(\xi) = \langle \xi, b \rangle$.

The duality between

$$\min\{f(x) + g(Ax) : x \in \mathcal{X}\} \quad \text{and} \quad \sup\{-g^*(\xi) - f^*(-A^t\xi) : \xi \in \mathcal{Y}'\}$$

is a classical object in convex analysis and a theorem guaranteeing, under some conditions, that the values are actually equal is known as the Fenchel–Rockafellar theorem. We will see in Sect. 4.5 a proof of this theorem, inspired by the precise proof of a concrete duality result presented in Sect. 4.3.

4.3 An Example of Convex Duality: Minimal Flows and Optimal Compliance

We start this section with two classical examples of convex optimization problems, which have a strong connection with PDEs, one from traffic congestion modeling and one from the mechanics of deformation.

The first problem consists in finding a flow transporting some mass from an original configuration f^+ (to be intended as a distribution of mass on a given compact domain Ω, so, for instance, one could take $f^+ \in \mathcal{P}(\Omega)$, the set of probability measures on Ω) to a target configuration f^-. The flow will be a vector field $\mathbf{v}$ which describes, in Eulerian variables, the motion: $\mathbf{v}(x)$ stands for the intensity and the direction of the movement of the mass at the point x (more precisely, $|\mathbf{v}|$ is the intensity and $\mathbf{v}/|\mathbf{v}|$ the direction); the motion is assumed to be stationary (i.e. the mass follows a permanent movement, and new mass is always injected at the points belonging to the support of f^+, distributed according to f^+, and withdrawn according to f^-) so that there is no explicit time dependence. The condition to move mass from f^+ to f^- can be written in terms of $\nabla \cdot \mathbf{v}$, thanks to the following heuristic consideration: for every small subdomain $A \subset \Omega$, the conservation of the mass imposes that the mass exiting A, which equals $\int_{\partial A} \mathbf{v} \cdot \mathbf{n}$ since mass can only leave A through the boundary (and the tangential movement on the boundary does not contribute to this), should be equal to the mass which has been inserted minus that which has been removed, i.e. to $\int_A f^+ - f^-$; writing $\int_{\partial A} \mathbf{v} \cdot \mathbf{n} = \int_A \nabla \cdot \mathbf{v}$ thanks to the divergence theorem and assuming that this equality holds for any A means $\nabla \cdot \mathbf{v} = f^+ - f^-$. Note that, if Ω itself has a boundary and no mass is injected/removed at points of the boundary, one should also impose $\mathbf{v} \cdot \mathbf{n} = 0$ on $\partial \Omega$ so that no mass leaves Ω. A convenient way to express this set of constraints is the weak formulation

$$\int_\Omega (\nabla \phi \cdot \mathbf{v} + \phi f) \, dx = 0 \text{ for all } \phi \in C^\infty(\Omega),$$

where $f = f^+ - f^-$. This includes the possibility that $f^\pm$ are singular measures with a part on the boundary, which should compensate $\mathbf{v} \cdot \mathbf{n}$.

Among all the possible flows $\mathbf{v}$ satisfying these constraints, we minimize a cost which penalizes the total movement which is realized, i.e. an integral cost which is increasing in $|\mathbf{v}|$. We will discuss in Sect. 4.6 the connection of this problem with optimal transport; here we just say that a natural choice would be to minimize $\int |\mathbf{v}| \, dx$ or $\int K(x)|\mathbf{v}| \, dx$, where $K > 0$ is a weight (exactly as we did in Sect. 1.4.3 for weighted geodesics); on the other hand, when we want to model traffic congestion, we can assume that the weight k is not given a priori but depends on the traffic itself and, in a very simplified setting, on $|\mathbf{v}|$. The simplest choice is to choose $K = |\mathbf{v}|$ and hence solve (adding, in order to simplify the computations, a factor $\frac{1}{2}$ in front of the integral)

$$\min \left\{ \int_\Omega \frac{1}{2}|\mathbf{v}|^2 \, dx \ : \ \nabla \cdot \mathbf{v} = f \right\}. \tag{4.4}$$

As we said, the second problem comes from mechanics, and is much more standard in its modeling. Assume that a membrane is originally at rest in horizontal position, which we model via the constant function $u = 0$ on Ω: the domain Ω stands for the projection on the horizontal plan of the membrane, and the value of $u(x) \in \mathbb{R}$ stands for its vertical displacement. If no force acts on the membrane, the configuration $u = 0$ is stable and is what can be observed in reality. Assume now that at some points the membrane is pushed down, and at others it is pushed up, thus applying a force $f = f^+ - f^-$ with positive and negative parts. If the two parts equilibrate (i.e. $\int f^+ = \int f^-$) we can imagine that the barycenter of the membrane does not move, but the shape of the membrane is deformed. The configuration which is realized by the membrane is the one which minimizes a sum of a potential energy given by the work of the force f (i.e. $\int uf$) and an elastic energy related to the deformation. The simplest choice is to choose the Dirichlet energy $\int \frac{1}{2}|\nabla u|^2$ as an elastic energy. One of the reason for this choice is that it is equivalent to the Taylor expansion (stopped at the first non-zero term) of the energy $\int \sqrt{1 + |\nabla u|^2}$, which represents the surface area of the deformed membrane (and hence approximates well such an area if u is small in C^1 norm). Alternatively, one can also describe the equilibrium shape of the membrane in PDE terms through $\Delta u = f$. We will discuss later how to insert Dirichlet data on u (i.e. regions where the displacement u is prescribed, for instance $u = 0$), and in particular we will see in Chap. 6 that an interesting question is how to optimize the Dirichlet regions, which can be considered as reinforcements for the membrane, under possible constraints or penalizations on their size. So far, we have stuck to the case where no value is prescribed for u, which corresponds to Neumann boundary conditions for the PDE $\Delta u = f$ if Ω has a boundary. The variational problem reads

$$\min \left\{ \int_\Omega \frac{1}{2}|\nabla u|^2 \, dx + \int_\Omega uf \, dx \ : \ u \in H^1(\Omega) \right\}. \tag{4.5}$$

In the case where f is a function instead of a more general distribution in the dual of H^1 (in which case one should replace $\int uf$ with $\langle f, u \rangle$), this is a particular case of the problems presented in Sect. 2.1. The minimal value is necessarily negative (since $u = 0$ is a competitor) and one can see that, using the optimality condition given by the Euler–Lagrange equation $\Delta u = f$, it equals $-\int \frac{1}{2}|\nabla u|^2$. Hence, if we consider that this minimal value depends on Ω and on the force f we see that how much it differs from 0 is a measure of how stiff the configuration is, since it measures the effective total deformation (in terms of elastic energy). The negative of this minimal value is called the *compliance* and it can be optimized in order to find the stiffest configuration (usually among domains Ω for given f, and usually adding Dirichlet boundary data).

The main goal of this section is to show that the two problems of the form (4.4) and (4.5) are actually dual to each other, and that the same holds for a large class of variants of the same problems, replacing the quadratic costs with other convex functions.

To be more precise we will consider a function $H : \Omega \times \mathbb{R}^d \to \mathbb{R}$ which is convex in the second variable

$$(Hyp1) \quad \text{for every } x \quad v \mapsto H(x, v) \text{ is convex}$$

and satisfying the following uniform bounds:

$$(Hyp2) \quad \frac{c_0}{q}|v|^q - h_0(x) \leq H(x, v) \leq \frac{c_1}{q}|v|^q + h_1(x),$$

where h_0, h_1 are L^1 functions on Ω, $c_0, c_1 > 0$ are given finite constants, and $q \in (1, +\infty)$ is a given exponent. For functions of this form, when we write $H^*(x, w)$ we mean the Legendre transform in the second variable, i.e. $H^*(x, w) = \sup_v w \cdot v - H(x, v)$.

We consider the problem

$$\min \left\{ \int_\Omega H(x, \mathbf{v}(x)) \, dx \; : \; \nabla \cdot \mathbf{v} = f \right\}.$$

Before giving rigorous results, we will formally build its dual problem, with the same informal derivation as we did in Sect. 4.2.3. This can be done in the following way: the constraint $\nabla \cdot \mathbf{v} = f$ can be written, in weak form, as $-\int \mathbf{v} \cdot \nabla u = \int fu$ for every u in a suitable space of test functions This means that we can rewrite the above problem in the min-max form

$$\min \left\{ \int_\Omega H(x, \mathbf{v}) \, dx + \sup_u \left(-\int_\Omega fu \, dx - \int_\Omega \mathbf{v} \cdot \nabla u \, dx \right) \right\},$$

since the last sup is 0 if the constraint is satisfied and $+\infty$ if not. Now, we have a min-max problem and the dual problem can be obtained just by inverting inf and sup. In this case we get

$$\sup\left\{-\int_\Omega fu\,dx + \inf_{\mathbf{v}}\left(\int_\Omega H(x,\mathbf{v})\,dx - \int_\Omega \mathbf{v}\cdot\nabla u\,dx\right)\right\}.$$

Since $\inf_{\mathbf{v}}\int_\Omega H(x,\mathbf{v})-\int_\Omega \mathbf{v}\cdot\nabla u = -\sup_{\mathbf{v}}\int_\Omega \nabla u\cdot\mathbf{v}-\int_\Omega H(x,\mathbf{v}) = \int_\Omega H^*(x,\nabla u)$, the problem becomes

$$\sup\left\{-\int_\Omega fu\,dx - \int_\Omega H^*(x,\nabla u)\,dx\right\}.$$

In the following, we will see precise statements about the duality between the two problems. The duality proof, based on the above convex analysis tools, is essentially inspired by the method used in [36] . Other proofs are obviously possible.

4.3.1 Rigourous Duality with No-Flux Boundary Conditions

We define the space $W_\diamond^{1,p}(\Omega)$ as the vector subspace of $W^{1,p}(\Omega)$ composed by functions with zero mean and the space $(W^{1,p})_\diamond'(\Omega)$ as the subspace of the dual of $W^{1,p}$ composed by those f such that $\langle f, 1\rangle = 0$ (i.e. those f with zero mean as well).

Note that for every $\mathbf{v} \in L^{p'}(\Omega;\mathbb{R}^d)$, the distribution $\nabla \cdot \mathbf{v}$, defined by

$$\langle \nabla \cdot \mathbf{v}, \phi\rangle := -\int_\Omega \mathbf{v}\cdot\nabla\phi\,dx,$$

naturally belongs to $(W^{1,p})_\diamond'(\Omega)$. Incidentally, this will be the definition we will use of the divergence operator (in weak form), and it includes a natural Neumann (no-flux) boundary condition on $\partial\Omega$. However, note that we will sometimes use the flat torus as a domain Ω, which gets rid of many boundary issues.

We will prove the following duality result.

Theorem 4.33 *Assume that Ω is smooth enough and that H satisfies Hyp1 and Hyp2 with $q = p' = p/(p-1)$. Then, for any $f \in (W^{1,p})_\diamond'(\Omega)$, we have*

$$\min\left\{\int_\Omega H(x,v(x))\,dx \ : \ v \in L^{p'}(\Omega;\mathbb{R}^d), \ \nabla\cdot v = f\right\}$$

$$= \max\left\{-\int_\Omega H^*(x,\nabla u(x))\,dx - \langle f,u\rangle \ : \ u \in W^{1,p}(\Omega)\right\}.$$

Proof We will define a function $\mathcal{F}: (W^{1,p})' \to \mathbb{R}$ in the following way

$$\mathcal{F}(f) := \min\left\{\int_\Omega H(x, \mathbf{v}(x))\, dx \; : \; \mathbf{v} \in L^{p'}(\Omega; \mathbb{R}^d), \nabla \cdot \mathbf{v} = f\right\}.$$

Note that if $f \notin (W^{1,p})'_\diamond \subset (W^{1,p})'$, then $\mathcal{F}(f) = +\infty$, as there is no $\mathbf{v} \in L^{p'}$ with $\nabla \cdot \mathbf{v} = f$. On the other hand, if $f \in (W^{1,p})'_\diamond$, then $\mathcal{F}(f)$ is well-defined and real-valued since $\int_\Omega H(x, \mathbf{v}(x))\, dx$ is comparable to the $L^{p'}$ norm of $\mathbf{v}$, and we use the following fact: for every $f \in (W^{1,p})'_\diamond$ there exists a $\mathbf{v} \in L^{p'}$ such that $f = \nabla \cdot \mathbf{v}$ and $\|\mathbf{v}\|_{L^{p'}} \leq \|f\|_{(W^{1,p})'_\diamond}$ (see the next lemma).

We now compute $\mathcal{F}^*: W^{1,p} \to \mathbb{R}$:

$$\mathcal{F}^*(u) = \sup_f \langle f, u\rangle - \mathcal{F}(f) = \sup_{f, \mathbf{v}: \nabla\cdot\mathbf{v}=f} \langle f, u\rangle - \int_\Omega H(x, \mathbf{v}(x))\, dx.$$

We can then go on using only $\mathbf{v}$ as a variable, thus obtaining

$$\mathcal{F}^*(u) = \sup_{\mathbf{v}} \langle \nabla \cdot \mathbf{v}, u\rangle - \int_\Omega H(x, \mathbf{v}(x))\, dx$$

$$= \sup_{\mathbf{v}} - \int (\mathbf{v} \cdot \nabla u)\, dx - \int_\Omega H(x, \mathbf{v}(x))\, dx = \int_\Omega H^*(x, -\nabla u(x))\, dx.$$

In the very last equality we used the fact that the vector field $\mathbf{v}$ can be chosen point by point as there are no longer any constraints on its divergence. The reader may wonder how we can restrict to $\mathbf{v} \in L^{p'}$, but this can be done in the following way. Of course we have

$$\sup_{\mathbf{v}\in L^{p'}} - \int (\mathbf{v} \cdot \nabla u)\, dx - \int_\Omega H(x, \mathbf{v}(x))\, dx \leq \int_\Omega H^*(x, -\nabla u(x))\, dx,$$

because of the pointwise inequality defining H^*. On the other hand, for every $M > 0$ we have

$$\sup_{\mathbf{v}\in L^{p'}} - \int (\mathbf{v}\cdot\nabla u)\, dx - \int_\Omega H(x, \mathbf{v}(x))\, dx \geq \sup_{\mathbf{v}:\,|\mathbf{v}|\leq M} - \int (\mathbf{v}\cdot\nabla u)\, dx - \int_\Omega H(x, \mathbf{v}(x))\, dx.$$

The sup in the right-hand side can be realized by choosing for every x a vector $v(x)$ which realizes the max (or almost the sup) in $-v \cdot \nabla u(x) - H(x, v)$ among vectors v with $|v| \leq M$. This can be done in a measurable way. If we set $H^{*,M}(x, w) := \sup_{v:\,|v|\leq M} v \cdot w - H(x, v)$ we then obtain

$$\sup_{\mathbf{v}\in L^{p'}} - \int (\mathbf{v} \cdot \nabla u)\, dx - \int_\Omega H(x, \mathbf{v}(x))\, dx \geq \int_\Omega H^{*,M}(x, -\nabla u(x))\, dx.$$

The desired equality can be obtained taking the limit $M \to \infty$, using the monotone increasing convergence of $H^{*,M}(x,w)$ to $H^*(x,w)$.

Now we want to make use of the double transform $\mathcal{F}^{**}$. By definition we have

$$\mathcal{F}^{**}(f) := \sup_u \langle f, u \rangle - \mathcal{F}^*(u) = \sup_u \langle f, u \rangle - \int_\Omega H^*(x, -\nabla u(x))\,dx.$$

By taking the sup on $-u$ instead of u we also have

$$\mathcal{F}^{**}(f) = \sup_u -\langle f, u \rangle - \int_\Omega H^*(x, \nabla u(x))\,dx.$$

Finally, if we prove that $\mathcal{F}$ is convex and l.s.c., then we also have $\mathcal{F}^{**} = \mathcal{F}$, and the desired duality result is proven.

The convexity of $\mathcal{F}$ is easy. We just need to take $f_0, f_1 \in (W^{1,p})'_\diamond(\Omega)$ and define $f_t := (1-t)f_0 + tf_1$. Let v_0, v_1 be optimal in the definition of $\mathcal{F}(f_0)$ and $\mathcal{F}(f_1)$, i.e. $\int H(x, v_i(x))\,dx = \mathcal{F}(f_i)$ and $\nabla \cdot v_i = f_i$. Let $v_t := (1-t)v_0 + tv_1$. Of course we have $\nabla \cdot v_t = f_t$ and, by convexity of $H(x, \cdot)$ we have

$$\mathcal{F}(f_t) \leq \int H(x, v_t(x))\,dx \leq (1-t) \int H(x, v_0(x))\,dx + t \int H(x, v_1(x))\,dx$$

$$\leq (1-t)\mathcal{F}(f_0) + t\mathcal{F}(f_1),$$

and the convexity is proven.

For the semicontinuity, we take a sequence $f_n \to f$ in $(W^{1,p})'$. We can assume that $\mathcal{F}(f_n) < +\infty$ otherwise there is nothing to prove, hence $f_n \in (W^{1,p})'_\diamond(\Omega)$. Take the corresponding optimal vector fields $v_n \in L^{p'}$, i.e. $\int H(x, v_n(x))\,dx = \mathcal{F}(f_n)$. We can extract a subsequence such that $\lim_k \mathcal{F}(f_{n_k}) = \liminf_n \mathcal{F}(p_n)$. Moreover, from the bound on H we can see that the $L^{p'}$ norm of v_n is bounded in terms of the values of $\mathcal{F}(f_n)$, which are (use Lemma 4.34) bounded by the $(W^{1,p})'_\diamond$ norms of f_n. Since f_n converges, we get a bound on $||v_n||_{L^{p'}}$. Hence, up to an extra subsequence extraction, we can assume $v_{n_k} \rightharpoonup v$. Obviously we have $\nabla \cdot v = f$ and, by semicontinuity of the integral functional $v \mapsto \int H(x, v)\,dx$, we get

$$\mathcal{F}(f) \leq \int H(x, v(x))\,dx \leq \liminf_k \int H(x, v_{n_k}(x))\,dx = \lim_k \mathcal{F}(f_{n_k}) = \liminf_n \mathcal{F}(f_n),$$

which gives the desired result. $\qquad\qquad\qquad\qquad\qquad\qquad\qquad\qquad\qquad\square$

The duality result that we proved can be written in the following form

$$\min\{A(v)\} + \min\{B(u)\} = 0, \tag{4.6}$$

where A is defined on $L^{p'}(\Omega; \mathbb{R}^d)$ and B on $W^{1,p}(\Omega)$ via

$$A(\mathbf{v}) := \begin{cases} \int_\Omega H(x, \mathbf{v}(x))\,dx & \text{if } \nabla \cdot \mathbf{v} = f, \\ +\infty & \text{otherwise,} \end{cases}$$

and

$$B(u) = \int_\Omega H^*(x, \nabla u(x))\,dx + \langle f, u \rangle.$$

Lemma 4.34 *Given $f \in (W^{1,p})'_\diamond(\Omega)$ there exists a $\mathbf{v} \in L^{p'}(\Omega; \mathbb{R}^d)$ such that $f = \nabla \cdot \mathbf{v}$ and $\|\mathbf{v}\|_{L^{p'}} \leq C\|f\|_{(W^{1,p})'}$.*

Proof Consider the minimization problem

$$\min \left\{ \frac{1}{p} \int_\Omega |\nabla \phi|^p\,dx + \langle f, \phi \rangle \;:\; \phi \in W^{1,p}(\Omega) \right\}.$$

It is easy to prove that this problem admits a solution, as the minimization can be restricted to the set $W^{1,p}_\diamond$. This solution ϕ satisfies[4] (see Sect. 2.4)

$$- \int_\Omega (\nabla \phi)^{p-1} \cdot \nabla \psi\,dx = \langle f, \psi \rangle$$

for all $\psi \in W^{1,p}(\Omega)$. This means precisely $\nabla \cdot \mathbf{v} = f$ for $\mathbf{v} = (\nabla \phi)^{p-1}$. Moreover, testing against ϕ, we get

$$\|\mathbf{v}\|_{L^{p'}}^{p'} = \int_\Omega |\mathbf{v}|^{p'}\,dx = \int_\Omega |\nabla \phi|^p\,dx = \langle f, \phi \rangle \leq \|f\|_{(W^{1,p})'}\|\phi\|_{W^{1,p}}$$

$$\leq C\|f\|_{(W^{1,p})'}\|\nabla \phi\|_{L^p} = C\|f\|_{(W^{1,p})'}\|\mathbf{v}\|_{L^{p'}}^{p'-1},$$

which gives the desired bound on $\|\mathbf{v}\|_{L^{p'}}$. $\qquad\square$

4.4 Regularity via Duality

In this section we will use the relation (4.6) to produce Sobolev regularity results for solutions of the minimization problems min A or min B.

---

[4] Pay attention to the notation: for every vector w and $\alpha > 0$ we denote by w^α the vector with modulus equal to $|w|^\alpha$, and same direction as w, i.e. $w^\alpha := |w|^{\alpha-1}w$.

We will start by describing the general strategy. We consider a function H not explicitly depending on x, and we assume that an inequality of the following form is true

$$(Hyp3) \quad H(v) + H^*(w) \geq v \cdot w + c|j(v) - j_*(w)|^2$$

for some given functions $j, j_* : \mathbb{R}^d \to \mathbb{R}^d$. This is an improvement of the Young inequality $H(v) + H^*(w) \geq v \cdot w$ (which is just a consequence of the definition of H^*). Of course this is always true taking $j = j_* = 0$, but the interesting cases are the ones where j and j_* are non-trivial.

To simplify the computations, we will assume that Ω is the flat d-dimensional torus $\mathbb{T}^d$. We start from the following observations, which we collect in a lemma. For the sake of the notation, we call $\bar{v}$ and $\bar{u}$ the minimizers (or some minimizers, in case there is no uniqueness) of A and B, respectively, and we denote by u_h the function $u_h(x) := \bar{u}(x + h)$. We define a function $g : \mathbb{R}^d \to \mathbb{R}$ given by

$$g(h) := \int_{\mathbb{T}^d} f(x)\bar{u}(x + h)\,dx - \int_{\mathbb{T}^d} f(x)\bar{u}(x)\,dx.$$

Lemma 4.35 *Assume H satisfies Hyp1, 2, 3 with $q = p'$, and let $\bar{v}$ and $\bar{u}$ be optimal. Then we have*

1. $j(\bar{v}) = j_*(\nabla\bar{u})$.
2. $c \int_{\mathbb{T}^d} |j_*(\nabla u_h) - j_*(\nabla\bar{u})|^2\,dx \leq g(h)$.
3. *If $g(h) = O(|h|^2)$, then $j_*(\nabla\bar{u}) \in H^1$.*
4. *If g is $C^{1,1}$, then $g(h) = O(|h|^2)$ and $j_*(\nabla\bar{u}) \in H^1$.*
5. *If $f \in W^{1,p'}_\diamond(\Omega)$, then $g \in C^{1,1}$ and hence $j_*(\nabla\bar{u}) \in H^1$.*

Proof First, we compute for arbitrary v and u admissible in the primal and dual problems (i.e. we need $\nabla \cdot v = f$), the sum $A(v) + B(u)$:

$$A(v) + B(u) = \int_{\mathbb{T}^d} (H(v) + H^*(\nabla u) + fu)\,dx = \int_{\mathbb{T}^d} (H(v) + H^*(\nabla u) - v \cdot \nabla u)\,dx$$

$$\geq c \int_{\mathbb{T}^d} |j(v) - j_*(\nabla u)|^2\,dx.$$

If we take $v = \bar{v}$ and $u = \bar{u}$, then $A(v) = \min A$, $B(u) = \min B$ and $A(v) + B(u) = 0$. Hence, we deduce $j(\bar{v}) = j_*(\nabla\bar{u})$, i.e. Point 1. in the statement.

Now, let us fix $v = \bar{v}$ but $u = u_h$. We obtain

$$\int_{\mathbb{T}^d} |j_*(\nabla\bar{u}) - j_*(\nabla u_h)|^2\,dx = \int_{\mathbb{T}^d} |j(\bar{v}) - j_*(\nabla u_h)|^2\,dx \leq \frac{A(\bar{v}) + B(u_h)}{c}$$

$$= \frac{B(u_h) - B(\bar{u})}{c}.$$

In computing $B(u_h) - B(\bar{u})$, we see that the terms $\int_{\mathbb{T}^d} H^*(\nabla u_h)$ and $\int_{\mathbb{T}^d} H^*(\nabla \bar{u})$ are equal, as one can see from an easy change-of-variable $x \mapsto x + h$. Hence,

$$B(u_h) - B(\bar{u}) = \int_{\mathbb{T}^d} f u_h \, dx - \int_{\mathbb{T}^d} f \bar{u} \, dx = g(h),$$

which gives Point 2.

Point 3 of the statement is an easy consequence of the classical characterization of Sobolev spaces (see Box 2.1). Point 4 comes from the optimality of $\bar{u}$, which means that we have $g(0) = 0$ and $g(h) \geq 0$ for all h. This implies, as soon as $g \in C^{1,1}$, $\nabla g(0) = 0$ and hence $g(h) = O(|h|^2)$.

For Point 5, we first differentiate $g(h)$, thus getting

$$\nabla g(h) = \int_{\mathbb{T}^d} f(x) \nabla \bar{u}(x + h) \, dx.$$

If we want to differentiate once more, we use the regularity assumption on f: we write

$$\int_{\mathbb{T}^d} f(x) \nabla \bar{u}(x + h) \, dx = \int_{\mathbb{T}^d} f(x - h) \nabla \bar{u}(x) \, dx$$

and then

$$D^2 g(h) = \int_{\mathbb{T}^d} \nabla f(x - h) \otimes \nabla \bar{u}(x) \, dx,$$

which also gives $|D^2 g| \leq ||\nabla f||_{L^{p'}} ||\nabla \bar{u}||_{L^p}$. Note that $\bar{u}$ naturally belongs to $W^{1,p}$, hence the integral is finite and bounded, and $g \in C^{1,1}$. □

Unfortunately, the last assumption ($f \in W^{1,p'}$) is quite restrictive, but we want to provide a case where it is reasonable to use it. First, we find interesting cases of functions H and H^* for which we can provide non-trivial functions j and j_*.

4.4.1 Pointwise Vector Inequalities

The first interesting case is the quadratic case. Take $H(v) = \frac{1}{2}|v|^2$ with $H^*(w) = \frac{1}{2}|w|^2$. In this case we have easily

$$H(v) + H^*(w) = \frac{1}{2}|v|^2 + \frac{1}{2}|w|^2 = v \cdot w + \frac{1}{2}|v - w|^2,$$

hence one can take $j(v) = v$ and $j_*(w) = w$.

Now we move to the general case.

Lemma 4.36 *Consider a function H which is uniformly convex, i.e. $v \mapsto H(v) - \frac{c}{2}|v|^2$ is convex for some $c > 0$ (or, equivalently, $D^2 H \geq cI$). In this case we have*

$$H(v) + H^*(w) \geq v \cdot w + \frac{c}{2}|v - \nabla H^*(w)|^2.$$

Proof We start from the equality

$$H(\nabla H^*(w)) + H^*(w) = \nabla H^*(w) \cdot w \qquad (4.7)$$

together with

$$H(v) \geq H(\nabla H^*(w)) + w \cdot (v - \nabla H^*(w)) + \frac{c}{2}|v - \nabla H^*(w)|^2, \qquad (4.8)$$

which is a consequence of $w \in \partial H(\nabla H^*(w))$ (note that H is not necessarily C^1, while H^* is indeed C^1 because H is strictly convex). If we sum (4.7) and (4.8) we obtain the claim. □

Remark 4.37 If we use Lemma 4.35 for a function H satisfying the assumptions of Lemma 4.36 we can use $j(v) = v$ and $j_*(w) = \nabla H^*(w)$. We obtain the equality $\mathbf{v} = \nabla H^*(\nabla \bar{u})$ for the optimizers $\mathbf{v}$ and $\bar{u}$. We deduce the condition $\nabla \cdot (\nabla H^*(\nabla \bar{u})) = f$, which is exactly the Euler–Lagrange equation for the variational problem solved by $\bar{u}$.

We now move to another interesting, though more difficult case, the case of powers. Take $H(v) = \frac{1}{p}|v|^p$ with $H^*(w) = \frac{1}{p'}|w|^{p'}$. We claim that in this case we can take $j(v) = v^{p/2}$ and $j_*(w) = w^{p'/2}$ (remember the notation for powers of vectors).

Lemma 4.38 *For any $v, w \in \mathbb{R}^d$ we have*

$$\frac{1}{p}|v|^p + \frac{1}{p'}|w|^{p'} \geq v \cdot w + \frac{1}{2\max\{p, p'\}}|v^{p/2} - w^{p'/2}|^2.$$

Proof First we write $a = v^{p/2}$ and $b = w^{p'/2}$ and we express the inequality in terms of a, b. Hence we try to prove $\frac{1}{p}|a|^2 + \frac{1}{p'}|b|^2 \geq a^{2/p} \cdot b^{2/p'} + \frac{1}{2\max\{p,p'\}}|a - b|^2$. In this way the inequality is more homogeneous, as it is of order 2 in all its terms (remember $1/p + 1/p' = 1$). Then we notice that we can also write the expression in terms of $|a|, |b|$ and $\cos\theta$, where θ is the angle between a and b (which is the same as the one between $v = a^{2/p}$ and $w = b^{2/p'}$). Hence, we want to prove

$$\frac{1}{p}|a|^2 + \frac{1}{p'}|b|^2 \geq \cos\theta \left(|a|^{2/p}|b|^{2/p'} - \frac{1}{\max\{p, p'\}}|a||b| \right)$$

$$+ \frac{1}{2\max\{p, p'\}}(|a|^2 + |b|^2).$$

Since this depends linearly on $\cos\theta$, it is enough to prove the inequality in the two limit cases $\cos\theta = \pm 1$.

For simplicity, due to the symmetry in p and p' of the claim, we assume $p \leq 2 \leq p'$. We start from the case $\cos\theta = 1$, i.e. $b = ta$, with $t \geq 0$ (the case $a = 0$ is trivial). In this case the l.h.s. of the inequality becomes

$$|a|^2\left(\frac{1}{p}+\frac{1}{p'}t^2\right) = |a|^2\left(\frac{1}{p}+\frac{1}{p'}(1+(t-1))^2\right)$$

$$= |a|^2\left(1+\frac{2}{p'}(t-1)+\frac{1}{p'}(t-1)^2\right)$$

$$\geq |a|^2\left(t^{2/p'}+\frac{1}{p'}(t-1)^2\right),$$

where we used the concavity of $t \mapsto t^{2/p'}$, which provides $1+\frac{2}{p'}(t-1) \geq t^{2/p'}$. This inequality is even stronger than the one we wanted to prove, as we get a factor $1/p'$ instead of $1/(2p')$ in the r.h.s.

The factor $1/(2p')$ appears in the case $\cos\theta = -1$, i.e. $b = -ta$, $t \geq 0$ (we do not claim that this coefficient is optimal). In this case we start from the r.h.s.

$$|a|^2\left(\frac{1}{2p'}(1+t)^2-t^{2/p'}\right)\leq|a|^2\frac{1}{2p'}(1+t)^2\leq|a|^2\frac{2}{2p'}(1+t^2)\leq|a|^2\left(\frac{1}{p}+\frac{1}{p'}t^2\right),$$

which gives the claim.   □

As a more involved variant of the power cost functions, we also consider some other convex functions which will be natural in the study of very degenerate elliptic PDEs and in congestion models.

Consider the case $H(v) = |v| + \frac{1}{p}|v|^p$. In this case, we can use $j(v) = v^{p/2}$ and $j_*(w) = (w-1)_+^{p'/2}$ (again, we use this weird notation: the vector $(w-1)_+^\alpha$ is the vector with norm equal to $(|w|-1)_+^\alpha$ and the same direction as w, i.e. $j_*(w) = (|w|-1)_+^\alpha w/|w|)$.

Indeed, we have

$$H^*(w) = \sup_v v\cdot w - |v| - \frac{1}{p}|v|^p = \frac{1}{p'}(|w|-1)_+^{p'}$$

and

$$H(v) + H^*(w) = |v| + \frac{1}{p}|v|^p + \frac{1}{p'}(|w|-1)_+^{p'}$$

$$\geq |v| + v\cdot(w-1)_+ + c|v^{p/2}-(w-1)_+^{p'/2}|^2.$$

We only need to prove $|v| + v \cdot (w - 1)_+ \geq v \cdot w$. This can be done by writing

$$|v| + v \cdot (w - 1)_+ = |v|(1 + (|w| - 1)_+ \cos \theta).$$

If $|w| \geq 1$ then we go on with

$$|v|(1 + (|w| - 1)_+ \cos \theta) \geq |v| \cos \theta (1 + (|w| - 1)_+) = |v| \cos \theta |w| = v \cdot w.$$

If $|w| \leq 1$ then we simply use

$$|v|(1 + (|w| - 1)_+ \cos \theta) = |v| \geq v \cdot w.$$

4.4.2 Applications to (Degenerate) Elliptic PDEs

If we look at the case $H(v) = \frac{1}{p'}|v|^{p'}$, we have $H^*(w) = \frac{1}{p}|w|^p$ and the solutions of $\Delta_p u = f$ (where $\Delta_p u := \nabla \cdot ((\nabla u)^{p-1})$) are the minimizers of $\int \frac{1}{p}|\nabla u|^p + fu$. We already mentioned in Sect. 2.4 some regularity issues about p-harmonic functions, and the same classical references [30, 131] also provide many results about solutions of $\Delta_p u = f$. In this section we first underline the simplest result that we can obtain from the consideration of the previous sections and these duality methods. It is indeed easy to obtain the following.

Proposition 4.39 *Assume that Ω is the flat torus and $\Delta_p \bar{u} = f \in W^{1,p'}(\Omega)$. Then $(\nabla \bar{u})^{p/2} \in H^1$.*

Proof This statement can be proven by combining Lemma 4.35 and Lemma 4.38.

$\square$

Remark 4.40 The above result is very classical (see for instance [140]), even if usually obtained using a slightly different technique. In particular, the pointwise inequality of Lemma 4.38 replaces, in this duality-based approach, the usual vector inequality that PDE methods require to handle equations involving Δ_p, i.e.

$$(w_0^{p-1} - w_1^{p-1}) \cdot (w_0 - w_1) \geq c|w_0^{p/2} - w_1^{p/2}|^2,$$

which is an improved version of the monotonicity of the gradient of $w \mapsto \frac{1}{p}|w|^p$.

Instead, if we consider $H(v) = |v| + \frac{1}{p'}|v|^{p'}$, we get the following result.

Proposition 4.41 *Let H be given by $H(v) = |v| + \frac{1}{p'}|v|^{p'}$ and $H^*(w) = \frac{1}{p}(|w| - 1)_+^p$. Assume that Ω is the flat torus and $f \in W^{1,p'}(\Omega)$. Let $\bar{v}$ be a solution of $\min A$ and $\bar{u}$ a solution of $\min B$ (equivalently, assume that $\bar{u}$ solves $\nabla \cdot ((\nabla \bar{u} - 1)_+^{p-1}) = f$). Then $\bar{v}^{p'/2} = (\nabla \bar{u} - 1)_+)^{p/2} \in H^1$.*

This result is the same as that proven in [44], which used PDE methods, and does not seem easy to improve. The equation $\nabla \cdot ((\nabla u - 1)_+^{p-1}) = f$, which can be written

$$\nabla \cdot \left((|\nabla u| - 1)_+^{p-1} \frac{\nabla u}{|\nabla u|} \right) = f,$$

is very degenerate in the sense that the coefficient $(|\nabla u| - 1)_+^{p-1}/|\nabla u|$ vanishes on the whole set where $|\nabla u| \leq 1$.

This equation and the corresponding minimization problems arise in traffic congestion (see [22, 44, 63] and Sect. 4.6) and the choice of the function H is very natural: we need a superlinear function of the form $H(v) = |v|h(|v|)$, with $h \geq 1$. This automatically implies the degeneracy.

We now move to the Poisson equation $\Delta u = f$, corresponding to the minimization of $\int \frac{1}{2}|\nabla u|^2 + fu$, and hence to $H(v) = \frac{1}{2}|v|^2$ and $H^*(w) = \frac{1}{2}|w|^2$. It is possible to treat this case by the same techniques as in the degenerate case above, but the result is disappointing. Indeed, from these techniques we just obtain $f \in H^1 \Rightarrow \nabla u \in H^1$, while it is well-known that $f \in L^2$ should be enough for the same result. Yet, with some more attention it is also possible to treat the L^2 case.

Proposition 4.42 *Assume that Ω is the flat torus and $\Delta \bar{u} = f \in L^2(\Omega)$. Then $\nabla \bar{u} \in H^1$.*

Proof We use the variational framework we presented before, with $H(v) = \frac{1}{2}|v|^2$. We then obtain

$$\frac{1}{2}\|\nabla u_h - \nabla \bar{u}\|_{L^2}^2 \leq g(h). \tag{4.9}$$

Now, set $\omega_t := \sup\{\|\nabla u_h - \nabla u\|_{L^2} : |h| \leq t\}$. From (4.9) we have

$$\omega_t^2 \leq \sup_{h:|h| \leq t} 2g(h) \leq 2t \sup_{h:|h| \leq t} |\nabla g(h)|.$$

From $\nabla g(h) = \nabla g(h) - \nabla g(0) = \int f(\nabla u_h - \nabla \bar{u})$ we deduce

$$|\nabla g(h)| \leq \|f\|_{L^2}\|\nabla u_h - \nabla \bar{u}\|_{L^2} \leq \|f\|_{L^2}\omega_t,$$

hence $\omega_t^2 \leq 2t\|f\|_{L^2}\omega_t$, which implies $\omega_t \leq 2t\|f\|_{L^2}$ and hence $\nabla \bar{u} \in H^1$. □

As we already pointed out, the result stating that solutions u of $\Delta_p \bar{u} = f \in W^{1,p'}(\Omega)$ satisfy $(\nabla \bar{u})^{p/2} \in H^1$ is very classical but not very satisfactory in the limit case $p = 2$. This is why we also look at the following other classical result. Before stating it, we recall some useful definitions of fractional Sobolev spaces.

Box 4.3 Important Notion—*Fractional Sobolev Spaces*
When Ω is bounded and its diameter is R, if $1 < p < +\infty$ and $0 < s < 1$, the space $W^{s,p}(\Omega)$ is defined as

$$W^{s,p}(\Omega) = \left\{ u \in L^p(\Omega) : [u]_{s,p}^p := \int_{B(0,R)} \frac{\|u_\delta - u\|_{L^p(\Omega_\delta)}^p}{\delta^{d+sp}} \, d\delta < +\infty \right\},$$

where $\Omega_\delta = \{x \in \Omega : x + \delta \in \Omega\}$. The norm on $W^{s,p}(\Omega)$ is given by $\|u\|_{L^p} + [u]_{s,p}$. The space H^s is defined as $W^{s,2}$.

Note that an inequality of the form $\|u_\delta - u\|_{L^p} \leq C|\delta|^s$ implies $u \in W^{s',p}$ for every $s' < s$. Also note that the Hilbert case $p = 2$ also enjoys an alternative definition in terms of the Fourier transform. Indeed, we have $u \in H^s$ if and only if $\xi \mapsto |\xi|^s \hat{u}(\xi) \in L^2$ and $[u]_{s,2}$ is equivalent to the L^2 norm of $\xi \mapsto |\xi|^s \hat{u}(\xi)$.

We refer to [1] or [77] for more details.

Proposition 4.43 *Assume that Ω is the flat torus $\mathbb{T}^d$ and $\Delta_p \bar{u} = f \in L^{p'}(\Omega)$, with $p > 2$. Then $\|(\nabla u_h)^{p/2} - (\nabla \bar{u})^{p/2}\|_{L^2} \leq C|h|^{p'/2}$, which implies in particular $(\nabla \bar{u})^{p/2} \in H^s$ for $s < p'/2 < 1$.*

Proof We use the same strategy as in Proposition 4.42. For simplicity, we set $G := (\nabla u)^{p/2}$. As in Proposition 4.42, we set $\omega_t := \sup_{h:|h| \leq t} \|G_h - G\|_{L^2}$. We have $\|G_h - G\|_{L^2}^2 \leq Cg(h)$, which implies

$$\omega_t^2 \leq Ct \sup_{h:|h| \leq t} |\nabla g(h) - \nabla g(0)| \leq Ct\|f\|_{L^{p'}} \sup_{h:|h| \leq t} \|\nabla u_h - \nabla \bar{u}\|_{L^p}.$$

From the α-Hölder behavior of the vector map $w \mapsto w^\alpha$ in $\mathbb{R}^d$ (see Lemma 4.44 below), with $\alpha = 2/p < 1$, we deduce, using $\nabla u = G^\alpha$,

$$\|\nabla u_h - \nabla \bar{u}\|_{L^p}^p = \int_{\mathbb{T}^d} |\nabla u_h - \nabla \bar{u}|^p \, dx \leq \int_{\mathbb{T}^d} |G_h - G|^2 \, dx = \|G_h - G\|_{L^2}^2.$$

Hence, we have

$$\omega_t^2 \leq Ct\|f\|_{L^{p'}} \omega_t^{2/p},$$

which implies

$$\omega_t^{2/p'} \leq Ct\|f\|_{L^{p'}},$$

i.e. the claim $\qquad\qquad\qquad\qquad\qquad\qquad\qquad\qquad\qquad\qquad\qquad\quad\square$

Lemma 4.44 *For $0 < \alpha < 1$, the map $w \mapsto w^\alpha$ is α-Hölder continuous in $\mathbb{R}^d$.*

Proof Let $a, b \in \mathbb{R}^d$. We write

$$|a^\alpha - b^\alpha| = \left| |a|^\alpha \frac{a}{|a|} - |a|^\alpha \frac{b}{|b|} + |a|^\alpha \frac{b}{|b|} - |b|^\alpha \frac{b}{|b|} \right| \leq |a|^\alpha \left| \frac{a}{|a|} - \frac{b}{|b|} \right| + \left| |a|^\alpha - |b|^\alpha \right|.$$

For the second term in the r.h.s., we use the α-Hölder behavior of $t \mapsto t^\alpha$ in $\mathbb{R}_+$ and get

$$\left| |a|^\alpha - |b|^\alpha \right| \leq \left| |a| - |b| \right|^\alpha \leq |a - b|^\alpha.$$

For the first term in the r.h.s., we use the inequality

$$\left| \frac{a}{|a|} - \frac{b}{|b|} \right| = \left| \frac{a}{|a|} - \frac{b}{|a|} + \frac{b}{|a|} - \frac{b}{|b|} \right| \leq \frac{|a - b|}{|a|} + |b| \frac{\left| |b| - |a| \right|}{|a||b|} \leq 2 \frac{|a - b|}{|a|}$$

and get

$$|a|^\alpha \left| \frac{a}{|a|} - \frac{b}{|b|} \right| \leq 2|a|^{\alpha-1}|a - b|.$$

If we choose a to be such $|a| \geq |b|$ (which is possible w.l.o.g.), we have $2|a| \geq |a - b|$ and hence $2^{\alpha-1}|a|^{\alpha-1} \leq |a - b|^{\alpha-1}$, i.e. $2|a|^{\alpha-1}|a - b| \leq 2^{2-\alpha}|a - b|^\alpha$.

Summing up, we have (without claiming that this constant is optimal)

$$|a^\alpha - b^\alpha| \leq (2^{2-\alpha} + 1)|a - b|^\alpha.$$

$\square$

Remark 4.45 Note that the result of Proposition 4.43 is also classical, and quite sharp. Indeed, one can informally consider the following example. Take $u(x) \approx |x|^r$ as $x \approx 0$ (and then multiply by a cut-off function out of 0). In this case we have

$$\nabla u(x) \approx |x|^{r-1}, \quad (\nabla u(x))^{p-1} \approx |x|^{(r-1)(p-1)}, \quad f(x) := \Delta_p u(x) \approx |x|^{(r-1)(p-1)-1}.$$

Hence, $f \in L^{p'}$ if and only if $((r-1)(p-1)-1)p' > -d$, i.e. $(r-1)p - p' > -d$. On the other hand, the fractional Sobolev regularity can be observed by considering that "differentiating s times" means subtracting s from the exponent, hence

$$(\nabla u(x))^{p/2} \approx |x|^{\frac{p(r-1)}{2}} \Rightarrow (\nabla u)^{\frac{p}{2}} \in H^s \Leftrightarrow |x|^{\frac{p(r-1)}{2}-s} \in L^2 \Leftrightarrow p(r-1)-2s > -d.$$

If we want this last condition to be true for arbitrary $s < p'/2$, then it amounts to $p(r-1) - p' > -d$, which is the same condition as above.

4.4.3 Variants: Local Regularity and Dependence on x

In the previous section we only provided global Sobolev regularity results on the torus. This guaranteed that we could do translations without boundary problems, and that by change-of-variable, the term $\int H(\nabla u_h)\,dx$ did not actually depend on h. We now provide a result concerning local regularity. As the result is local, boundary conditions should not be very important. Yet, as the method itself is global, we need to fix them and be precise about the variational problems that we use. We will consider, as usual, variational problems without any boundary constraint, which correspond to PDEs with no-flux boundary conditions. Since, as we said, boundary conditions should play no role, with some work it is also possible to modify them, but we will not develop these considerations here.

We will only provide the following result, in the easiest case $p = p' = 2$.

Theorem 4.46 *Let H, H^*, j and j_* satisfy Hyp1, 2, 3 with $q = 2$. Assume $f \in H^1$. Assume also $H^* \in C^{1,1}$ and $j_* \in C^{0,1}$. Assume $\nabla \cdot (\nabla H^*(\nabla \bar{u})) = f$ in Ω with no-flux boundary conditions on $\partial \Omega$. Then, $j_*(\nabla \bar{u}) \in H^1_{loc}$.*

Proof The condition $\nabla \cdot \nabla H^*(\nabla \bar{u}) = f$ is equivalent to the fact that $\bar{u}$ is solution of

$$\min \left\{ \int_\Omega H^*(\nabla u)\,dx + \int_\Omega fu\,dx \ : \ u \in W^{1,p}(\Omega) \right\}.$$

We will follow the usual duality strategy as in the rest of the section. Yet, in order not to have boundary problems, we need to use a cut-off function $\eta \in C_c^\infty(\Omega)$ and define

$$u_h(x) = \bar{u}(x + h\eta(x))$$

(note that for small h we have $x + h\eta(x) \in \Omega$ for all $x \in \Omega$). In this case it is no longer true that we have $\tilde{g}(h) := \int_\Omega H^*(\nabla u_h)\,dx = \int_\Omega H^*(\nabla \bar{u})\,dx$. If this term is not constant in h, then we need to prove that it is a $C^{1,1}$ function of h. To do this, and to avoid differentiating $\nabla \bar{u}$, we use a change-of-variable. Set $y = x + h\eta(x)$. We have $\nabla(u_h)(x) = (\nabla \bar{u})(y)(I + h \otimes \nabla \eta(x))$, hence

$$\tilde{g}(h) = \int_\Omega H^*(\nabla u_h)\,dx = \int_\Omega H^*(\nabla \bar{u}(y) + (\nabla \bar{u}(y) \cdot h)\nabla \eta(x)) \frac{1}{1 + h \cdot \nabla \eta(x)}\,dy,$$

where $x = X(h, y)$ is a function of h and y obtained by inverting $x \mapsto x + h\eta(x)$ and we used $\det(I + h \otimes \nabla \eta(x)) = 1 + h \cdot \nabla \eta(x)$. The function X is C^∞ by the implicit function theorem, and all the other ingredient of the above integral are at least $C^{1,1}$ in h. This proves that $\tilde{g}$ is $C^{1,1}$. The regularity of the term $g(h) = \int_\Omega fu_h$ should also be considered. Differentiating once we get $\nabla g(h) = \int_\Omega f(x)\nabla \bar{u}(x + h\eta(x))\eta(x)\,dx$. To differentiate once more, we use the same change-of-variable, thus getting

$$\nabla g(h) = \int_\Omega f(X(h, y)) \nabla \bar{u}(y) \eta(X(h, y)) \frac{1}{1 + h \cdot \nabla \eta(x)} \, dy.$$

From $y = X(h, y) + h\eta(X(h, y))$ we get a formula for $D_h X(h, y)$, i.e. $0 = D_h X(h, y) + \eta(X(h, y))I + h \otimes \nabla \eta(\eta(X(h, y)) D_h X(h, y)$. This allows us to differentiate once more the function g and proves $g \in C^{1,1}$.

Finally, we come back to the duality estimate. What we can easily get is

$$c\| j_*(\nabla(u_h)) - j_*(\nabla \bar{u}) \|_{L^2}^2 \le g(h) + \tilde{g}(h) = O(|h|^2).$$

A small difficulty remains in the fact that $j_*(\nabla(u_h))$ is not the translation of $j_*(\nabla \bar{u})$. Yet, if we take a subdomain $\Omega' \subset \Omega$ such that $(\Omega')_r \subset \{\eta = 1\}$, then for every h with $|h| < r$ it is true that, on Ω', the function $j_*(\nabla(u_h))$ is the translation of $j_*(\nabla \bar{u})$ by a vector h. This allows us to conclude, as in Lemma 4.35 (4), and obtain $j_*(\nabla \bar{u}) \in H^1_{loc}$. □

The duality theory has been presented in the case where H and H^* could also depend on x, while for the moment regularity results have only be presented under the assumption that they do not. We now consider a small variant and look at how to handle the following particular case, corresponding to the minimization problem

$$\min \left\{ \frac{1}{p} \int_\Omega a(x) |\nabla u(x)|^p \, dx + \int_\Omega f(x) u(x) \, dx \; : \; u \in W^{1,p}(\Omega) \right\}. \tag{4.10}$$

We will use $\Omega = \mathbb{T}^d$ to avoid accumulating difficulties (boundary issues and dependence on x). Note that the PDE corresponding to the above minimization problem is $\nabla \cdot (a(\nabla u)^{p-1}) = f$.

First, we need to compute the transform of $w \mapsto H^*(w) := \frac{a}{p}|w|^p$. Set $b = a^{1/(p-1)}$. It is easy to obtain $H(v) = \frac{1}{bp'}|v|^{p'}$. Also, we can check (just by scaling the inequality of Lemma 4.38), that we have

$$\frac{1}{bp'}|v|^{p'} + \frac{b^{p-1}}{p}|w|^p \ge v \cdot w + b^{p-1} \left| w^{p/2} - \frac{v^{p'/2}}{b^{p/2}} \right|^2.$$

In particular, if we assume that $a(x)$ is bounded from below by a positive constant, and we set $H^*(x, w) = \frac{a(x)}{p}|w|^p$, then we get

$$H(x, v) + H^*(x, w) \ge v \cdot w + c|j(x, v) - j_*(w)|^2$$

where $j_*(w) = w^{p/2}$.

We can now prove the following theorem.

Theorem 4.47 *Assume* $f \in W^{1,p'}(\mathbb{T}^d)$ *and* $a \in \text{Lip}, a \geq a_0 > 0,$ *and let* $\bar{u}$ *be the minimizer of (4.10). Then* $G := \nabla \bar{u}^{p/2} \in H^1(\mathbb{T}^d)$.

Proof Our usual computations show that we have

$$c\|G_h - G\|_{L^2}^2 \leq g(h) + \tilde{g}(h),$$

where $g(h) = \int f u_h - \int f \bar{u}$ and $\tilde{g}(h) = \int \frac{a(x)}{p} |\nabla u_h|^p - \int \frac{a(x)}{p} |\nabla \bar{u}|^p$. With our assumptions, $g \in C^{1,1}$. As for $\tilde{g}(h)$, we write

$$\int_{\mathbb{T}^d} \frac{a(x)}{p} |\nabla u_h|^p = \int_{\mathbb{T}^d} \frac{a(x-h)}{p} |\nabla \bar{u}|^p$$

and hence

$$\nabla \tilde{g}(h) = \int_{\mathbb{T}^d} \frac{\nabla a(x-h)}{p} |\nabla \bar{u}|^p = \int_{\mathbb{T}^d} \frac{\nabla a(x)}{p} |\nabla u_h|^p.$$

Hence,

$$|\nabla \tilde{g}(h) - \nabla \tilde{g}(0)| \leq \int_{\mathbb{T}^d} \frac{|\nabla a(x)|}{p} \left| |\nabla u_h|^p - |\nabla \bar{u}|^p \right|$$

$$\leq C \int_{\mathbb{T}^d} \left| |G_h|^2 - |G|^2 \right| \leq C\|G_h - G\|_{L^2}\|G_h + G\|_{L^2}.$$

Here we used the L^∞ bound on $|\nabla a|$. Then, from the lower bound on a, we also know $G \in L^2$, hence we get $|\nabla \tilde{g}(h) - \nabla \tilde{g}(0)| \leq C\|G_h - G\|_{L^2}$.

Now, we define as usual $\omega_t := \sup_{h=|h| \leq t} \|G_h - G\|_{L^2}$ and we get

$$\omega_t^2 \leq C \sup_{h=|h| \leq t} g(h) + \tilde{g}(h) \leq Ct \sup_{h=|h| \leq t} |\nabla g(h) + \nabla \tilde{g}(h)|$$

$$= Ct \sup_{h=|h| \leq t} |\nabla g(h) - \nabla g(0) + \nabla \tilde{g}(h) - \nabla \tilde{g}(0)| \leq Ct^2 + Ct\omega_t,$$

which allows us to deduce $\omega_t \leq Ct$ and hence $G \in H^1$. $\qquad\square$

We also provide the following theorem, which is also interesting for $p = 2$.

Theorem 4.48 *Assume* $p \geq 2, f \in L^{p'}(\mathbb{T}^d)$ *and* $a \in \text{Lip}, a \geq a_0,$ *and let* $\bar{u}$ *be the minimizer of (4.10). Then* $G = \nabla \bar{u}^{p/2}$ *satisfies* $\|G_h - G\|_{L^2} \leq C|h|^{p'/2}$. *In particular,* $G \in H^1(\mathbb{T}^d)$ *if* $p = 2$ *and* $G \in H^s(\mathbb{T}^d)$ *for all* $s < p'/2$ *if* $p > 2$.

Proof The only difference with the previous case is that we cannot say g is $C^{1,1}$ but we should stick to the computation of ∇g. We use as usual

$$|\nabla g(h) - \nabla g(0)| \leq ||f||_{L^{p'}} ||\nabla u_h - \nabla \bar{u}||_{L^p}.$$

As we are forced to let the norm $||\nabla u_h - \nabla \bar{u}||_{L^p}$ appear, we will use it also in $\tilde{g}$. Indeed, observe that we can estimate

$$|\nabla \tilde{g}(h) - \nabla \tilde{g}(0)| \leq \int_{\mathbb{T}^d} \frac{|\nabla a(x)|}{p} \left| |\nabla u_h|^p - |\nabla \bar{u}|^p \right|$$

$$\leq C \int_{\mathbb{T}^d} (|\nabla u_h|^{p-1} + |\nabla \bar{u}|^{p-1}) |\nabla u_h - \nabla \bar{u}| \leq C ||\nabla \bar{u}^{p-1}||_{L^{p'}} ||\nabla u_h - \nabla \bar{u}||_{L^p}.$$

We then use $||\nabla \bar{u}^{p-1}||_{L^{p'}} = ||\nabla \bar{u}||_{L^p}^{p-1}$ and conclude

$$|\nabla \tilde{g}(h) - \nabla \tilde{g}(0)| \leq C ||\nabla u_h - \nabla \bar{u}||_{L^p}.$$

This gives, defining ω_t as usual,

$$\omega_t \leq Ct \sup_{h=|h|\leq t} |\nabla g(h) - \nabla g(0) + \nabla \tilde{g}(h) - \nabla \tilde{g}(0)| \leq Ct \sup_{h=|h|\leq t} ||\nabla u_h - \nabla \bar{u}||_{L^p}$$

and hence

$$\omega_t^2 \leq Ct\omega_t^{2/p}$$

as in Proposition 4.43.   □

4.5 A Proof of the Fenchel–Rockafellar Duality Theorem

In this section we want to take advantage of the technique presented in Sect. 4.3 for the precise case of minimal flow problems in order to prove a general abstract version of the Fenchel–Rockafellar duality theorem. For simplicity and in order to understand the technique, we will start from the case where all spaces are reflexive. In the following theorem we consider two reflexive spaces X and $\mathcal{Y}$ and use duality pairs including them and their duals.

Theorem 4.49 *Let X and $\mathcal{Y}$ be two reflexive normed vector spaces, $f : X \to \mathbb{R} \cup \{+\infty\}$ and $g : \mathcal{Y} \to \mathbb{R} \cup \{+\infty\}$ two convex and lower semicontinuous functions, and $A : X \to \mathcal{Y}$ a continuous linear mapping. Assume that g is bounded from below* *and f is coercive. Then we have*

$$\min \{f(x) + g(Ax) \, : \, x \in X\} = \sup \{-g^*(\xi) - f^*(-A^t\xi) \, : \, \xi \in Y'\},$$

where the existence of the minimum on the left-hand side is part of the claim.

Proof We define a function $\mathcal{F} : Y \to \mathbb{R} \cup \{+\infty\}$ by

$$\mathcal{F}(p) := \min \{f(x) + g(Ax + p) \, : \, x \in X\}.$$

The existence of the minimum is a consequence of the following fact: for any sequence (x_n, p_n) with $f(x_n) + g(Ax_n + p_n) \leq C$ the sequence x_n is bounded. This boundedness comes from the lower bound on g and from the coercive behavior of f. Once we know this, we can use $p_n = p$, take a minimizing sequence x_n for fixed p, and extract a weakly convergent subsequence $x_n \rightharpoonup x$ using the Banach–Alaoglu theorem in reflexive spaces. We also have $Ax_n + p \rightharpoonup Ax + p$ and the semicontinuity of f and g provide the minimality of x (since, being convex, f and g are both l.s.c. for the strong and the weak convergence, in X and Y, respectively).

We now compute $\mathcal{F}^* : Y' \to \mathbb{R} \cup \{+\infty\}$:

$$\mathcal{F}^*(\xi) = \sup_p \, \langle \xi, p \rangle - \mathcal{F}(p)$$

$$= \sup_{p,x} \langle \xi, p \rangle - f(x) - g(Ax + p)$$

$$= \sup_{y,x} \langle \xi, y - Ax \rangle - f(x) - g(y)$$

$$= \sup_y \langle \xi, y \rangle - g(y) + \sup_x \langle -A^t\xi, x \rangle - f(x)$$

$$= g^*(\xi) + f^*(-A^t\xi).$$

Now we use, as we did in Sect. 4.3, the double transform $\mathcal{F}^{**}$. In this case, as the variable in the function $\mathcal{F}$ is a perturbation p which has been added to the problem and we are interested in the case $p = 0$, we will use $\mathcal{F}^{**}(0) = \sup -\mathcal{F}^*$. The claim is proven as soon as we prove that $\mathcal{F}$ is convex and l.s.c. so that $\mathcal{F}^{**}(0) = \mathcal{F}(0)$.

The convexity of $\mathcal{F}$ is easy. We just need to take $p_0, p_1 \in Y$, and define $p_t := (1 - t)p_0 + tp_1$. Let x_0, x_1 be optimal in the definition of $\mathcal{F}(p_0)$ and $\mathcal{F}(p_1)$, i.e. $f(x_i) + g(Ax_i + p_i) = \mathcal{F}(p_i)$, and set $x_t := (1 - t)x_0 + tx_1$. We have

$$\mathcal{F}(p_t) \leq f(x_t) + g(Ax_t + p_t) \leq (1 - t)\mathcal{F}(p_0) + t\mathcal{F}(p_1),$$

and the convexity is proven.

For the semicontinuity, we take a sequence $p_n \to p$ in Y. We can assume $\mathcal{F}(p_n) \leq C$ otherwise there is nothing to prove. Take the corresponding optimal points x_n and, applying the very first observation of this proof, we obtain $\|x_n\| \leq C$. We can extract a subsequence such that $\lim_k \mathcal{F}(p_{n_k}) = \liminf_n \mathcal{F}(p_n)$ and $x_{n_k} \rightharpoonup x$.

The semicontinuity of f and g provides

$$\mathcal{F}(p) \leq f(x) + g(Ax + p) \leq \liminf_k f(x_{n_k}) + g(Ax_{n_k} + p_{n_k})$$

$$= \lim_k \mathcal{F}(p_{n_k}) = \liminf_n \mathcal{F}(p_n),$$

which gives the desired result.   □

Remark 4.50 The use of a perturbation p as a tool to prove duality results is nowadays classical, and very useful as it allows the use of min-max exchange theorems to be avoided, which in general require some compactness assumptions which are difficult to prove. In Theorem 4.33 instead of using a perturbation p we directly defined a function on the value of the constraint f, but when the function g is not an indicator function this is not a doable procedure. Another interesting feature of the perturbation approach is that the set of solutions of the dual problem is equal to the set $\partial \mathcal{F}(0)$ (simply because the subdifferential at 0 of a convex function $\mathcal{F}$ coincides with the set of points where $\partial \mathcal{F}^*$ contains 0, i.e. of the minimizers of $\mathcal{F}^*$). This is the point of view of [62, 81], for instance.

We now note that, if g is not bounded from below, it is always possible to make it so by removing a suitable linear function from it, since all convex and l.s.c. functions are bounded from below by an affine function. We can then define $\tilde{g}$ by $\tilde{g}(y) = g(y) - \langle \xi_0, y \rangle$ for a suitable ξ_0, and guarantee $\inf \tilde{g} > -\infty$. In order not to change the value of the primal problem we also need to modify f into $\tilde{f}$ defined by $\tilde{f}(x) := f + \langle A^t \xi_0, x \rangle$, so that

$$\tilde{f}(x) + \tilde{g}(Ax) = f(x) + \langle A^t \xi_0, x \rangle + g(Ax) - \langle \xi_0, Ax \rangle = f(x) + g(Ax).$$

Moreover, we can compute what changes in the dual problem. Is it true that we have $\sup_\xi -g^*(\xi) - f^*(-A^t \xi) = \sup_\xi -\tilde{g}^*(\xi) - \tilde{f}^*(-A^t \xi)$?

In order to do this, we need to compute the Legendre transform of $\tilde{f}$ and $\tilde{g}$. A general and easy fact, which is proposed as an exercise (see Exercise 4.9), states that subtracting a linear function translates into a translation on the Legendre transform. We then have

$$\tilde{g}^*(\xi) = g^*(\xi + \xi_0); \qquad \tilde{f}^*(\zeta) = f^*(\zeta - A^t \xi_0)$$

and then

$$\tilde{g}^*(\xi) + \tilde{f}^*(-A^t \xi) = g^*(\xi + \xi_0) + f^*(-A^t(\xi + \xi_0))$$

and a simple change of variable $\xi \mapsto \xi + \xi_0$ shows that the sup has not changed. This shows that the duality result is not affected by this reformulation in terms of $\tilde{f}$ and $\tilde{g}$. It is then enough, for the duality to hold, that the assumptions of Theorem 4.49 are satisfied by $(\tilde{f}, \tilde{g})$ instead of (f, g). Since we chose ξ_0 on purpose in order to have $\tilde{g}$ lower bounded, we only need now to require that $\tilde{f}$ is coercive. Not that this would be the case if f was superlinear, as it would stay superlinear after adding any linear function, but it is not automatic when speaking of a generic coercive function.

The condition on ξ_0 such that, at the same time, $\tilde{g}$ is bounded from below and $\tilde{f}$ superlinear can be more easily translated in terms of f^* and g^*.

Yet, we will directly state the main result of the section, which is the most standard formulation of the Fenchel–Rockafellar duality theorem. This requires us to switch the roles of f and f^*, g and g^*, and of the spaces. In this statement we also remove the reflexivity assumption that we used, for simplicity, in the previous statement.

Theorem 4.51 *Let X and Y be two normed vector spaces, and suppose that Y is separable. Let $f : X \to \mathbb{R} \cup \{+\infty\}$ and $g : Y \to \mathbb{R} \cup \{+\infty\}$ be two convex and lower semicontinuous functions, and $A : X \to Y$ a continuous linear mapping. Assume that there exists an $x_0 \in X$ such that $f(x_0) < +\infty$ and that g is continuous and finite at Ax_0. Then we have*

$$\inf\{f(x) + g(Ax) : x \in X\} = \max\left\{-g^*(\xi) - f^*(-A^t\xi) : \xi \in Y'\right\},$$

where the existence of the maximum on the right-hand side is part of the claim.

Proof First of all, we note that the assumption on x_0 allows us to write

$$f^*(\eta) \geq \langle \eta, x_0 \rangle - f(x_0),$$

as well as

$$g^*(\xi) \geq \langle \xi, Ax_0 \rangle + c_0\|\xi\| - c_1,$$

for suitable constants $c_0, c_1 > 0$. This second inequality comes from the fact that $\xi \mapsto g^*(\xi) - \langle \xi, Ax_0 \rangle$ is coercive, as a consequence of the fact that its Fenchel–Legendre transform is locally bounded around 0. We then have

$$g^*(\xi) + f^*(-A^t\xi + \eta) \geq \langle \xi, Ax_0 \rangle + c_0\|\xi\| - c_1 + \langle -A^t\xi, x_0 \rangle + \langle \eta, x_0 \rangle - f(x_0)$$
$$\geq c_0\|\xi\| - C(\|\eta\|).$$

This proves that, for every sequence (ξ_n, η_n) such that $g^*(\xi_n) + f^*(-A^t\xi_n + \eta_n) \leq C$, if $\|\eta_n\|$ is bounded then $\|\xi_n\|$ is also bounded. Using the Banach–Alaoglu theorem, since the sequence ξ_n lies in the dual Y' of a separable space, we can also extract a weakly-$*$ convergent subsequence from ξ_n.

We now consider the function $\mathcal{F}$ defined on X' via

$$\mathcal{F}(\eta) := \min\{g^*(\xi) + f^*(-A^t\xi + \eta) : \xi \in Y'\}.$$

The minimum in this definition exists because every minimizing sequence is compact for the weak-* convergence, so that we have $\xi_{n_k} \overset{*}{\rightharpoonup} \xi$ and $-A^t \xi_{n_k} + \eta) \overset{*}{\rightharpoonup} -A^t \xi + \eta)$, and both f^* and g^* are weakly-* lower semicontinuous, as is always the case for Fenchel–Legendre transforms defined on dual spaces.

We then compute the Fenchel–Legendre transform of $\mathcal{F}$ using the duality pair with X (and not X''):

$$\mathcal{F}^*(x) := \sup_{\xi} \langle x, \eta \rangle - \mathcal{F}(\eta)$$

$$= \sup_{\xi, \eta} \langle x, \eta \rangle - g^*(\xi) - f^*(-A^t \xi + \eta)$$

$$= \sup_{\xi, \zeta} \langle x, (\zeta + A^t \xi) \rangle - g^*(\xi) - f^*(\zeta)$$

$$= f^{**}(x) + g^{**}(Ax)$$

$$= f(x) + g(Ax),$$

where we used the change of variable $\zeta = \eta - A^t \xi$ and then the equalities $f^{**} = f$ and $g^{**} = g$ (because f and g are convex and l.s.c.). We then conclude by using $\mathcal{F}^{**}(0) = \mathcal{F}(0)$ as usual, but we need to prove that $\mathcal{F}$ is convex and lower semicontinuous.

The proof of convexity follows the same lines as in Theorem 4.49. For the semicontinuity, we take $\eta_n \overset{*}{\rightharpoonup} \eta$ and we observe that we have $\|\eta_n\| \leq C$. We can assume, up to extracting a subsequence, that $\lim_n \mathcal{F}(\eta_n)$ exists and is finite. If we call ξ_n some minimizers in the definition of $\mathcal{F}(\eta_n)$ we have $g^*(\xi_n) + f^*(-A^t \xi_n + \eta_n) \leq C$. We deduce that $\|\xi_n\|$ is bounded and we can extract a weakly-* convergent subsequence $\xi_{n_k} \overset{*}{\rightharpoonup} \xi$. We then have

$$\mathcal{F}(\eta) \leq g^*(\xi) + f^*(-A^t \xi + \eta) \leq \liminf_k g^*(\xi_{n_k}) + f^*(-A^t \xi_{n_k} + \eta_{n_k})$$

$$= \lim_n \mathcal{F}(\eta_n),$$

which proves the lower semicontinuity.   □

4.6   Discussion: From Optimal Transport to Congested Traffic and Mean Field Games

In this section we want to underline the connections of some of the problems studied in this chapter with problems from optimal transport theory, and with their variants involving traffic congestion.

We start with a very brief presentation of the optimal transport problem, for which we refer for instance to the books [195] and [177].

The starting point is the following problem proposed by the French mathematician Gaspard Monge in 1781, [149], that we present here in modern language. Given two metric spaces X and Y, two probability measures $\mu \in \mathcal{P}(X)$, $\nu \in \mathcal{P}(Y)$, and a cost function $c : X \times Y \to \mathbb{R}$, we look for a map $T : X \to Y$ pushing the first one onto the other, i.e. satisfying $T_{\#}\mu = \nu$, and minimizing the integral

$$M(T) := \int_X c(x, T(x)) \, d\mu(x).$$

Box 4.4 Memo—*Image Measures*

An important notion in measure theory is that of image measure. Given a measure $\mu \in \mathcal{M}(X)$ on a space X and a measurable function $T : X \to Y$, we define the *image measure* or *push-forward* of μ through T as a measure $T_{\#}\mu$ on Y characterized by

$$(T_{\#}\mu)(A) := \mu(T^{-1}(A))$$

for every measurable set $A \subset Y$. Equivalently, we can require

$$\int_Y \phi \, d(T_{\#}\mu) := \int_X \phi \circ T \, d\mu$$

for every bounded and measurable function (or, by approximation, for every bounded and continuous function) $\phi : Y \to \mathbb{R}$. The case of sets can be recovered using $\phi = \mathbb{1}_A$. Note that image measures pass to the limit through a.e. convergence (if $T_n \to T$ μ-a.e. then $(T_n)_{\#}\mu \overset{*}{\rightharpoonup} T_{\#}\mu$) but not through weak convergence (for instance consider $X = [0, 2\pi]$, $\mu = \mathcal{L}^1$, $T_n(x) = \sin(nx)$ and $T(x) = 0$).

In the problem considered by Monge X and Y were subsets of Euclidean space, μ and ν were absolutely continuous, and $c(x, y) = |x - y|$ was the Euclidean distance. Roughly speaking, this means that we have a collection of particles, we know how they are distributed (this is the measure μ), and they have to be moved so that they will be arranged according to a new prescribed distribution ν. The displacement has to be chosen so as to minimize the average displacement, and the map T describes the movement.

Monge proved many properties of the optimal transport map T in the case he was interested in, but never cared to prove that such a map existed. In this sense, the problem remained unsolved until the reformulation that Leonid Kantorovich gave in 1942, [124]. His formulation consists in the problem

$$(K) \qquad \min\left\{ \int_{X\times Y} c \, d\gamma \; : \; \gamma \in \Pi(\mu, \nu) \right\}. \qquad (4.11)$$

Here $\Pi(\mu, \nu)$ is the set of the so-called *transport plans*, i.e. $\Pi(\mu, \nu) = \{\gamma \in \mathcal{P}(X \times Y) : (\pi_x)_\#\gamma = \mu, (\pi_y)_\#\gamma = \nu, \}$ where π_x and π_y are the two projections of $X \times Y$ onto the two factors X and Y. These probability measures over $X \times Y$ are an alternative way to describe the displacement of the particles of μ: instead of saying, for each x, which is the destination $T(x)$ of the particle originally located at x, we say for each pair (x, y) how many particles go from x to y. It is clear that this description allows for more general movements, since from a single point x particles can a priori move to different destinations y. If multiple destinations really occur, then this movement cannot be described through a map T.

If we define the map $(id, T) : X \to X \times Y$ by $(id, T)(x) := (x, T(x))$, it can be easily checked that $\gamma_T := (id, T)_\#\mu$ belongs to $\Pi(\mu, \nu)$ if and only if T pushes μ onto ν and the functional $\int c \, d\gamma_T$ takes the form $\int c(x, T(x)) \, d\mu(x)$, thus generalizing Monge's problem.

Kantorovich's generalized problem is much easier to handle than the original one proposed by Monge and the direct method of the calculus of variations proves that a minimum does exist (at least when c is l.s.c. and bounded from below). This is not the case for the original Monge problem, since in general we can only obtain weakly convergent minimizing sequences $T_n \rightharpoonup T$ but the limit T could have a different image measure than ν. Hence, the general strategy to prove the existence of an optimizer in the Monge problem consists in first considering the minimizer γ of the Kantorovich problem and then trying to prove that it is actually of the form $(id, T)_\#\mu$. This is possible (see, for instance, [177, Chapter 1]) under some conditions on μ and on the cost c. Anyway, we will not consider this question here. Another important fact which makes the Kantorovich problem easier to consider is that it is convex (it is a linear optimization problem under linear constraints) and hence an important tool will be the theory of convex duality.

As we already did in this chapter, we first find a formal dual problem, by means of an inf-sup exchange. With this aim, we first express the constraint $\gamma \in \Pi(\mu, \nu)$ in the following way: notice that, if γ is a non-negative measure on $X \times Y$, then we have

$$\sup_{\substack{\phi \in C(X), \\ \psi \in C(Y)}} \int_X \phi \, d\mu + \int_Y \psi \, d\nu - \int_{X\times Y} (\phi(x) + \psi(y)) \, d\gamma = \begin{cases} 0 & \text{if } \gamma \in \Pi(\mu, \nu) \\ +\infty & \text{otherwise} \end{cases}.$$

Hence, we can remove the constraints on γ if we add the previous sup. Then, we look at the problem we get by interchanging the inf in γ and the sup in ϕ, ψ:

$$\sup_{\phi,\psi} \int_X \phi \, d\mu + \int_Y \psi \, d\nu + \inf_{\gamma} \int_{X \times Y} (c(x, y) - (\phi(x) + \psi(y))) \, d\gamma.$$

We can then re-write the inf in γ as a constraint on ϕ and ψ, since we have

$$\inf_{\gamma \geq 0} \int_{X \times Y} (c(x, y) - (\phi(x) + \psi(y))) \, d\gamma = \begin{cases} 0 & \text{if } \phi \oplus \psi \leq c, \\ -\infty & \text{otherwise} \end{cases},$$

where $\phi \oplus \psi$ denotes the function $(x, y) \mapsto \phi(x) + \psi(y)$.

The validity of the exchange of inf and sup can be proven with the techniques that we developed in Sects. 4.3 and 4.5. A precise proof can be found, for instance, in Sect. 1.6 of [177], based on the fact that it is (magically) possible to prove a compactness result for the functions ϕ and ψ when X and Y are compact and c is continuous (indeed, it is possible to restrict to functions sharing the same modulus of continuity as c). A different proof, based on the reduction to a particular case of the Fenchel–Rockafellar duality theorem, is presented in [195].

We then have the following dual optimization problem:

$$(\text{D}) \qquad \max \left\{ \int_X \phi \, d\mu + \int_Y \psi \, d\nu \; : \; \phi \in C(X), \psi \in C(Y), \phi \oplus \psi \leq c \right\}.$$

$$(4.12)$$

We now look in particular at the case where c is the Euclidean distance (i.e. $X = Y \subset \mathbb{R}^d$ and $c(x, y) = |x - y|$; for the case where the cost is a strictly convex function of $x - y$ we refer to Box 4.5 below). A first observation concerning the dual problem is that for any ϕ one can choose the best possible (the largest) function ψ which satisfies, together with ϕ, the constraint $\phi(x) + \psi(y) \leq c(x, y)$. Such a function is given by

$$\psi(y) = \inf_x |x - y| - \phi(x)$$

and it belongs to Lip_1. This means that we can restrict the maximization in the dual problem to $\psi \in \text{Lip}_1$ and, analogously, to $\phi \in \text{Lip}_1$. Furthermore, if we now know $\phi \in \text{Lip}_1$, there is an easy expression for the function $\psi(y) = \inf_x |x - y| - \phi(x)$, which is just $\psi = -\phi$ (see Exercise 4.22). In this case the dual problem becomes

$$(\text{D} - \text{dist}) \qquad \max \left\{ \int_X \phi \, d(\mu - \nu) \; : \; \phi \in \text{Lip}_1(X) \right\}.$$

Box 4.5 Good to Know!—*The Brenier Theorem in Optimal Transport*

The use of primal-dual optimality conditions in optimal transport allows us to say that for any optimal γ in (K) and any optimal pair (ϕ, ψ) in (D) we have $\phi(x) + \psi(y) = c(x, y)$ for all $(x, y) \in \text{spt}(\gamma)$, while in the other points we have an inequality. When $\nabla\phi(x)$ and $\nabla_x c(x, y)$ exist, this implies $\nabla\phi(x) = \nabla_x c(x, y)$. If c is of the form $c(x, y) = h(x - y)$ for a function h which is strictly convex and C^1, this gives $y = x - \nabla h^*(\nabla\phi(x)) := T(x)$, which proves the existence of a transport map T. The only delicate assumption is the differentiability of ϕ, but ϕ can be assumed to share the modulus of continuity of c, which is Lipschitz continuous on compact sets, hence $\nabla\phi$ exists a.e. This applies to all costs of the form $h(z) = c|z|^p$ for $p > 1$, for instance.

In the very particular case $h(z) = \frac{1}{2}|z|^2$ we have $\nabla h^* = id$. We then obtain $T(x) = x - \nabla\phi(x) = \nabla u(x)$, where $u(x) = \frac{1}{2}|x|^2 - \phi(x)$. Moreover, by the trick of choosing the best possible ϕ given ψ, we can assume that ϕ is of the form

$$\phi(x) = \inf_y \frac{1}{2}|x - y|^2 - \psi(y).$$

This can be written ad $u = v^*$, where $v(y) = \frac{1}{2}|y|^2 - \psi(y)$ and shows that the optimal transport map for the quadratic cost is the gradient of a convex function (see [45, 46]). This applies to the case where convex functions are differentiable μ-a.e., in particular if μ is absolutely continuous. Moreover, when μ and ν have densities f and g respectively, one can write the condition $T_\#\mu = \nu$ in the form of a change-of-variable formula involving the determinant of the Jacobian: $|\det(DT(x))| = f(x)/g(T(x))$ a.e. In the case $T = \nabla u$ and u convex this becomes the Monge–Ampère equation

$$\det(D^2 u) = \frac{f}{g(\nabla u)}.$$

This is a second-order non-linear elliptic PDE which would need to be completed by some boundary conditions. What replaces the boundary conditions is the condition $\nabla u(\text{spt } f) \subset \text{spt } g$.

Besides Brenier's original papers [45, 46] we refer to [177, Chapter 1].

Another important aspect of the optimal transport problem when the cost is given by the Euclidean distance is its connection with minimal-flow problems. Let us concentrate on the case of a compact connected domain $\Omega \subset \mathbb{R}^d$ and consider

$$\text{(B)} \qquad \min \left\{ ||\mathbf{v}|| \; : \; \mathbf{v} \in \mathcal{M}^d(\Omega); \; \nabla \cdot \mathbf{v} = \mu - \nu \right\}, \qquad (4.13)$$

where $||\mathbf{v}||$ denotes the mass of the vector measure $\mathbf{v}$ and the divergence condition is to be read in the weak sense, with no-flux boundary conditions, i.e. $-\int \nabla \phi \cdot d\mathbf{v} = \int \phi \, d(\mu - \nu)$ for any $\phi \in C^1(\Omega)$. Indeed, if Ω is convex then the value of this problem is equal to that of the optimal transport problem with $c(x, y) = |x - y|$ (if Ω is not convex then the Euclidean distance has to be replaced with the geodesic distance inside Ω)

This minimal flow problem was first proposed by Beckmann in [22], under the name of the *continuous transportation model*. This problem is actually strongly related to the Kantorovich problem with the distance cost $c(x, y) = |x - y|$, as we will show in a while, although Beckmann was not aware of this, the two theories being developed essentially at the same time.

In order to see the connection between the two problems, we start from a formal computation. We re-write the constraint on $\mathbf{v}$ in (4.13) by means of the equality

$$\sup_{\phi} \int_{\Omega} -\nabla \phi \cdot d\mathbf{v} + \int_{\Omega} \phi \, d(\mu - \nu) = \begin{cases} 0 & \text{if } \nabla \cdot \mathbf{v} = \mu - \nu \\ +\infty & \text{otherwise} \end{cases}.$$

Hence one can write (4.13) as

$$\min_{\mathbf{v}} ||\mathbf{v}|| + \sup_{\phi} \int_{\Omega} -\nabla \phi \cdot d\mathbf{v} + \int_{\Omega} \phi \, d(\mu - \nu)$$

$$= \sup_{\phi} \int_{\Omega} \phi \, d(\mu - \nu) + \inf_{\mathbf{v}} ||\mathbf{v}|| - \int_{\Omega} \nabla \phi \cdot d\mathbf{v},$$

where inf and sup have been exchanged formally as in the previous computations. Then, one notices that we have

$$\inf_{\mathbf{v}} ||\mathbf{v}|| - \int_{\Omega} \nabla \phi \cdot d\mathbf{v} = \inf_{\mathbf{v}} \int_{\Omega} d|\mathbf{v}| \left(1 - \nabla \phi \cdot \frac{d\mathbf{v}}{d|\mathbf{v}|} \right) = \begin{cases} 0 & \text{if } |\nabla \phi| \leq 1 \\ -\infty & \text{otherwise} \end{cases}$$

and this leads to the dual formulation for (B), which gives

$$\sup_{\phi \, : \, |\nabla \phi| \leq 1} \int_{\Omega} \phi \, d(\mu - \nu).$$

Since this problem is exactly the same as (D-dist) (a consequence of the fact that Lip_1 functions are exactly those functions whose gradient is smaller than 1, when the domain is convex), this gives the equivalence between (B) and (K) (when we use $c(x, y) = |x - y|$ in (K)).

The proof of the duality for (B) is proposed as an exercise (Exercise 4.23), in the more general case

$$\text{(B–K)} \qquad \min\left\{ \int_\Omega K\, d|\mathbf{v}| \; : \; \mathbf{v} \in \mathcal{M}^d(X); \; \nabla \cdot \mathbf{v} = \mu - \nu \right\},$$

where a given continuous weight $K : X \to \mathbb{R}_+$ is integrated against the total variation measure $|\mathbf{v}|$. In this case the dual is given by

$$\sup\left\{ \int_\Omega \phi\, d(\mu - \nu) \; : \; \phi \in \text{Lip}, \; |\nabla \phi| \leq K \right\}$$

and the corresponding Kantorovich problem is the one with cost given by $c = d_K$, the weighted distance with weight K (see Sect. 1.4.3), since the functions ϕ satisfying $|\nabla \phi| \leq K$ are exactly those which are Lip_1 w.r.t. this distance.

As a transition to congestion problems, we will now see how to provide a new equivalent formulation of (K) and (B) with geodesic cost functions.

We will use absolutely continuous curves $\omega : [0, 1] \mapsto \Omega$. Given $\omega \in \text{AC}([0, 1]; \Omega)$ and a continuous function ϕ, we write

$$L_\phi(\omega) := \int_0^1 \phi(\omega(t))|\omega'(t)|\, dt,$$

which coincides with the weighted length if $\phi \geq 0$. We will write C (the space of "curves") for $\text{AC}([0, 1]; \Omega)$, and consider probability measures Q on the space C, which is endowed with the uniform convergence. Note that the Ascoli–Arzelà theorem guarantees that the sets $\{\omega \in C : \text{Lip}(\omega) \leq \ell\}$ are compact for every ℓ. We will associate with Q two measures on Ω. The first is a scalar one, called the *traffic intensity* and denoted by $i_Q \in \mathcal{M}_+(\Omega)$; it is defined (see [63]) by

$$\int_\Omega \phi\, di_Q := \int_C \left(\int_0^1 \phi(\omega(t))|\omega'(t)|\, dt \right) dQ(\omega) = \int_C L_\phi(\omega)\, dQ(\omega),$$

for all $\phi \in C(\Omega, \mathbb{R}_+)$. The interpretation is the following: for a subregion A, $i_Q(A)$ represents the total cumulated traffic in A induced by Q, i.e. for every path we compute "how long" it stays in A, and then we average on paths.

We also associate with any traffic plan $Q \in \mathcal{P}(C)$ a vector measure $\mathbf{v}_Q$ via

$$\forall \xi \in C(\Omega; \mathbb{R}^d) \qquad \int_\Omega \xi \cdot d\mathbf{v}_Q := \int_C \left(\int_0^1 \xi(\omega(t)) \cdot \omega'(t)\, dt \right) dQ(\omega).$$

We will call $\mathbf{v}_Q$ the *traffic flow* induced by Q. Taking a gradient field $\xi = \nabla\phi$ in the previous definition yields

$$\int_\Omega \nabla\phi \cdot d\mathbf{v}_Q = \int_C [\phi(\omega(1)) - \phi(\omega(0))] \, dQ(\omega) = \int_\Omega \phi \, d((e_1)_\# Q - (e_0)_\# Q)$$

(e_t denotes the evaluation map at time t, i.e. $e_t(\omega) := \omega(t)$) so that, if we set $(e_0)_\# Q = \mu$ and $(e_1)_\# Q = \nu$, we have

$$\nabla \cdot \mathbf{v}_Q = \mu - \nu$$

in the usual sense with no-flux boundary conditions.

It is easy to check that we have $|\mathbf{v}_Q| \leq i_Q$, where $|\mathbf{v}_Q|$ is the total variation measure of the vector measure $\mathbf{v}_Q$. This last inequality is in general not an equality, since the curves of Q could produce some cancellations.

A re-formulation of the optimal transport problem with cost d_K can then be given in the following way:

$$\min\left\{ \int_C L_K(\omega) \, dQ(\omega) \; : \; (e_0)_\# Q = \mu, \; (e_1)_\# Q = \nu \right\}.$$

It it then possible to show that, given an optimal Q, the measure $\gamma := (e_0, e_1)_\# Q$ is an optimal transport plan in (K) for the cost $c = d_K$, and the vector measure $\mathbf{v}_Q$ is optimal in (B-K).

Yet, this language allows us to consider many other interesting situations, and in particular the case of congested traffic, where the weight function K is in general not given a priori but depends on the traffic intensity itself. We present here the extension to the continuous framework of the notion of Wardrop equilibrium (a very common definition of equilibrium in traffic problems, introduced in [196], and generally used on networks) proposed in [63].

Congestion effects are captured by the metric associated with Q via its traffic intensity: assume $i_Q \ll \mathcal{L}^d$ and set

$$K_Q(x) := g(x, i_Q(x))$$

for a given increasing function $g(x, .) : \mathbb{R}_+ \to \mathbb{R}_+$. We then consider the weighted length L_{K_Q}, as well as the corresponding weighted distance d_{K_Q}, and define a Wardrop equilibrium as a measure $Q \in \mathcal{P}(C)$ such that

$$Q(\{\omega \; : \; L_{K_Q}(\omega) = d_{K_Q}(\omega(0), \omega(1))\}) = 1. \tag{4.14}$$

Of course this requires some technicalities, to take into account the case where i_Q is not absolutely continuous or its density is not smooth enough.

This is a typical situation in game theory, and a particular case of *Nash equilibrium*: a configuration where every agent makes a choice, has a cost depending on his own choice and on the others' choices, and no agent will change his mind after knowing what the others chose. Here the choice of every agent consists of a trajectory ω, the configuration of choices is described via a measure Q on the set of choices (since all agents are indistinguishable, we only care about how many of them made each choice), the cost which is paid is $L_{K_Q}(\omega)$, which depends on the global configuration of choices Q and, of course, on ω, and we require that Q is concentrated on optimal trajectories, so that nobody will change their mind.

Box 4.6 Important Notion—*Nash Equilibria*

Definition Consider a game where several players $i = 1, \ldots, n$ must choose a strategy among a set of possibilities S_i and assume that the each player tries to optimize (say, minimize) a result depending on what everybody chooses. More precisely, for every i we have a function $f_i : S_1 \times \cdots \times S_n \to \mathbb{R}$ and we say that a configuration $(s_1, \ldots, s_n)$ (where $s_i \in S_i$) is an equilibrium (a *Nash equilibrium*) if, for every i, the choice s_i optimizes $S_i \ni s \mapsto f_i(s_1, \ldots, s_{i-1}, s, s_{i+1}, \ldots, s_n)$ (i.e. s_i is optimal for player i under the assumption that the other players freeze their choice).

Nash equilibria need not exist in all situations, but Nash proved (via convex analysis and fixed point arguments) that they always exist when we consider the so-called *mixed strategies*. This means that we accept that every player, instead of choosing an element $s_i \in S_i$, only chooses a probability on S_i and then randomly picks a strategy according to the law he has chosen.

The notion of Nash equilibrium, first introduced by J. Nash in [157, 158] in the case of a finite number of players, can be easily extended to a continuum of players where each one is negligible compared to the others (*non-atomic games*). Considering for simplicity the case of identical players, we have a common space S of possible strategies and we look for a measure $Q \in \mathcal{P}(S)$. This measure induces a payoff function $f_Q : S \to \mathbb{R}$ and we want the following condition to be satisfied: there exists a $C \in \mathbb{R}$ such that $f_Q(x) = C$ for Q-a.e. x, and $f_Q(x) \geq C$ everywhere (if the players want to minimize the payoff f_Q, otherwise, if it has to be maximized, we impose $f_Q(x) \leq C$), i.e. f_Q must be optimal Q-a.e.

For all notions related to game theory we refer for instance to the classical book [162].

A typical problem in road traffic is to find a Wardrop equilibrium Q with prescribed transport plan $\gamma = (e_0, e_1)_\# Q$ (what is usually called an *origin-destination* matrix in the engineering community, in the discrete case). Another possibility is to only prescribe its marginals $\mu = (e_0)_\# Q$ and $\nu = (e_1)_\# Q$. More generally, we impose a constraint $(e_0, e_1)_\# Q \in \Gamma \subset \mathcal{P}(\Omega \times \Omega)$. A way to find such equilibria is the following.

Let us consider the (convex) variational problem

$$(W) \qquad \min\left\{\int_\Omega H(x, i_Q(x)) \ dx \ : \ Q \in \mathcal{P}(C), (e_0, e_1)_\# Q \in \Gamma\right\} \qquad (4.15)$$

where $H'(x, \cdot) = g(x, \cdot)$, $H(x, 0) = 0$. Under some technical assumptions, the main result of [63] is that (W) admits at least one minimizer, and that such a minimizer is a Wardrop equilibrium (which is consistent with what proven in [23] in the discrete case on a network). Hence, the Wardop equilibrium problem is a *potential game* (see [148]) as one can find an equilibrium by minizing an overall functional. Moreover, $\gamma_Q := (e_0, e_1)_\# Q$ solves the optimization problem

$$\min\left\{\iint_{\Omega\times\Omega} d_{K_Q}(x, y) \ d\gamma(x, y) \ : \ \gamma \in \Gamma\right\}.$$

In particular, if Γ is a singleton, this last condition does not play any role (there is only one competitor) and we have the existence of a Wardrop equilibrium corresponding to any given transport plan γ. If, on the contrary, $\Gamma = \Pi(\mu, \nu)$, then the second condition means that γ solves a Monge–Kantorovich problem for a distance cost depending on Q itself, which is a new equilibrium condition.

In the case where $\Gamma = \Pi(\mu, \nu)$ it is possible to prove that (W) is indeed equivalent to a variational divergence constrained problem *à la* Beckmann, i.e.

$$(B - cong) \qquad \min\left\{\int_\Omega H(x, |\mathbf{v}(x)|) \ dx \ : \ \nabla \cdot \mathbf{v} = \mu - \nu\right\}, \qquad (4.16)$$

which is a variational problem of the form of those studied in Sects. 4.3 and 4.4. It is then possible to prove that the optimizers of this problem are the vector fields of the form $\mathbf{v}_Q$ where Q solves (W).

We now switch to optimal transport problems with other costs, and in particular $c(x, y) = |x - y|^p$. Since these costs do not satisfy the triangle inequality the dual formulation is more involved than that in (D-dist) and no Beckmann formulation exists.[5] Instead, there exists a dynamic formulation, known as the *Benamou–Brenier* problem, [24]. We can indeed consider the optimization problem

$$(BB_p) \qquad \min\left\{\int_0^1 \int_\Omega |v_t|^p \ d\varrho_t \ dt \ : \ \partial_t\varrho_t + \nabla \cdot (\varrho_t v_t) = 0, \ \varrho_0 = \mu, \ \varrho_1 = \nu\right\}$$

---

[5] We mention anyway [120] which reformulates optimal transport problems with costs which are convex but not 1-homogeneous in $x - y$ into transport costs with 1-homogeneous costs, for which a Beckmann formulation exists, after adding a time variable.

which consists in minimizing the integral in time and space of a p-variant of the kinetic energy among solutions of the continuity equation with prescribed initial and final data.

Box 4.7 Good to Know!—*The Continuity Equation in Mathematical Physics*

Assume that a family of particles moves according to a velocity field $\mathbf{v}$, depending on time and space: $\mathbf{v}(t, x)$ stands for the velocity at time t of any particle which is located at point x at such an instant of time. The position of the particle originally located at x will be given by the solution of the ODE

$$\begin{cases} y'_x(t) = \mathbf{v}_t(y_x(t)) \\ y_x(0) = x. \end{cases} \tag{4.17}$$

We then define the map Y_t through $Y_t(x) = y_x(t)$, and, if we are given the distribution of particles at $t = 0$, we look for the measure $\varrho_t := (Y_t)_{\#}\varrho_0$. We can then prove that ϱ_t and $\mathbf{v}_t$ solve together the so-called continuity equation

$$\partial_t \varrho_t + \nabla \cdot (\varrho_t \mathbf{v}_t) = 0$$

(as usual, with no-flux boundary condition on $\partial\Omega$ if we assume that we have $y_x(t) \in \Omega$ for all (t, x)). For the use of this equation in optimal transport theory and related topics we refer to [177, Chapter 4].

It is possible to prove that the minimal value in (BB_p) equals that of (K) with cost $|x - y|^p$ and that the optimal solutions of (BB_p) are obtained by taking an optimal transport map T for this same cost in the Monge problem, defining $T_t := (1-t)id+tT$, and setting $\varrho_t = (T_t)_{\#}\mu$ and $\mathbf{v}_t = (T-id) \circ (T_t)^{-1}$. A corresponding formula may be given using the optimizers of (K) if they are not given by transport maps.

Concerning Problem (BB_p), we observe that it is not a convex optimization problem in the variables $(\rho, \mathbf{v})$, because of the product term $\rho|\mathbf{v}|^p$ in the functional and the product $\rho\mathbf{v}$ in the differential constraint. But if one changes variables, defining $\mathbf{w} = \rho\mathbf{v}$, then using the variables $(\rho, \mathbf{w})$, the constraint becomes linear and the functional convex. The important point for convexity is that the function

$$\mathbb{R} \times \mathbb{R}^d \ni (s, w) \mapsto \begin{cases} \dfrac{|w|^p}{ps^{p-1}} & \text{if } s > 0, \\ 0 & \text{if } (s, w) = (0, 0), \\ +\infty & \text{otherwise} \end{cases} \tag{4.18}$$

is convex (and it is actually obtained as $\sup\{as + b \cdot w \; : \; a + \frac{1}{p'}|b|^{p'} \leq 0\}$; see Exercise 4.8).

Some analogies and some differences may be underlined between the optimal flow formulation à la Beckmann of the OT problem with $p = 1$ and the dynamic formulation à la Benamou–Brenier of the case $p > 1$. Both involve differential constraints on the divergence, and actually we can also look at the continuity equation as a time-space constraint on $\nabla_{t,x} \cdot (\rho, \mathbf{w})$ making the analogy even stronger. Yet, there is no time in the Beckmann problem, and even when time appears in the Lagrangian formulation with measures Q on curves it is a fictitious time parameter since everything is invariant under reparameterization. On the other hand, time plays an important role in the case $p > 1$. This is also the reason why the models that we will present in a while, for congestion problems where the congestion effect is evaluated at each time, will be so close to the Benamou–Brenier problem.

The models that we want to introduce are known as Mean Field Games (MFG for short). This theory was introduced around 2006 at the same time by Lasry and Lions, [132–134], and by Caines, Huang and Malhamé, [117], in order to describe the evolution of a population in a game with a continuum of players where collective effect of the players' presence on each player recalls what is called in physics a *mean field*. These games are very particular examples of differential games: typically, in a differential game the role of the time variable is crucial since if a player decides to deviate from a given strategy (a notion which is at the basis of the Nash equilibrium definition), the others can react to this change, so that the choice of a strategy is usually not defined as the choice of a path, but of a function selecting a path according to the information the player has at each given time. Yet, when each player is considered as negligible, any deviation he/she performs will have no effect on the other players, so that they will not react. In this way we have a static game where the space of strategies is a space of paths.

We can give a Lagrangian description of the equilibria by using again measures on paths, or an Eulerian description through a system of PDEs, where the key ingredients are the density ρ and the value function φ of the control problem solved by each player, the velocity $\mathbf{v}(t, x)$ of the agents at (t, x) being, by optimality, related to the gradient $\nabla\varphi(t, x)$. For a general overview of MFG theory, see the lecture notes by P. Cardaliaguet [60], based on the course by P.-L. Lions at Collège de France [141].

Let us describe in a more precise way the simplest MFG models. We look at a population of agents moving inside a compact connected domain Ω and assume that every agent chooses his own trajectory ω solving a minimization problem

$$\min \int_0^T \left(\frac{|\omega'(t)|^2}{2} + h[\varrho_t](\omega(t)) \right) dt + \Psi(\omega(T)),$$

with given initial point $x(0)$. Here Ψ stands for the preferences of the agents in terms of their final position, and $h[\varrho](x)$ is the cost of moving around x: the mean-field effect is modeled through the fact that the function h depends on the density ϱ_t of the agents at time t. The dependence of the cost on the velocity ω' could of course be more general than a simple quadratic function.

For now, we consider the evolution of the density ϱ_t as an input, i.e. we assume that agents know it. Hence, we can assume the function h to be given, and we want to study the above optimization problem. The main tool to analyze it, coming from optimal control theory, is the value function φ (see Sect. 1.7): we know that it solves the *Hamilton–Jacobi equation* $-\partial_t \varphi + \frac{1}{2}|\nabla\varphi|^2 = h$ with final datum $\varphi(T, x) = \Psi(x)$ and that the optimal trajectories $\omega(t)$ can be computed using φ, since they are the solutions of

$$\omega'(t) = -\nabla\varphi(t, \omega(t)).$$

This means that the velocity field which advects the particles when each agent follows the optimal curves will be given by $\mathbf{v} = -\nabla\varphi$. If we want to find an equilibrium, the density ϱ_t that we fixed at the beginning should also coincide with that which is obtained by following this optimal velocity field $\mathbf{v}$ and so ϱ should evolve according to the continuity equation with such a velocity field. This means solving the following coupled (HJ)+(CE) system:

$$\begin{cases} -\partial_t\varphi + \frac{|\nabla\varphi|^2}{2} = h[\rho] \\ \partial_t\rho - \nabla \cdot (\rho\nabla\varphi) = 0, \\ \varphi(T, x) = \Psi(x), \quad \rho(0, x) = \rho_0(x). \end{cases} \qquad (4.19)$$

The above system is a PDE, and Eulerian, description of the equilibrium we were looking for. In Lagrangian terms we could express the equilibrium condition in a similar way to what we did for Wardrop equilibria. Consider again the space of curves C and define the kinetic energy functional $K : C \to \mathbb{R}$ given by

$$K(\omega) = \frac{1}{2} \int_0^T |\omega'|^2(t)\, dt.$$

For notational simplicity, we write K_Ψ for the kinetic energy augmented by a final cost: $K_\Psi(\omega) := K(\omega) + \Psi(\omega(T))$; similarly, we denote by $K_{\Psi,h}$ the same quantity when also a running cost h is included: $K_{\Psi,h}(\omega) := K_\Psi(\omega) + \int_0^T h(t, \omega(t))\, dt$.

We now define an MFG equilibrium as a measure $Q \in \mathcal{P}(C)$ such that Q-a.e. the curve ω minimizes $K_{\Psi,h}$ with given initial point when $h(t, \cdot) := h[(e_t)_\# Q]$. Whenever $(x, \varrho) \mapsto h[\varrho](x)$ has some continuity properties it is possible to prove the existence of an equilibrium by the Kakutani fixed point method (see, for instance, [107] or [144], where such a theorem is used to prove the existence of an MFG equilibrium in a slightly different setting).

On the other hand, the assumption that h is continuous does not cover a very natural case, which is the local one, where $h[\rho](x)$ directly depends on the density of ρ at the point x. We can focus for instance on the case $h[\rho](x) = V(x)+g(\rho(x))$, where we identify the measure ρ with its density w.r.t. the Lebesgue measure on Ω. The function $V : \Omega \to \mathbb{R}$ is a potential taking into account different local costs of different points in Ω and the function $g : \mathbb{R}_+ \to \mathbb{R}$ is assumed to be nondecreasing and models congestion effects.

In this case it is not possible to prove the existence of an equilibrium using a fixed point theorem, but luckily there exists a variational formulation. Indeed, we can consider all the possible population evolutions, i.e. pairs $(\rho, \mathbf{v})$ satisfying $\partial_t \rho + \nabla \cdot (\rho \mathbf{v}) = 0$, and we minimize the following energy

$$\mathcal{A}(\rho, \mathbf{v}) := \int_0^T \int_\Omega \left(\frac{1}{2} \rho_t |\mathbf{v}_t|^2 + \rho_t V + G(\rho_t) \right) \, dx \, dt + \int_\Omega \Psi \, d\rho_T,$$

where G is the anti-derivative of g, i.e. $G'(s) = g(s)$ for $s \in \mathbb{R}^+$ with $G(0) = 0$. We fix by convention $G(s) = +\infty$ for $\rho < 0$. Note in particular that G is convex, as its derivative is the nondecreasing function g.

The above minimization problem recalls, in particular when $V = 0$, the Benamou–Brenier dynamic formulation for optimal transport. The main difference with the Benamou–Brenier problem is that here we add to the kinetic energy a congestion cost G; also note that usually in optimal transport the target measure ρ_T is fixed, and here it is part of the optimization (but this is not a crucial difference).

As is often the case in congestion games, the quantity $\mathcal{A}(\rho, \mathbf{v})$ is not the total cost for all the agents. Indeed, the term $\int \int \frac{1}{2} \rho |\mathbf{v}|^2$ is exactly the total kinetic energy, the last term $\int \Psi \, d\rho_T$ is the total final cost, and the cost $\int V \, d\rho_t$ exactly coincides with the total cost induced by the potential V; yet, the term $\int G(\rho)$ is not the total congestion cost, which should instead be $\int \rho g(\rho)$. This means that the equilibrium minimizes an overall energy (we have what is called a *potential game*), but not the total cost; this gives rise to the so-called *price of anarchy*.

In order to convince the reader of the connection between the minimization of $\mathcal{A}(\rho, \mathbf{v})$ and the equilibrium system (4.19), we will use a formal argument from convex duality. A rigorous proof of duality would, by the way, require the problem to be rewritten as a convex optimization problem, with a change of variables, using again $\mathbf{w} = \rho \mathbf{v}$.

We will formally produce a dual problem to $\min \mathcal{A}$ via a min-max exchange procedure. First, we write the constraint $\partial_t \rho + \nabla \cdot (\rho \mathbf{v}) = 0$ in weak form, i.e.

$$\int_0^T \int_\Omega (\rho \partial_t \phi + \nabla \phi \cdot \rho \mathbf{v}) \, dx \, dt + \int_\Omega \phi_0 \, d\rho_0 - \int_\Omega d\phi_T \rho_T = 0 \qquad (4.20)$$

for every function $\phi \in C^1([0, T] \times \Omega)$ (note that we do not impose conditions on the values of ϕ on $\partial \Omega$, hence this is equivalent to completing the continuity equation with a no flux boundary condition $\rho \mathbf{v} \cdot n = 0$).

Using (4.20), we can re-write our problem as

$$\min_{\rho, \mathbf{v}} \mathcal{A}(\rho, \mathbf{v}) + \sup_{\phi} \int_0^T \int_\Omega (\rho \partial_t \phi + \nabla \phi \cdot \rho \mathbf{v}) \, dx \, dt + \int_\Omega \phi_0 \, d\rho_0 - \int_\Omega d\phi_T \rho_T,$$

since the sup in ϕ takes the value 0 if the constraint is satisfied and $+\infty$ if not. We now switch the inf and the sup and look at the maximization over ϕ of

$$\int_\Omega \phi_0 d\rho_0 + \inf_{\rho, \mathbf{v}} \int_\Omega (\Psi - \phi_T) d\rho_T$$

$$+ \int_0^T \int_\Omega \left(\frac{1}{2} \rho_t |\mathbf{v}_t|^2 + \rho_t V + G(\rho_t) + \rho \partial_t \phi + \nabla \phi \cdot \rho \mathbf{v} \right) dx dt.$$

First, we minimize w.r.t. $\mathbf{v}$, thus obtaining $\mathbf{v} = -\nabla \phi$ (on $\{\rho_t > 0\}$) and we replace $\frac{1}{2} \rho |\mathbf{v}|^2 + \nabla \phi \cdot \rho \mathbf{v}$ with $-\frac{1}{2} \rho |\nabla \phi|^2$. Then we get, in the double integral,

$$\inf_\rho \left\{ G(\rho) - \rho(-V - \partial_t \phi + \frac{1}{2} |\nabla \phi|^2) \right\} = -\sup_\rho \{p\rho - G(\rho)\} = -G^*(p),$$

where we set $p := -V - \partial_t \phi + \frac{1}{2} |\nabla \phi|^2$ and G^* is defined as the Legendre transform of G. Then, we observe that the minimization in the final cost simply gives as a result 0 if $\Psi \geq \phi_T$ (since the minimization is only performed among non-negative ρ_T) and $-\infty$ otherwise. Hence we obtain a dual problem of the form

$$\sup \left\{ -\mathcal{B}(\phi, p) := \int_\Omega \phi_0 \, d\rho_0 - \int_0^T \int_\Omega G^*(p) \, dx \, dt \ : \ \begin{aligned} \phi_T &\leq \Psi, \\ -\partial_t \phi + \frac{1}{2} |\nabla \phi|^2 &= V + p \end{aligned} \right\}.$$

Note that the condition $G(s) = +\infty$ for $s < 0$ implies $G^*(p) = 0$ for $p \leq 0$. This in particular means that in the above maximization problem one can assume $p \geq 0$ (indeed, replacing p with p_+ does not change the G^* part, but improves the value of ϕ_0, considered as a function depending on p). The choice of using two variables (ϕ, p) connected by a PDE constraint instead of only ϕ is purely conventional, and it allows for a dual problem which has a sort of symmetry w.r.t. the primal one.

The arguments from convex duality developed in Sect. 4.2.4 would allow us to say that optimal pairs $(\rho, \mathbf{v})$ are obtained by looking at saddle points $((\rho, \mathbf{v}), (\phi, p))$, provided that there is no duality gap between the primal and the dual problems, and that both problems admit a solution. This would mean that, whenever $(\rho, \mathbf{v})$ minimizes $\mathcal{A}$, then there exists a solution (ϕ, p) of the dual problem, such that

- $\mathbf{v}$ minimizes $\frac{1}{2} \rho |\mathbf{v}|^2 + \nabla \phi \cdot \rho \mathbf{v}$, i.e. $\mathbf{v} = -\nabla \phi$ ρ-a.e. This gives (CE): $\partial_t \rho - \nabla \cdot (\rho \nabla \phi) = 0$.
- ρ minimizes $G(\rho) - p\rho$, i.e. $g(\rho) = p$ if $\rho > 0$ or $g(\rho) \geq p$ if $\rho = 0$ (in particular, when we have $g(0) = 0$, we can write $g(\rho) = p_+$); this gives (HJ):

$-\partial_t \phi + \frac{1}{2}|\nabla \phi|^2 = V + g(\rho)$ on $\{\rho > 0\}$ (as the reader can see, there are some subtleties where the mass ρ vanishes).

- ρ_T minimizes $(\Psi - \phi_T)\rho_T$ among $\rho_T \geq 0$. This is not a condition on ρ_T, but rather on ϕ_T: we must have $\phi_T = \Psi$ on $\{\rho_T > 0\}$, otherwise there is no minimizer. This gives the final condition in (HJ).

This provides an informal justification for the equivalence between the equilibrium and the global optimization. It is only informal because we have not discussed if we have duality gaps and whether or not the maximization in (ϕ, p) admits a solution. Moreover, once these issues are clarified, we will still only get a very weak solution to the coupled system (CE)+(HJ). Nothing guarantees that this solution actually encodes the individual minimization problem of each agent.

It is also possible to provide a Lagrangian version of the variational problem, which then has the following form:

$$\min \left\{ J(Q) := \int_C K \, dQ + \int_0^T \mathcal{G}((e_t)_\# Q) \, dt + \int_\Omega \Psi \, d(e_T)_\# Q, \atop Q \in \mathcal{P}(C), (e_0)_\# Q = \rho_0 \right\}, \tag{4.21}$$

where $\mathcal{G} : \mathcal{P}(\Omega) \to \overline{\mathbb{R}}$ is defined by

$$\mathcal{G}(\rho) = \begin{cases} \int_\Omega (V(x)\rho(x) + G(\rho(x))) \, dx & \text{if } \rho \ll \mathcal{L}^d, \\ +\infty & \text{otherwise.} \end{cases}$$

The functional $\mathcal{G}$ is a typical local functional defined on measures (see [35] and Sect. 3.3). It is lower semicontinuous w.r.t. the weak-* convergence of probability measures provided $\lim_{s \to \infty} G(s)/s = +\infty$ (which is the same as $\lim_{s \to \infty} g(s) = +\infty$).

In the Lagrangian language, it is possible to prove that the optimal $\bar{Q}$ satisfies the following optimality conditions: setting $\rho_t = (e_t)_\# \bar{Q}$ and $h(t, x) = g(\rho_t(x))$ (identifying as usual measures and densities), if we take another competitor $\tilde{Q}$, we have

$$J_h(\tilde{Q}) \geq J_h(\bar{Q}),$$

where J_h is the linear functional

$$J_h(Q) = \int_C K \, dQ + \int_0^T \int_\Omega h(t, x)(e_t)_\# Q + \int_\Omega \Psi d(e_T)_\# Q.$$

Since we only consider absolutely continuous $(e_t)_\# Q$, the functional J_h is well-defined for any measurable $h \geq 0$. Formally, we can also write $\int_0^T \int_\Omega h(t,x)(e_t)_\# Q = \int_C dQ \int_0^T h(t,\gamma(t))\, dt$ and hence we get that

$$Q \mapsto \int_C dQ(\gamma) \left(K(\gamma) + \int_0^T h(t,\gamma(t))\, dt + \Psi(\gamma(T)) \right)$$

is minimal for $Q = \bar{Q}$. This corresponds to saying that $\bar{Q}$ is concentrated on curves minimizing $K_{\psi,h}$, hence it is a Lagrangian MFG equilibrium. Unfortunately, this argument is not rigorous at all because of two main difficulties: first, integrating h on a path has no meaning if h is a function which is only defined a.e.; second, in order to prove that $\bar{Q}$ is concentrated on optimal curves one needs to compare it to a competitor $\widetilde{Q}$ which is indeed concentrated on optimal curves, but at the same time we need $(e_t)_\# \widetilde{Q} \ll \mathcal{L}^d$, and these two conditions can be incompatible with each other. These difficulties could be fixed by a density argument if h is continuous; another, more technical approach requires $h \in L^\infty$ by choosing a precise representative. This is explained for instance in [178] and is based on techniques from incompressible fluid mechanics (see [13]). In any case, it becomes crucial to prove regularity results on h or, equivalently, on ρ.

A general L^∞ regularity result is presented in [135] and is based on optimal transport techniques. Here we want, instead, to briefly mention that this question can be attacked using the techniques for regularity via convex duality of Sect. 4.4. By the way, the first use of duality-based methods as a way to prove regularity can indeed be found in time-dependent problems, in [47] (later improved by [12]), in the study of variational models for the incompressible Euler equation. This was later adapted in [61] and [171] to density-constrained or density-penalized mean-field games.

We can then come back to the primal and the dual problems in variational MFG and do our usual computation taking arbitrary $(\rho, \mathbf{v})$ and (ϕ, p) admissible in the primal and dual problem. We compute

$$\mathcal{A}(\rho, \mathbf{v}) + \mathcal{B}(\phi, p) = \int_\Omega (\Psi - \phi_T) d\rho_T$$

$$+ \int_0^T \int_\Omega \left(G(\rho) + G^*(p) - p\rho \right) dx\, dt$$

$$+ \frac{1}{2} \int_0^T \int_\Omega |\mathbf{v} + \nabla\phi|^2 d\rho. \qquad (4.22)$$

Note that we have $(G(\rho) + G^*(p) - p\rho) \geq \frac{\lambda}{2}|\rho - g^{-1}(p)|^2$ where $g = G'$ and $\lambda = \inf G''$. We assume $\lambda > 0$ and, for simplicity, $\Omega = \mathbb{T}^d$. Using

$$\mathcal{A}(\rho, \mathbf{v}) + \mathcal{B}(\phi, p) \geq c \int_0^T \int_\Omega |\rho - g^{-1}(p)|^2\, dx\, dt$$

we can deduce, with the same technique as in Sect. 4.4, $\rho \in H^1$ (we can get both regularity in space and local in time). By the way, using the last term in (4.22), we can also get $\iint \rho |D^2 \phi|^2 < \infty$.

4.7 Exercises

Exercise 4.1 Assume that a functional $F : W^{1,p}(\Omega) \to \mathbb{R} \cup \{+\infty\}$ of the form $F(u) = \int_\Omega L(x, u(x), \nabla u(x)) \, dx$ is convex and that the function $L : \Omega \times \mathbb{R} \times \mathbb{R}^d \to \mathbb{R}$ is continuous in all variables and C^2 in its last variable. Prove that $L(x, s, \cdot)$ is convex for every (x, s).

Exercise 4.2 Given a function $f : X \to \mathbb{R} \cup \{+\infty\}$ and a function $h : X \to \mathbb{R}_+$, prove that we have $(f + \varepsilon h)^* \to f^*$ in the sense of pointwise convergence, as $\varepsilon \to 0^+$.

Exercise 4.3 Prove that the function $f : \mathbb{R}_+ \to \mathbb{R} \cup \{+\infty\}$ given by $f(x) = x \log x$ admits a unique extension to $\mathbb{R}$ which is convex and l.s.c., and compute its Legendre transform.

Exercise 4.4 If $f : \mathbb{R}^N \to \mathbb{R}$ is given by $f(x) = |x| \log |x|$, compute f^* and f^{**}.

Exercise 4.5 Given $p, r \in (1, \infty)$, let $||\cdot||_r$ be the norm on $\mathbb{R}^N$ defined via $||x||_r := (\sum_i |x_i|^r)^{1/r}$, and f the function given by $f(x) = \frac{1}{p}||x||_r^p$. Compute f^*.

Exercise 4.6 Let $f : \mathbb{R}^N \to \mathbb{R}$ be convex. Prove that f is $C^{1,1}$ if and only if f^* is elliptic (meaning that there exists a $c > 0$ such that $f(x) - c|x|^2$ is convex).

Exercise 4.7 Prove that a convex and l.s.c. function $f : X \to \mathbb{R} \cup \{+\infty\}$ is 1-homogeneous (i.e. $f(tx) = tf(x)$ for every $t \geq 0$ and every $x \in X$) if and only if f^* takes its values in $\{0, +\infty\}$.

Exercise 4.8 Let $f : \mathbb{R}^N \to \mathbb{R}$ be a finite-valued superlinear convex function. Prove that the function $h : \mathbb{R}^N \times \mathbb{R} \to \mathbb{R} \cup \{+\infty\}$ defined via

$$h(x, t) := \begin{cases} tf\left(\frac{x}{t}\right) & \text{if } t > 0, \\ 0 & \text{if } t = 0, x = 0, \\ +\infty & \text{if } t = 0, x \neq 0, \\ +\infty & \text{if } t < 0 \end{cases}$$

is convex and l.s.c. The function h is called the *perspective function* of f.

Exercise 4.9 Given $f : X \to \mathbb{R} \cup \{+\infty\}$ and $\xi_0 \in X'$, define $f_{\xi_0} : X \to \mathbb{R} \cup \{+\infty\}$ via $f_{\xi_0}(x) := f(x) - \langle \xi_0, x \rangle$. Prove that we have $f_{\xi_0}^*(\xi) = f^*(\xi + \xi_0)$ for every ξ.

Exercise 4.10 In dimension $N = 1$, prove the equivalence between these two facts:

1. f^* is strictly convex;
2. f satisfies the following property: it is C^1 in the interior Ω of its domain, and it coincides with $x \mapsto \sup_{x' \in \Omega} f(x') + \nabla f(x') \cdot (x - x')$.

Exercise 4.11 Let X be a normed space with the following property: all convex and l.s.c. functions f such that ∂f is surjective are superlinear. Prove that X' has this other property: all convex and weakly-* l.s.c. functions are bounded on bounded sets.

Exercise 4.12 Discuss what does not work in the following (wrong) counter-example to the statement of Exercise 4.11. Consider $X = L^2(\Omega)$ and $F(u) := \int |\nabla u|^2$ (a functional set to $+\infty$ if $u \notin H^1(\Omega)$). This functional is not superlinear because it vanishes on constant functions, but for every $u \in H^2(\Omega)$ we do have $\partial F(u) = \{-\Delta u\}$, which shows that any L^2 function (the elements of the dual of the Hilbert space on which the functional is defined) is a subgradient, so the subgradient is surjective.

Also discuss what would change if we took $F(u) := \int |\nabla u|^2$ if $u \in H_0^1(\Omega)$ (and $F = +\infty$ outside of $H_0^1(\Omega)$).

Exercise 4.13 Given a smooth open and connected domain $\Omega \subset \mathbb{R}^d$, and $f \in L^2(\Omega)$, set $H_{div}(\Omega) = \{\mathbf{v} \in L^2(\Omega; \mathbb{R}^d) : \nabla \cdot \mathbf{v} \in L^2(\Omega)\}$ and consider the minimization problems

$$(P) := \min \left\{ F(u) := \int_\Omega \left(\frac{1}{2}|\nabla u|^2 + \frac{1}{2}|u|^2 + f(x)u \right) dx \; : \; u \in H^1(\Omega) \right\}$$

$$(D) := \min \left\{ G(\mathbf{v}) := \int_\Omega \left(\frac{1}{2}|\mathbf{v}|^2 + \frac{1}{2}|\nabla \cdot \mathbf{v} - f|^2 \right) dx \; : \; \mathbf{v} \in H_{div}(\Omega) \right\},$$

1. Prove that (P) admits a unique solution;
2. Prove $\min(P) + \inf(D) \geq 0$;
3. Prove that there exist $\mathbf{v} \in H_{div}(\Omega)$ and $u \in H^1(\Omega)$ such that $F(u) + G(v) = 0$;
4. Deduce that $\min(D)$ is attained and $\min(P) + \inf(D) = 0$;
5. Justify by a formal inf-sup exchange the duality $\min F(u) = \sup -G(v)$;
6. Prove $\min F(u) = \sup -G(v)$ via a duality proof based on convex analysis.

Exercise 4.14 Consider the problem

$$\min \left\{ A(\mathbf{v}) := \int_{\mathbb{T}^d} H(\mathbf{v}(x)) \, dx \; : \; \mathbf{v} \in L^2, \nabla \cdot \mathbf{v} = f \right\}$$

for a function H which is elliptic. Prove that the problem has a solution, provided there exists at least an admissible $\mathbf{v}$ with $A(\mathbf{v}) < +\infty$. Prove that, if f is an H^1 function with zero mean, then the optimal $\mathbf{v}$ is also H^1.

Exercise 4.15 Let $H : \mathbb{R}^d \to \mathbb{R}$ be given by $H(z) = \sqrt{1 + |z|^2 + |z|^4}$ and Ω be the d-dimensional torus. Consider the equation

$$\nabla \cdot \left(\frac{(1 + 2|\nabla u|^2)\nabla u}{H(\nabla u)} \right) = f.$$

1. Given $f \in H^{-1}(\Omega)$ such that $\langle f, 1 \rangle = 0$ (i.e. f has zero average), prove that there exists a solution $u \in H^1(\Omega)$ to this equation, which is unique up to additive constants.
2. If $f \in H^1$, prove that the solution u is H^2.
3. What can we say if $f \in L^2$?

Exercise 4.16 Let $f \in L^1([0, 1])$ and $F \in W^{1,1}([0, 1])$ be such that $F' = f$ and $F(0) = F(1) = 0$ (in particular, f has zero mean). Let $1 < p < \infty$ be a given exponent, and q be its conjugate exponent. Prove

$$\min \left\{ \int_0^1 \frac{1}{p} |u'(t)|^p \, dt + \int_0^1 f(t)u(t) \, dt \; : \; u \in W^{1,p}([0, 1]) \right\} = -\frac{1}{q} \|F\|_{L^q([0,1])}^q.$$

Exercise 4.17 Let $H : \mathbb{R}^d \to \mathbb{R}$ be given by

$$H(v) = \frac{(4|v| + 1)^{3/2} - 6|v| - 1}{12}.$$

1. Prove that H is C^1 and strictly convex. Is it $C^{1,1}$? Is it elliptic?
2. Compute H^*. Is it C^1, strictly convex, $C^{1,1}$ and/or elliptic?
3. Consider the problem $\min\{ \int H(\mathbf{v}) \; : \; \nabla \cdot \mathbf{v} = f \}$ (on the d-dimensional torus, for simplicity) and find its dual.
4. Supposing $f \in L^2$, prove that the optimal u in the dual problem is H^2.
5. Under the same assumption, prove that the optimal $\mathbf{v}$ in the primal problem belongs to $W^{1,p}$ for every $p < 2$ if $d = 2$, for $p = d/(d - 1)$ if $3 \le d \le 5$, and for $p = 6/5$ if $d \ge 3$.

Exercise 4.18 Given a function $H : \mathbb{R}^d \to \mathbb{R}$ which is both elliptic and $C^{1,1}$ (i.e. there are two positive constants c_0, c_1 such that $c_0 \mathbf{I} \le D^2 H \le c_1 \mathbf{I}$), let $u \in H^1_{loc}(\Omega)$ be a weak solution in Ω of $\nabla \cdot (\nabla H(\nabla u)) = 0$.

1. Prove that we have $u \in H^2_{loc}$.
2. Prove the same result for solutions of $\nabla \cdot (\nabla H(\nabla u)) = f$, $f \in L^2_{loc}$.
3. If Ω is the torus, prove that the same result is global and only uses the lower bound on $D^2 H$.

Exercise 4.19 Given a function $g \in L^2([0, L])$, consider the problem

$$\min\left\{\int_0^L \frac{1}{2}|u(t) - g(t)|^2 \, dt \; : \; u(0) = u(L) = 0, u \in \text{Lip}([0, L]), |u'| \leq 1 \text{ a.e.}\right\}.$$

1. Prove that this problem admits a solution.
2. Prove that the solution is unique.
3. Find the optimal solution in the case where g is the constant function $g = 1$ in terms of the value of L, distinguishing $L > 2$ and $L \leq 2$.
4. Computing the value of

$$\sup\left\{-\int_0^L (u(t)z'(t) + |z(t)|) \, dt \; : \; z \in H^1([0, L])\right\}$$

    find the dual of the previous problem by means of a formal inf-sup exchange.
5. Assuming that the equality $\inf \sup = \sup \inf$ in the duality is satisfied, write the necessary and sufficient optimality conditions for the solutions of the primal and dual problem. Check that these conditions are satisfied by the solution found in the case $g = 1$.
6. Prove the equality $\inf \sup = \sup \inf$.

Exercise 4.20 Given $u_0 \in C^1([0, 1])$ consider the problem

$$\min\left\{\int_0^1 \frac{1}{2}|u(t) - u_0(t)|^2 \, dt \; : \; u' \geq 0\right\},$$

which consists in the projection of u_0 onto the set of monotone increasing functions (where the condition $u' \geq 0$ is intended in the weak sense).

1. Prove that this problem admits a unique solution.
2. Write the dual problem
3. Prove that the solution is actually the following: define U_0 through $U_0' = u_0$, set $U_1 := (U_0)^{**}$ to be the largest convex and l.s.c. function smaller than U_0, take $u = U_1'$.

Exercise 4.21 Consider the optimal control problem (1.20): write the constraint $y' = f(t, y; \alpha)$ as a sup over vector test functions p and, via a formal inf-sup inversion as is usually done in convex duality, formally recover the Pontryagin Maximum Principle.

Exercise 4.22 Given a space X let us fix a cost function $c : X \times X \to \mathbb{R}_+$ which is symmetric ($c(x, y) = c(y, x)$ for all x, y) and satisfies the triangle inequality $c(x, y) \leq c(x, z) + c(z, y)$, as well as $c(x, x) = 0$. For $\psi : X \to \mathbb{R}$ define $\psi^c : X \to \mathbb{R}$ via $\psi^c(x) := \inf_y c(x, y) - \psi(y)$. Prove that a function ϕ is of the form

ψ^c if and only if it satisfies $|\phi(x) - \phi(y)| \leq c(x, y)$ for all x, y. Also prove that for functions satisfying this condition we have $\phi^c = -\phi$.

Exercise 4.23 Prove the equality

$$\min \left\{ \int_\Omega K \, d|\mathbf{v}| \; : \; \mathbf{v} \in \mathcal{M}^d(\Omega); \; \nabla \cdot \mathbf{v} = \mu - \nu \right\}$$

$$= \max \left\{ \int_\Omega \phi \, d(\mu - \nu) \; : \; |\nabla \phi| \leq K \right\}$$

using the duality methods inspired by those developed in Sect. 4.3.

Hints

Hint to Exercise 4.1 If F is convex, then it is weakly l.s.c.

Hint to Exercise 4.2 Prove that the limit is actually a sup and do a sup-sup exchange.

Hint to Exercise 4.3 The only extension takes the value $+\infty$ on $(-\infty, 0)$ and $f^*(y) = e^{y-1}$.

Hint to Exercise 4.4 The function f is no longer convex, and $f^*(y) = e^{|y|-1}$.

Hint to Exercise 4.5 Check that we have $f^*(y) = \frac{1}{p'} ||y||_{r'}^{p'}$.

Hint to Exercise 4.6 For elliptic functions f we have an improved monotonicity of ∂f, and, applying this to f^*, if we assume it to be elliptic, we obtain $c|\xi - \xi_0|^2 \leq (\xi - \xi_0) \cdot (x - x_0)$ whenever $x \in \partial f^*(\xi)$ and $x_0 \in \partial f^*(\xi_0)$. This implies that f is C^1 and ∇f is c^{-1}-Lipschitz. For the converse implication, we want to prove the inequality $f^*(\xi) \geq f^*(\xi_0) + x_0 \cdot (\xi - \xi_0) + \frac{1}{2L}|\xi - \xi_0|^2$, where $L = \text{Lip}(\nabla f)$ and $x_0 \in \partial f^*(\xi_0)$. To prove this, first consider $\min_\xi f^*(\xi) - f^*(\xi_0) - x_0 \cdot (\xi - \xi_0) + \frac{1}{2L+\varepsilon}|\xi - \xi_0|^2 + \delta|\xi - \xi_0|^4$ for $\varepsilon, \delta > 0$, and prove that the minimum is attained at $\xi = \xi_0$.

Hint to Exercise 4.7 When f is 1-homogeneous, use $f^*(y) = \sup_{t>0,x}(tx) \cdot y - f(tx) = \sup_t tf^*(y)$.

Hint to Exercise 4.8 Consider $\sup_{a,b:a+f^*(b)\leq 0} at + b \cdot x$.

Hint to Exercise 4.9 Use the definition of f^* and a change of variables.

Hint to Exercise 4.10 Use Point 3. of Proposition 4.23.

Hint to Exercise 4.11 Use the fact that the superlinearity of f is equivalent to the boundedness of f^* on each ball, and that the surjectivity of ∂f is equivalent to $\partial f^*(\xi) \neq \emptyset$ for every $\xi \in \mathcal{X}'$.

Hint to Exercise 4.12 Even for smooth functions it is not true that we have $F(v) \geq F(u) + \int (-\Delta u)(v - u)$ because of boundary terms. Taking into account these terms, the subdifferentials of F always have zero-mean.

Hint to Exercise 4.13 In point 3. use $\mathbf{v} = \nabla u$ where u solves (P).

Hint to Exercise 4.14 Use the results on regularity via duality, after proving that we can use $j(v) = v$.

Hint to Exercise 4.15 Note that the equation is just $\nabla \cdot (\nabla H(\nabla u)) = f$ and associate with it a variational problem.

Hint to Exercise 4.16 Use the duality of Sect. 4.3 in 1D, where only one vector field $\mathbf{v} = F$ is admissible.

Hint to Exercise 4.17 H is not elliptic (since it does not grow as $|v|^2$) but $C^{1,1}$, while H^* is elliptic and not $C^{1,1}$.

Hint to Exercise 4.18 Use the regularity via duality and adapt the case of the Laplacian when $f \in L^2$.

Hint to Exercise 4.19 The sup in the last point gives the constraint $|u'| \leq 1$. The solution in point 3. is given by $u(t) = \min\{t, L - t, 1\}$.

Hint to Exercise 4.20 The dual is $\max_{\varphi \in H_0^1, \varphi \geq 0} -\frac{1}{2} \int |\varphi'|^2 + \int \varphi' u_0$. Use the function $\varphi = U_0 - U_1$.

Hint to Exercise 4.21 Write $\min_{y,\alpha} \sup_p \int L(t, y, \alpha) - p' \cdot y + p \cdot f(t, y, \alpha)$.

Hint to Exercise 4.22 When ϕ satisfies $|\phi(x) - \phi(y)| \leq c(x, y)$, use this inequality to obtain $\phi^c \geq -\phi$ (the opposite inequality is trivial).

Hint to Exercise 4.23 Consider $\mathcal{F}(\mathbf{w}) := \max\{\int \phi \, \mathrm{d}(\mu - v) : |\nabla \phi + \mathbf{w}| \leq K\}$.

Chapter 5
Hölder Regularity

This chapter presents some classical results from the regularity theory of uniformly elliptic PDEs and in particular those results related to Hölder regularity (i.e. proving that the weak solutions of some elliptic PDEs, and/or their derivatives, belong to $C^{0,\alpha}$). It is a priori less variational in spirit than the rest of the book, since it concerns solutions of some PDEs, which may or may not be obtained as minimizers of some variational problems, but its connections with the calculus of variations are crucial. In particular, the whole modern approach to the calculus of variations consists in relaxing the problems to large functional spaces, where compactness results are available, but where functions are not smooth enough to provide a "classical" formulation of the problem: it is then important to be able to come back, at least in some cases, to classical solutions, i.e. to prove that the solution of a minimization problem in H^1 is actually C^1 (or $C^{1,\alpha}$, or C^∞). Moreover, the very theory that we present in this chapter is the core argument in the solution of Hilbert's 19th problem, which is a problem in the calculus of variations: is it true that the solution of smooth variational problems are smooth functions? A historical discussion of this problem will be presented, after Sects. 5.1 and 5.2. These two sections are devoted to the classical Schauder estimates (prove that the solution gains in regularity compared to the data, if the data are themselves in $C^{k,\alpha}$; these sections essentially follow what is done in many classical books on the topic, such as [95, 97]). We then move to Sect. 5.4 which is devoted to the most difficult case: proving at least some regularity without any assumption on the coefficients of the equation besides their uniform ellipticity. This was the key ingredient in the solution given by De Giorgi to Hilbert's 19th problem. The chapter closes with a discussion section on other regularity results (in $W^{k,p}$ spaces instead of $C^{k,\alpha}$ and including boundary regularity) followed by a list of exercises.

© The Author(s), under exclusive license to Springer Nature Switzerland AG 2023
F. Santambrogio, *A Course in the Calculus of Variations*, Universitext,
https://doi.org/10.1007/978-3-031-45036-5_5

5.1   Morrey and Campanato Spaces

We will define and analyze in this section some functional spaces defined by integral inequalities which will turn out to coincide, in some cases, with the space of Hölder continuous functions, and which will be very useful in proving Hölder regularity results.

We consider an open domain Ω with the following regularity property: there exists a constant $A \in (0, 1)$ such that

$$|\Omega \cap B(x_0, R)| \geq A|B(x_0; R)| \tag{5.1}$$

for every $x_0 \in \Omega$ and every $R < \text{diam}(\Omega)$. We denote by $\Omega_{x_0,R}$ the set $\Omega \cap B(x_0, R)$. Every Lipschitz domain (in particular, every convex domain) satisfies this property while a typical example of a domain not satisfying it is a domain with a cusp such as $\{(x_1, x_2) \in \mathbb{R}^2 : x_2^2 \leq x_1^4, x_1 \in [0, 1]\}$.

Given $p \in [1, \infty)$ and $\lambda \in \mathbb{R}_+$ we define the Morrey space

$$L^{p,\lambda}(\Omega) := \left\{ u \in L^p(\Omega) : \sup_{x_0,R} R^{-\lambda} \int_{\Omega_{x_0,R}} |u|^p \, dx < +\infty \right\},$$

where the sup in the definition is taken over $x_0 \in \Omega$ and $R < \text{diam}(\Omega)$. We define a norm on this vector space via

$$\|u\|_{L^{p,\lambda}} := \left(\sup_{x_0,R} R^{-\lambda} \int_{\Omega_{x_0,R}} |u|^p \, dx \right)^{1/p}.$$

We then define the Campanato space

$$\mathcal{L}^{p,\lambda}(\Omega) := \left\{ u \in L^p(\Omega) : \sup_{x_0,R} R^{-\lambda} \int_{\Omega_{x_0,R}} |u - \langle u \rangle_{x_0,R}|^p \, dx < +\infty \right\},$$

where the sup in the definition is taken over the same set of pairs (x_0, R) and $\langle u \rangle_{x_0,R} := \fint_{\Omega_{x_0,R}} u$. We also define a seminorm on this vector space by

$$[u]_{\mathcal{L}^{p,\lambda}} := \left(\sup_{x_0,R} R^{-\lambda} \int_{\Omega_{x_0,R}} |u - \langle u \rangle_{x_0,R}|^p \, dx \right)^{1/p}$$

and a norm by setting

$$\|u\|_{\mathcal{L}^{p,\lambda}} := \|u\|_{L^p} + [u]_{\mathcal{L}^{p,\lambda}}.$$

We summarize in the table below the different spaces that one obtains for $L^{p,\lambda}$ and $\mathcal{L}^{p,\lambda}$ depending on p and λ. This will depend on the dimension d. We consider for simplicity scalar-valued functions (otherwise, the same would be true componentwise).

	$\lambda < d$	$\lambda = d$	$\lambda \in (d, d+p]$	$\lambda > d+p$
$L^{p,\lambda}$	$= \mathcal{L}^{p,\lambda}$	L^∞	$\{0\}$	$\{0\}$
$\mathcal{L}^{p,\lambda}$	$= L^{p,\lambda}$	"BMO"	$C^{0,\frac{\lambda-d}{p}}$	$\mathbb{R}$

What we mean is

- If $\lambda < d$ the two spaces $L^{p,\lambda}$ and $\mathcal{L}^{p,\lambda}$ coincide and their norms are equivalent. The only non-trivial fact to prove is the inequality $||u||_{L^{p,\lambda}} \leq C||u||_{\mathcal{L}^{p,\lambda}}$, which is done in Proposition 5.3.
- If $\lambda = d$ the spaces $L^{p,\lambda}$ all coincide with $L^\infty(\Omega)$. This can be easily seen considering that we have a.e. $|u(x_0)|^p = \lim_{R \to 0} f_{\Omega_{x_0,R}} |u|^p$ and this quantity is bounded whenever $u \in L^{p,d}$ (on the other hand, if $u \in L^\infty$ then we easily have $||u||_{L^{p,d}} \leq C(A)||u||_{L^\infty}$).
- If $\lambda = d$ the spaces $\mathcal{L}^{p,\lambda}$ do not coincide with $L^\infty(\Omega)$ but with the spaces where a suitable mean oscillation of u is bounded. This is called the space of *Bounded Mean Oscillation* (BMO) functions when $p = 1$ (this explains the quotation marks in the table above, since calling it BMO when $p > 1$ is an abuse of language) and was introduced in [121], but we will not discuss it further.
- If $\lambda > d$ the space $L^{p,\lambda}$ only contains the zero function. This can also be easily seen by considering the formula $|u(x_0)|^p = \lim_{R \to 0} f_{\Omega_{x_0,R}} |u|^p$ together with $f_{\Omega_{x_0,R}} |u|^p \leq C(A)R^{-d} \int_{\Omega_{x_0,R}} |u|^p \leq R^{\lambda-d}||u||_{L^{p,\lambda}}$ and taking $R \to 0$.
- If $\lambda > d$ the space $\mathcal{L}^{p,\lambda}$ coincides with the space of functions which admit a representative satisfying

$$|u(x) - u(y)| \leq C|x - y|^\alpha$$

for $\alpha = \frac{\lambda-d}{p}$ and $C \leq C(A)||u||_{\mathcal{L}^{p,\lambda}}$, i.e. $C^{0,\alpha}$ functions; on the other hand, it is clear that $C^{0,\alpha}$ functions belong to $\mathcal{L}^{p,d+\alpha p}$ functions and $||u||_{\mathcal{L}^{p,\lambda}}$ can be estimated in terms of the $C^{0,\alpha}$ norm. This will be proven in Theorem 5.4 and is the main result of this section. Once we identify $\mathcal{L}^{p,\lambda}$ with $C^{0,\alpha}$ functions it is clear that for $\alpha > 1$ we only have constant functions.

In order to prove the results leading to the above characterizations of the spaces $L^{p,\lambda}$ and $\mathcal{L}^{p,\lambda}$ we first need the following lemmas, where we assume that Ω satisfies (5.1) and we set $\lambda = d + p\alpha$, for α which can be positive or negative.

Lemma 5.1 *Take $u \in \mathcal{L}^{p,\lambda}(\Omega)$ and two numbers $r, R < \mathrm{diam}(\Omega)$ with $r \leq R \leq 2r$. Then we have*

$$|\langle u \rangle_{x_0,R} - \langle u \rangle_{x_0,r}| \leq C[u]_{\mathcal{L}^{p,\lambda}} r^\alpha,$$

where the constant C only depends on A, p and d.

Proof For every point $x \in \Omega_{x_0,r}$ we have $|\langle u \rangle_{x_0,R} - \langle u \rangle_{x_0,r}| \leq |\langle u \rangle_{x_0,R} - u(x)| + |u(x) - \langle u \rangle_{x_0,r}|$ so that we have

$$|\langle u \rangle_{x_0,R} - \langle u \rangle_{x_0,r}|^p \leq 2^{p-1}(|\langle u \rangle_{x_0,R} - u(x)|^p + |u(x) - \langle u \rangle_{x_0,r}|^p)$$

and then

$$|\langle u \rangle_{x_0,R} - \langle u \rangle_{x_0,r}|^p$$

$$\leq 2^{p-1}\left(\fint_{\Omega_{x_0,r}} |\langle u \rangle_{x_0,R} - u(x)|^p \, \mathrm{d}x + \fint_{\Omega_{x_0,r}} |u(x) - \langle u \rangle_{x_0,r}|^p \, \mathrm{d}x \right).$$

Then we use

$$\fint_{\Omega_{x_0,r}} |\langle u \rangle_{x_0,R} - u(x)|^p \, \mathrm{d}x \leq C \fint_{\Omega_{x_0,R}} |\langle u \rangle_{x_0,R} - u(x)|^p \, \mathrm{d}x,$$

where the constant C is an upper bound to $\frac{|\Omega_{x_0,R}|}{|\Omega_{x_0,r}|} \leq \frac{|B(x_0,R)|}{A|B(x_0,r)|} = \frac{R^d}{Ar^d} \leq \frac{2^d}{A}$. Then, using $u \in \mathcal{L}^{p,\lambda}(\Omega)$ and $\lambda = d + p\alpha$, we obtain

$$|\langle u \rangle_{x_0,R} - \langle u \rangle_{x_0,r}|^p \leq C(R^{\alpha p} + r^{\alpha p}) \leq Cr^{\alpha p}.$$

The result follows.   □

Lemma 5.2 *Take two points $x_0, y_0 \in \Omega$, a function $u \in \mathcal{L}^{p,\lambda}(\Omega)$, and set $R = |x_0 - y_0|$. There exists a constant C only depending on A, p, d such that*

$$|\langle u \rangle_{x_0,2R} - \langle u \rangle_{y_0,2R}| \leq C[u]_{\mathcal{L}^{p,\lambda}} R^\alpha,$$

provided $2R < \mathrm{diam}(\Omega)$.

Proof Consider the set $\Omega_{x_0,2R} \cap \Omega_{y_0,2R}$. Since it contains $\Omega_{x_0,R}$ its measure is at least $A|B(x_0, R)|$, so that we have

$$\frac{|\Omega_{x_0,2R}|}{|\Omega_{x_0,2R} \cap \Omega_{y_0,2R}|}, \quad \frac{|\Omega_{y_0,2R}|}{|\Omega_{x_0,2R} \cap \Omega_{y_0,2R}|} \leq \frac{2^d}{A}.$$

We then take $x \in \Omega_{x_0,2R} \cap \Omega_{y_0,2R}$ and write

$$|\langle u \rangle_{x_0,2R} - \langle u \rangle_{y_0,2R}|^p \le 2^{p-1}(|\langle u \rangle_{x_0,2R} - u(x)|^p + |u(x) - \langle u \rangle_{y_0,2R}|^p),$$

which gives, averaging over x,

$$|\langle u \rangle_{x_0,2R} - \langle u \rangle_{y_0,2R}|^p$$

$$\le C \left(\fint_{\Omega_{x_0,2R} \cap \Omega_{y_0,2R}} |\langle u \rangle_{x_0,2R} - u(x)|^p dx + \fint_{\Omega_{x_0,2R} \cap \Omega_{y_0,2R}} |u(x) - \langle u \rangle_{y_0,2R}|^p dx \right)$$

$$\le C \left(\fint_{\Omega_{x_0,2R}} |\langle u \rangle_{x_0,2R} - u(x)|^p dx + \fint_{\Omega_{y_0,2R}} |u(x) - \langle u \rangle_{y_0,2R}|^p dx \right) \le C R^{\alpha p},$$

which concludes the proof. □

We now prove the following.

Proposition 5.3 Assume $\lambda < d$. Then we have $L^{p,\lambda}(\Omega) = \mathcal{L}^{p,\lambda}(\Omega)$ and the two norms are equivalent (up to a multiplicative constant depending on Ω).

Proof Since we can cover Ω with a ball of the form $B(x_0, \mathrm{diam}(\Omega))$ it is clear that we have $\|u\|_{L^p} \le C \|u\|_{L^{p,\lambda}}$ and by using

$$|\langle u \rangle_{x_0,R}|^p \le \fint_{\Omega_{x_0,R}} |u|^p \, dx$$

we have

$$R^{-\lambda} \int_{\Omega_{x_0,R}} |u - \langle u \rangle_{x_0,R}|^p \, dx = R^{-\alpha p} \fint_{\Omega_{x_0,R}} |u - \langle u \rangle_{x_0,R}|^p \, dx$$

$$\le C R^{-\alpha p} \fint_{\Omega_{x_0,R}} |u|^p \, dx + C R^{-\alpha p} |\langle u \rangle_{x_0,R}|^p$$

$$\le C R^{-\alpha p} \fint_{\Omega_{x_0,R}} |u|^p \, dx \le \|u\|_{L^{p,\lambda}},$$

which implies $[u]_{\mathcal{L}^{p,\lambda}} \le C \|u\|_{L^{p,\lambda}}$.

We now need to obtain the opposite inequality. Take $u \in \mathcal{L}^{p,\lambda}$ with $\lambda = d + \alpha p$, $\alpha < 0$. We fix $R < \mathrm{diam}(\Omega)$ and consider a sequence $2^k R$ for $k = 0, \ldots, k_0$ such that $2^{k_0} R \in (\text{diam}(\Omega)/2, \text{diam}(\Omega))$. Using Lemma 5.1 we obtain

$$|\langle u \rangle_{x_0,R}| \leq |\langle u \rangle_{x_0,2^{k_0}R}| + \sum_{k=0}^{k_0-1} |\langle u \rangle_{x_0,2^kR} - \langle u \rangle_{x_0,2^{k+1}R}|$$

$$\leq C\|u\|_{L^1} + C[u]_{\mathcal{L}^{p,\lambda}} R^\alpha \sum_{k=0}^{\infty} 2^{k\alpha}.$$

Since $\alpha < 0$, the last sum is convergent, so that we obtain $|\langle u \rangle_{x_0,R}| \leq C\|u\|_{L^p} + CR^\alpha[u]_{\mathcal{L}^{p,\lambda}}$. Using again $\alpha < 0$ together with $R < \text{diam}(\Omega)$ we obtain $|\langle u \rangle_{x_0,R}| \leq C\|u\|_{\mathcal{L}^{p,\lambda}} R^\alpha$.

Then we use

$$R^{-\lambda} \int_{\Omega_{x_0,R}} |u|^p \leq CR^{-\alpha p} \fint_{\Omega_{x_0,R}} |u - \langle u \rangle_{x_0,R}|^p + CR^{-\alpha p}|\langle u \rangle_{x_0,R}|^p$$

$$\leq C[u]^p_{\mathcal{L}^{p,\lambda}} + C\|u\|_{\mathcal{L}^{p,\lambda}},$$

which proves the opposite inequality. $\square$

The most interesting case is the following

Theorem 5.4 *Assume $\lambda > d$. Then we have $u \in \mathcal{L}^{p,\lambda}(\Omega)$ if and only if u admits a continuous representative satisfying $|u(x) - u(y)| \leq C|x - y|^\alpha$. Moreover, the best constant C in this inequality and the seminorm $[u]_{\mathcal{L}^{p,\lambda}}$ are comparable up to multiplicative constants only depending on A, p, d.*

Proof First of all, it is clear that every u satisfying $|u(x) - u(y)| \leq C|x - y|^\alpha$ belongs to $\mathcal{L}^{p,\lambda}(\Omega)$, since in this case we have

$$|u(x) - \langle u \rangle_{x_0,R}| \leq CR^\alpha \quad \text{for all } x \in \Omega_{x_0,R}$$

and integrating this inequality over $\Omega_{x_0,R}$ after raising it to the power p provides $u \in \mathcal{L}^{p,\lambda}(\Omega)$ with a seminorm bounded by C.

We now have to find a Hölder continuous representative for functions $u \in \mathcal{L}^{p,\lambda}(\Omega)$. We fix a point $x_0 \in \Omega$ and look at $R \mapsto \langle u \rangle_{x_0,R}$. We consider the sequence $\langle u \rangle_{x_0,2^{-k}R}$ and use (Lemma 5.1)

$$|\langle u \rangle_{x_0,2^{-k}R} - \langle u \rangle_{x_0,2^{-(k+1)}R}| \leq C[u]_{\mathcal{L}^{p,\lambda}} R^\alpha 2^{-k\alpha}.$$

This proves that this sequence is Cauchy, and that it admits a limit. Moreover, for every $r > 0$ we can find k such that $2^{-(k+1)}R \leq r \leq 2^{-k}R$ and $|\langle u \rangle_{x_0,2^{-k}R} - \langle u \rangle_{x_0,r}| \leq C[u]_{\mathcal{L}^{p,\lambda}} r^\alpha$. This shows that $R \mapsto \langle u \rangle_{x_0,R}$ has a limit $\ell(x_0)$ as $R \to 0$

and that we have

$$|\ell(x_0) - \langle u \rangle_{x_0,R}| \leq C[u]_{\mathcal{L}^{p,\lambda}} R^\alpha. \tag{5.2}$$

We observe that the function $x_0 \mapsto \ell(x_0)$ is a representative of u since it coincides with u at every Lebesgue point of u. Up to changing u on a negligible set, we will assume $\ell(x_0) = u(x_0)$ for every x_0. We now consider $x_0, y_0 \in \Omega$, set $R = |x_0 - y_0|$, and assume $2R \leq \mathrm{diam}(\Omega)$. We then consider

$$|u(x_0) - u(y_0)| \leq |u(x_0) - \langle u \rangle_{x_0,2R}| + |\langle u \rangle_{x_0,2R} - \langle u \rangle_{y_0,2R}| + |\langle u \rangle_{y_0,2R} - u(y_0)|$$

$$\leq C[u]_{\mathcal{L}^{p,\lambda}} R^\alpha,$$

where we estimate the first and last term thanks to (5.2) and the second one using Lemma 5.2. $\qquad\square$

Box 5.1 Memo—*Lebesgue Points of L^1 Functions*
A very useful notion concerning L^1_{loc} functions is that of Lebesgue points, as it allows us, for instance, to select a precise representative.

Definition Given a function $u \in L^1_{loc}(\Omega)$ a point x is said to be a Lebesgue point of u if we have

$$\lim_{r \to 0} \fint_{B(x,r)} |u(y) - u(x)|\, dy = 0.$$

In particular, this condition also implies $u(x) = \lim_{r \to 0} \fint_{B(x,r)} u(y)\, dy$, but is stronger. Note that on Lebesgue points we also have $u_n(x) \to u(x)$ every time that u_n is defined as a convolution $u_n = \eta_n * u$ for a rescaling $\eta_n \overset{*}{\rightharpoonup} \delta_0$ of a fixed convolution kernel $\eta \in C_c^\infty$ (actually, bounded kernels with bounded supports are enough).

A well-known theorem in geometric measure theory is the following:

Theorem *For any $u \in L^1_{loc}(\Omega)$ a.e. point $x \in \Omega$ is a Lebesgue point.*

The set of Lebesgue points depends on the representative of u (since $u(x)$ appears in the definition). Yet, it is possible to define $\ell(x) := \liminf_{r \to 0} \fint_{B(x,r)} u(y)\, dy$; the value of ℓ does not depend on the representative of u. We then consider the set of points x where $\lim_{r \to 0} \fint_{B(x,r)} |u(y) - \ell(x)|\, dy = 0$; this set is independent of the representative; moreover ℓ is a representative of u. It is also possible to define ℓ as the limsup, of course.

Theorem 5.4 provides an extremely useful characterization of Hölder continuous functions. It allows us, for instance, to recover the well-known injection of Sobolev spaces into Hölder continuous functions, due to Morrey. Indeed, we have:

Proposition 5.5 *The space* $W^{1,p}(\Omega)$ *continuously injects into* $\mathcal{L}^{1,\lambda}(\Omega)$ *for* $\lambda = 1 + \frac{d}{p'} = d + 1 - \frac{d}{p}$. *Whenever* $p > d$, *we then have a continuous injection of* $W^{1,p}(\Omega)$ *into* $C^{0,\alpha}(\Omega)$ *for* $\alpha = 1 - \frac{d}{p}$.

Proof We start from the Poincaré–Wirtinger inequality followed by a Hölder inequality with exponents p and p':

$$\int_{\Omega_{x_0,R}} |u - \langle u \rangle_{x_0,R}| \, dx \leq C R \int_{\Omega_{x_0,R}} |\nabla u| \, dx$$

$$\leq C R |\Omega_{x_0,R}|^{1/p'} \left(\int_{\Omega_{x_0,R}} |\nabla u|^p \, dx \right)^{1/p} \leq C R^{1+\frac{d}{p'}}.$$

We used the fact that the constant in all forms of the Poincaré inequality is proportional to the diameter of the domain, here R (see Box 5.2). The above inequality proves

$$[u]_{\mathcal{L}^{1,\lambda}} \leq C \|\nabla u\|_{L^p}$$

and from Proposition 5.4 we conclude the Hölder regularity whenever $1 - \frac{d}{p} > 0$.

$\square$

Box 5.2 Memo—*Dependence of the Constants on the Domain in Poincaré and Sobolev Inequalities*

It is well-known that for every open and connected domain Ω there exist constants $C_p(\Omega)$ such that we have

$$\|u\|_{L^p} \leq C_p(\Omega)\|\nabla u\|_{L^p}$$

for functions u which belong either to $W_0^{1,p}(\Omega)$ or to $W^{1,p}(\Omega)$ and have zero mean. These inequalities are called the *Poincaré inequality* in the case of zero boundary conditions, or the *Poincaré–Wirtinger* inequality for zero-mean functions. We also saw in Proposition 2.2 that this also holds when u vanishes on a subset of measure at least $|\Omega|/2$ (or on a subset of the boundary with $\mathcal{H}^{d-1}$ measure at least $\mathcal{H}^{d-1}(\partial\Omega)/2$; the ratio $1/2$ could be replaced with other ratios, but then the constant would also depend on this ratio). How do these constants C_p depend on the domain Ω? In the Poincaré case (in $W_0^{1,p}$), it is easy to see that the constant is monotone by inclusion (since

(continued)

Box 5.2 (continued)

if $\Omega \subset \Omega'$ and $u \in W_0^{1,p}(\Omega)$ then the extension of u to 0 outside Ω also belongs to $W_0^{1,p}(\Omega')$). Hence, one can estimate the value of the constant C_p just by looking at it when Ω is a ball. For the case of the Poincaré–Wirtinger inequality or of functions vanishing on a certain proportion of the domain this is not true, and we will only concentrate on the case of balls.

When $\Omega = B(0, R)$ a simple scaling argument shows $C_p(\Omega) = RC_p(B(0, 1))$. Indeed, if $u \in W^{1,p}(B(0, R))$ then u_R defined via $u_R(x) = u(Rx)$ belongs to $W^{1,p}(B(0, 1))$ (and this transformation preserves the boundary value, the mean, and the proportion of the vanishing set) and we have

$$R^{-d/p}\|u\|_{L^p(B(0,R))} = \|u_R\|_{L^p(B(0,1))} \leq C_p(B(0,1))\|\nabla u_R\|_{L^p(B(0,1))}$$

$$= C_p(B(0,1))R\|\nabla u(Rx)\|_{L^p(B(0,1))} = C_p(B(0,1))R^{1-d/p}\|\nabla u\|_{L^p(B(0,R))}.$$

The scaling is different when the norms on the sides of the inequality do not share the same exponent. This is possible thanks to the Sobolev injection : if $q \leq p^* := \frac{dp}{d-p}$ (when $p < d$, otherwise one just needs $q < \infty$) we have indeed, under the same assumptions on u,

$$\|u\|_{L^q} \leq C_{p,q}(\Omega)\|\nabla u\|_{L^p}.$$

Here for $\Omega = B(0, R)$ the same scaling argument shows $C_{p,q}(\Omega) = R^{1-\frac{d}{p}+\frac{d}{q}}C_p(B(0, 1))$. This also shows that only the critical injection for $q = p^*$ can be true in the whole space as it is the only case where $1 - \frac{d}{p} + \frac{d}{q} = 0$.

When different exponents are involved it is more convenient to express the norms in terms of averages instead of integrals: the Sobolev inequality for zero-mean or $W_0^{1,p}$ functions becomes

$$\left(\fint_{B(0,R)} |u|^q \, dx \right)^{1/q} \leq CR \left(\fint_{B(0,R)} |\nabla u|^p \, dx \right)^{1/p}$$

and the general case for functions in $W^{1,p}$ becomes

$$\left(\fint_{B(0,R)} |u|^q \, dx \right)^{1/q} \leq \left(R^p \fint_{B(0,R)} |\nabla u|^p \, dx + \fint_{B(0,R)} |u|^p \, dx \right)^{1/p}.$$

5.2 Regularity for Elliptic Equations in Divergence Form and Smooth Coefficients

5.2.1 Preliminaries

We start with a very useful algebraic lemma, which can be found in any classical book on elliptic regularity.

Lemma 5.6 *Assume that an increasing function* $\phi : [0, R_0] \to \mathbb{R}_+$ *satisfies, for every* $r < R/2$,

$$\phi(r) \leq A \left(\varepsilon + \left(\frac{r}{R} \right)^\alpha \right) \phi(R) + B R^\beta$$

for some constants $A, B, \varepsilon \geq 0$ *and exponents* $\alpha > \beta > 0$. *Then there exists an* $\varepsilon_0 = \varepsilon_0(A, \alpha, \beta)$ *such that, if* $\varepsilon < \varepsilon_0$, *we also have* $\phi(r) \leq C r^\beta$ *for all* $r \in [0, R_0]$ *for a constant* $C = C(A, B, R_0, \phi(R_0), \alpha, \beta)$.

Proof For every $\tau < 1/2$ we have $\phi(\tau R) \leq A(\varepsilon + \tau^\alpha)\phi(R) + B R^\beta$. Choosing $\gamma \in (\beta, \alpha)$, we can take τ such that $2A\tau^\alpha = \tau^\gamma$ and we can assume without loss of generality that A is large enough so that $\tau < 1/2$. We then take $\varepsilon_0 = \tau^\alpha$ so that, for $\varepsilon < \varepsilon_0$, we have

$$\phi(\tau R) \leq A(\varepsilon_0 + \tau^\alpha)\phi(R) + B R^\beta \leq 2A\tau^\alpha \phi(R) + B R^\beta \leq \tau^\gamma \phi(R) + B R^\beta. \tag{5.3}$$

We now iterate this estimate, thus obtaining

$$\phi(\tau^k R) \leq \tau^{k\gamma} \phi(R) + B R^\beta \tau^{(k-1)\beta} \sum_{j=0}^{k-1} \tau^{j(\gamma-\beta)}.$$

This inequality can be proven by induction. It is already proven for $k = 1$ as it coincides with (5.3). Assuming it is true for k we then obtain

$$\phi(\tau^{k+1} R) \leq \tau^\gamma \phi(\tau^k R) + B \tau^{k\beta} R^\beta$$

$$\leq \tau^{(k+1)\gamma} \phi(R) + B R^\beta \tau^{k\beta} \tau^{\gamma-\beta} \sum_{j=0}^{k-1} \tau^{j(\gamma-\beta)} + B \tau^{k\beta} R^\beta$$

$$= \tau^{(k+1)\gamma} \phi(R) + B R^\beta \tau^{k\beta} \sum_{j=0}^{k} \tau^{j(\gamma-\beta)}.$$

Using $\tau < 1$ and $\gamma > \beta$ we have $\sum_{j=0}^{k} \tau^{j(\gamma-\beta)} \leq C$ and then we obtain

$$\phi(\tau^k R) \leq \tau^{k\gamma} \phi(R) + B C R^\beta \tau^{(k-1)\beta} \leq C \tau^{k\beta},$$

where C is a new constant also depending on B, R and $\phi(R)$. We also used $\tau^\gamma \leq \tau^\beta$. Choosing $R = R_0$ and finding for every r a value of k such that $\tau^{(k+1)} R_0 \leq r \leq \tau^k R_0$ one obtains, using the monotonicity of ϕ and its boundedness, the estimate $\phi(r) \leq Cr^\beta$. $\qquad \square$

We now need some estimates on constant coefficient equations. In the following pages, when there is no ambiguity and a point x_0 is fixed, we write B_r for $B(x_0, r)$ for every radius $r > 0$.

Proposition 5.7 *Assume that u solves $\nabla \cdot (A\nabla u) = 0$ in B_R, where A is a fixed symmetric and positive definite matrix with $\Lambda_{min} I \leq A \leq \Lambda_{max} I$, and take $r < R$. Then there exists a constant C, only depending on $\Lambda_{min}, \Lambda_{max}$, such that we have*

$$\int_{B_r} |u|^2 \, dx \leq C \left(\frac{r}{R}\right)^d \int_{B_R} |u|^2 \, dx, \tag{5.4}$$

$$\int_{B_r} |u - \langle u \rangle_{x_0, r}|^2 \, dx \leq C \left(\frac{r}{R}\right)^{d+2} \int_{B_R} |u - \langle u \rangle_{x_0, R}|^2 \, dx. \tag{5.5}$$

Proof We can prove (5.4) with constant $C = \Lambda_{max}^d \Lambda_{min}^{-d}$. First, note that if u solves $\nabla \cdot (A\nabla u) = 0$ then $u_A(x) := u(A^{1/2}x)$ is harmonic. In particular u_A^2 is subharmonic and Remark 2.7 implies that its averages on balls are non-decreasing functions of the radius of the ball:

$$\int_{B_r} |u(A^{1/2}x)|^2 \, dx \leq \left(\frac{r}{R}\right)^d \int_{B_R} |u(A^{1/2}x)|^2 \, dx.$$

A change of variable $y = A^{1/2}x$ gives

$$\int_{B_{\Lambda_{min}r}} |u|^2 \, dx \leq \int_{A^{1/2}B_r} |u(y)|^2 \, dy \leq \left(\frac{r}{R}\right)^d \int_{A^{1/2}B_R} |u(y)|^2 \, dy \leq \left(\frac{r}{R}\right)^d \int_{B_{\Lambda_{max}R}} |u|^2 \, dx.$$

Setting $r' = \Lambda_{min}r$ and $R' = \Lambda_{max}R$, this provides

$$\int_{B_{r'}} |u|^2 \, dx \leq C \left(\frac{r'}{R'}\right)^d \int_{B_{R'}} |u|^2 \, dx$$

as soon as $r' < \frac{\Lambda_{min}}{\Lambda_{max}} R'$, with $C = \Lambda_{max}^d \Lambda_{min}^{-d}$. Since the inequality is true with constant $C = c^{-d}$ whenever restricted to $r \geq cR$, then (5.4) is also satisfied without the restriction $r' < \frac{\Lambda_{min}}{\Lambda_{max}} R'$.

In order to prove (5.5), we use the Poincaré Wirtinger inequality, then (5.4) applied to ∇u, which also solves the same constant-coefficient equations, and then the Caccioppoli inequality (see Proposition 5.8):

$$\int_{B_r} |u - \langle u \rangle_{x_0,r}|^2 \, dx \leq Cr^2 \int_{B_r} |\nabla u|^2 \, dx$$

$$\leq Cr^2 \left(\frac{r}{R}\right)^d \int_{B_R} |\nabla u|^2 \, dx$$

$$\leq Cr^2 \left(\frac{r}{R}\right)^d \frac{1}{R^2} \int_{B_{2R}} |u - \langle u \rangle_{x_0,2R}|^2 \, dx.$$

This proves (5.5) as soon as $r < R/2$. In the case $r \geq cR$ (here $c = 1/2$) we use the fact that $t \mapsto \int_E |u - t|^2 \, dx$ is minimal when $t = \fint_E u \, dx$, so that we have

$$\int_{B_r} |u - \langle u \rangle_{x_0,r}|^2 \, dx \leq \int_{B_r} |u - \langle u \rangle_{x_0,R}|^2 \, dx \leq \int_{B_R} |u - \langle u \rangle_{x_0,R}|^2 \, dx,$$

which yields the inequality with $C = c^{-(d+2)}$. □

We now finish this preliminary section by proving the Caccioppoli inequality, which we present in the general case, i.e. for non-constant coefficients, for future use.

Proposition 5.8 *Assume that u solves $\nabla \cdot (A(x)\nabla u) = 0$ in B_R, in the sense that we have $\int A(x)\nabla u \cdot \nabla \varphi = 0$ for every $\varphi \in H_0^1(B_R)$, or that u is non-negative and that it solves $\nabla \cdot (A(x)\nabla u) \geq 0$, in the sense that we have $\int A(x)\nabla u \cdot \nabla \varphi \leq 0$ for every $\varphi \in H_0^1(B_R)$ with $\varphi \geq 0$. Also assume that $A(x)$ is, for every x, a symmetric and positive definite matrix with $\Lambda_{min}I \leq A(x) \leq \Lambda_{max}I$. Then, for every $r < R$, we have*

$$\int_{B_r} |\nabla u|^2 \, dx \leq \frac{C}{(R-r)^2} \int_{B_R} |u|^2 \, dx,$$

where $C = 4\Lambda_{max}/\Lambda_{min}$.

Proof Let $\eta \in \text{Lip}(B_R)$ be a function such that $\eta = 0$ on ∂B_R, $\eta = 1$ on B_r and $|\nabla \eta| \leq (R - r)^{-1}$. The important fact is that, in both cases (when the equation is satisfied as an equality or as an inequality), we can use $\varphi = \eta^2 u$. Then we have $\nabla \varphi = 2u\eta\nabla\eta + \nabla u \eta^2$ and hence

$$\int_{B_R} A(x)\nabla u \cdot \nabla u \eta^2 \, dx \leq -2 \int_{B_R} A(x)\nabla u \cdot \nabla \eta \eta u \, dx.$$

It is possible to apply a Hölder inequality on the right-hand side, thus obtaining

$$\left| \int_{B_R} A(x)\nabla u \cdot \nabla \eta \eta u \right| dx \leq \left(\int_{B_R} A(x)\nabla u \cdot \nabla u \eta^2 dx \right)^{\frac{1}{2}} \left(\int_{B_R} A(x)\nabla \eta \cdot \nabla \eta u^2 dx \right)^{\frac{1}{2}},$$

so that we have

$$\int_{B_R} A(x)\nabla u \cdot \nabla u\, \eta^2\, dx \le 4\int_{B_R} A(x)\nabla \eta \cdot \nabla \eta\, u^2\, dx \le \frac{4\Lambda_{max}}{(R-r)^2}\int_{B_R} \eta u^2\, dx.$$

Using $A \ge \Lambda_{min}I$ and restricting the integral on the left-hand side to the set B_r where $\eta = 1$ provides the desired inequality.   □

5.2.2 Schauder Regularity

We prove here a sequence of three incremental theorems about the regularity of solutions of $\nabla \cdot (A(x)\nabla u) = \nabla \cdot F$ in Morrey and Campanato spaces. These results are part of what is usually called Schauder regularity theory, presented here using the powerful tools of Campanato spaces. The analysis requires a small preliminary lemma.

Lemma 5.9 *For every vector field $F \in L^2(B_R)$ and every matrix field $A(x)$ such that $A(x)$ is symmetric and satisfies $\Lambda_{min}I \le A(x) \le \Lambda_{max}I$ for every x for two positive constants $\Lambda_{min}, \Lambda_{max} > 0$, there exists a unique solution $w \in H_0^1(B_R)$ of $\nabla \cdot (A(x)\nabla w) = \nabla \cdot F$, and such a function w satisfies*

$$\Lambda_{min}^2 \int_{B_R} |\nabla w|^2\, dx \le \int_{B_R} |F - \langle F\rangle_{x_0,R}|^2\, dx \le \int_{B_R} |F|^2\, dx.$$

Proof The existence of such a w can be obtained by minimizing the functional $w \mapsto \int \left(\frac{1}{2}A(x)\nabla w \cdot \nabla w + F \cdot \nabla w\right) dx$ among $w \in H_0^1(B_R)$ and the uniqueness is due to the strict convexity of this functional. Testing the equation against w itself provides

$$\int_{B_R} A(x)\nabla w \cdot \nabla w\, dx = -\int_{B_R} F \cdot \nabla w\, dx = -\int_{B_R} (F - \langle F\rangle_{x_0,R}) \cdot \nabla w\, dx,$$

where we use for the last inequality the fact that we have $\int \nabla w = 0$ because $w \in H_0^1(B_R)$. Replacing A with $\Lambda_{min}I$ on the left-hand side and applying a Hölder inequality on the right-hand side provides

$$\Lambda_{min} \int_{B_R} |\nabla w|^2\, dx \le \left(\int_{B_R} |F - \langle F\rangle_{x_0,R}|^2\, dx\right)^{1/2} \left(\int_{B_R} |\nabla w|^2\, dx\right)^{1/2}.$$

Dividing by $\left(\int |\nabla w|^2\right)^{1/2}$ on both sides and taking the square gives the first part of the claim. We then conclude by using once more the fact that $v \mapsto \int |F - v|^2\, dx$ is minimized when v is the average of F.   □

Theorem 5.10 *Let u be a solution of $\nabla \cdot (A\nabla u) = \nabla \cdot F$ in Ω, where A is a constant symmetric and positive-definite matrix and $F \in \mathcal{L}^{2,\lambda}(\Omega)$ with $\lambda < d + 2$. Then $\nabla u \in \mathcal{L}^{2,\lambda}(\Omega')$ for every subdomain Ω' compactly contained in Ω.*

Proof Given a ball $B_R \subset \Omega$ we write $u = \tilde{u} + w$ in such a ball, where w is the solution of $\nabla \cdot (A\nabla w) = \nabla \cdot F$ with $w \in H_0^1(B_R)$. Then $\tilde{u}$ solves $\nabla \cdot (A\nabla \tilde{u}) = 0$. Note that we have $\langle \nabla w \rangle_{x_0,R} = 0$, hence $\langle \nabla u \rangle_{x_0,R} = \langle \nabla \tilde{u} \rangle_{x_0,R}$. We then define $\phi(r) := \int_{B_r} |\nabla u - \langle \nabla u \rangle_{x_0,r}|^2 \, dx$ and we have

$$
\begin{aligned}
\phi(r) &\leq \int_{B_r} |\nabla u - \langle \nabla \tilde{u} \rangle_{x_0,r}|^2 \, dx \leq 2 \int_{B_r} |\nabla \tilde{u} - \langle \nabla \tilde{u} \rangle_{x_0,r}|^2 \, dx + 2 \int_{B_r} |\nabla w|^2 \, dx \\
&\leq C \left(\frac{r}{R} \right)^{d+2} \int_{B_R} |\nabla \tilde{u} - \langle \nabla \tilde{u} \rangle_{x_0,R}|^2 \, dx + 2 \int_{B_R} |\nabla w|^2 \, dx \\
&\leq C \left(\frac{r}{R} \right)^{d+2} \int_{B_R} |\nabla u - \langle \nabla u \rangle_{x_0,R}|^2 \, dx + C \int_{B_R} |\nabla w|^2 \, dx \\
&\leq C \left(\frac{r}{R} \right)^{d+2} \phi(R) + C \int_{B_R} |F - \langle F \rangle_{x_0,R}|^2 \, dx \leq C \left(\frac{r}{R} \right)^{d+2} \phi(R) + C R^\lambda.
\end{aligned}
$$

In the inequalities above we used the formula (5.5) applied to $\nabla \tilde{u}$, which also solves the same constant-coefficient equation as $\tilde{u}$, and then Lemma 5.9, as well as the assumption $F \in \mathcal{L}^{2,\lambda}$. Finally, by Lemma 5.6 we obtain $\phi(r) \leq C r^\lambda$, which gives the claim. The restriction to a subdomain Ω' compactly contained in Ω is required to cover it with finitely many balls contained in Ω, with a uniform radius R_0. $\qquad \square$

Theorem 5.11 *Let u be a solution in Ω of $\nabla \cdot (A(x)\nabla u) = \nabla \cdot F$, where A is a symmetric and positive-definite matrix depending continuously on x and satisfying $\Lambda_{min} I \leq A(x) \leq \Lambda_{max} I$ for every x, for two positive constants $\Lambda_{min}, \Lambda_{max} > 0$. Assume $F \in L^{2,\lambda}(\Omega)$ with $\lambda < d$. Then $\nabla u \in L^{2,\lambda}(\Omega')$ for every subdomain Ω' compactly contained in Ω.*

Proof Given a ball $B_R \subset \Omega$ we write $u = \tilde{u} + w$ in such a ball, where w is the solution of $\nabla \cdot (A(x_0)\nabla w) = \nabla \cdot (F + (A(x_0) - A(x))\nabla u)$ with $w \in H_0^1(B_R)$. Then $\tilde{u}$ solves $\nabla \cdot (A(x_0)\nabla \tilde{u}) = 0$. This is a useful trick: if a function solves an equation with variable coefficients, we can "freeze" the coefficients at a point, and of course add on the other side of the equation the difference between the coefficients at x and at such a point, which will hopefully be small. We then define $\phi(r) := \int_{B_r} |\nabla u|^2 \, dx$ and we have

$$
\begin{aligned}
\phi(r) &\leq \int_{B_r} |\nabla u|^2 \, dx \leq 2 \int_{B_r} |\nabla \tilde{u}|^2 \, dx + 2 \int_{B_r} |\nabla w|^2 \, dx \\
&\leq C \left(\frac{r}{R} \right)^d \int_{B_R} |\nabla \tilde{u}|^2 \, dx + 2 \int_{B_R} |\nabla w|^2 \, dx
\end{aligned}
$$

$$\leq C \left(\frac{r}{R}\right)^d \int_{B_R} |\nabla u|^2 \, dx + C \int_{B_R} |\nabla w|^2 \, dx$$

$$\leq C \left(\frac{r}{R}\right)^d \phi(R) + C \int_{B_R} |G|^2 \, dx,$$

where $G := F + (A(x_0) - A(x))\nabla u$. Here we used formula (5.4) applied to $\nabla \tilde{u}$, and again Lemma 5.9. We now use

$$\int_{B_R} |G|^2 \, dx \leq 2 \int_{B_R} |F|^2 \, dx + 2\omega(R)^2 \int_{B_R} |\nabla u|^2 \, dx,$$

where ω is the modulus of continuity of A. Using the assumption $F \in L^{2,\lambda}$ we then obtain

$$\phi(r) \leq C \left(\frac{r}{R}\right)^d \phi(R) + C R^\lambda + C \varepsilon \phi(R)$$

as soon as $\omega(R)$ is small enough. Again, by Lemma 5.6 we obtain $\phi(r) \leq C r^\lambda$, which gives the claim (with the same need to restrict to compact subdomains as in Theorem 5.10). □

Theorem 5.12 *Let u be a solution of $\nabla \cdot (A(x)\nabla u) = \nabla \cdot F$ in Ω, where A is a symmetric and positive-definite matrix which is $C^{0,\alpha}$ w.r.t. x and satisfies $\Lambda_{min}I \leq A(x) \leq \Lambda_{max}I$ for every x, for two positive constants $\Lambda_{min}, \Lambda_{max} > 0$. Assume $F \in C^{0,\alpha}(\Omega)$. Then $\nabla u \in C^{0,\alpha}(\Omega')$ for every subdomain Ω' compactly contained in Ω.*

Proof We now combine the ideas of the two previous proofs. We fix again a ball $B_R \subset \Omega$ and write $u = \tilde{u} + w$, where w is the solution of $\nabla \cdot (A(x_0)\nabla w) = \nabla \cdot (F + (A(x_0) - A(x))\nabla u)$ with $w \in H_0^1(B_R)$. We then define, as in Theorem 5.10, $\phi(r) := \int_{B_r} |\nabla u - \langle \nabla u \rangle_{x_0,r}|^2 \, dx$, and we get

$$\phi(r) \leq C \left(\frac{r}{R}\right)^{d+2} \phi(R) + C \int_{B_R} |\nabla w|^2 \, dx.$$

Again as in Theorem 5.10 we bound $\int_{B_R} |\nabla w|^2$ by

$$\int_{B_R} |G - \langle G \rangle_{x_0,R}|^2 \, dx \leq \int_{B_R} |G - \langle F \rangle_{x_0,R}|^2 \, dx.$$

We then obtain

$$\int_{B_R} |\nabla w|^2 \, dx \le C \int_{B_R} |G - \langle F \rangle_{x_0, R}|^2 dx$$

$$\le C \int_{B_R} |F - \langle F \rangle_{x_0, R}|^2 dx + C \omega(R)^2 \int_{B_R} |\nabla u|^2 dx.$$

We use Theorem 5.11 to obtain $\nabla u \in L^{2,d-\varepsilon}$ for arbitrary $\varepsilon > 0$, so that $\omega(R)^2 \int |\nabla u|^2$ can be bounded by $C R^{2\alpha + d - \varepsilon}$ (we use $\omega(R) \le C R^\alpha$, by assumption). Then we get, using the assumption $F \in C^{0,\alpha} = \mathcal{L}^{2,d+2\alpha}$,

$$\phi(r) \le C \left(\frac{r}{R} \right)^{d+2} \phi(R) + C R^{d+2\alpha} + C R^{2\alpha + d - \varepsilon}.$$

Using Lemma 5.6 we obtain $\phi(r) \le Cr^{2\alpha + d - \varepsilon}$, i.e. $\nabla u \in \mathcal{L}^{2,d+2\alpha - \varepsilon}$. This means $\nabla u \in C^{0,\alpha - \varepsilon/2}$. In particular, ∇u is continuous and hence locally bounded. Up to restricting to a compact subdomain, we can use $\nabla u \in L^\infty$ and then the term $\omega(R)^2 \int |\nabla u|^2$ is now bounded by $C R^{2\alpha + d}$. This allows us to write

$$\phi(r) \le C \left(\frac{r}{R} \right)^{d+2} \phi(R) + C R^{d+2\alpha}$$

and the result follows by using once more Lemma 5.6.   □

We can then extend this to higher-order regularity:

Theorem 5.13 *Let u be a solution of $\nabla \cdot (A(x)\nabla u) = \nabla \cdot F$ in Ω where A is a symmetric and positive-definite matrix which is $C^{k,\alpha}$ w.r.t. x and satisfies $\Lambda_{min} I \le A(x) \le \Lambda_{max} I$ for every x, for two positive constants $\Lambda_{min}, \Lambda_{max} > 0$. Assume $F \in C^{k,\alpha}$. Then $\nabla u \in C^{k,\alpha}(\Omega')$ for every subdomain Ω' compactly contained in Ω.*

Proof This theorem can be proven by induction. The case $k = 0$ coincides with Theorem 5.12. Once we know it for a certain k, in order to prove it for $k + 1$ we differentiate the equation. Denoting by the superscript $'$ the derivatives w.r.t. a certain variable x_i we have

$$\nabla \cdot (A(x)\nabla u') = \nabla \cdot F' - \nabla \cdot (A'(x)\nabla u)$$

so that u' solves the same equation where in the right-hand side F is replaced by $F' - A'(x)\nabla u$. Since we assume $F \in C^{k+1,\alpha}$ we also have $F \in C^{k,\alpha}$ and, by the induction assumption, we know $\nabla u \in C^{k,\alpha}$. Moreover $F', A'(x) \in C^{k,\alpha}$ since $F, A \in C^{k+1,\alpha}$. Then using the equation solved by u' we obtain $\nabla u' \in C^{k,\alpha}$. Applying this to all partial derivatives of u this provides $\nabla u \in C^{k+1,\alpha}$.   □

It is also possible to obtain a result when the right-hand side is not expressed as a divergence.

Theorem 5.14 *Let u be a solution of $\nabla \cdot (A(x)\nabla u) = f$ in Ω, where A is a symmetric and positive-definite matrix which is $C^{k,\alpha}$ w.r.t. x and satisfies $\Lambda_{min}I \leq A(x) \leq \Lambda_{max}I$ for every x, for two positive constants $\Lambda_{min}, \Lambda_{max} > 0$. Assume $f \in C^{k-1,\alpha}(\Omega)$. Then $u \in C^{k+1,\alpha}(\Omega')$ for every subdomain Ω' compactly contained in Ω.*

Proof First we prove $u \in C^{k,\alpha}_{loc}$. Locally around a point with $x_1 = a$, consider the vector field $F(x_1,\ldots,x_n) = e_1 \int_a^{x_1} f(s,x_2,\ldots,x_n)\,ds$. This vector field satisfies $\nabla \cdot F = f$ and has at least the same regularity as f (indeed, we took the antiderivative in one variable, which improves the regularity, but in the other variables the regularity is not improved). We then have $\nabla \cdot (A(x)\nabla u) = \nabla \cdot F$ and this provides $\nabla u \in C^{k-1,\alpha}_{loc}$. We then differentiate the equation in a direction x_i and denote by the superscript $'$ the derivatives w.r.t. to x_i. We have

$$\nabla \cdot (A(x)\nabla u') = f' - \nabla \cdot (A'(x)\nabla u) = \nabla \cdot (fe_i - A'(x)\nabla u).$$

In the right-hand side the vector field $fe_i - A'(x)\nabla u$ is at least $C^{k-1,\alpha}$ because of the assumptions on A and f, and on what we proved on ∇u. Then we obtain $\nabla u' \in C^{k-1,\alpha}_{loc}$, which means $u \in C^{k+1,\alpha}_{loc}$. □

5.3 Hilbert's 19th Problem and the De Giorgi Theorem

Hilbert's nineteenth problem is one of the 23 Hilbert problems proposed in 1900 by David Hilbert (see [115]): the question is whether the solutions of "regular" problems in the calculus of variations are always analytic. Indeed, Hilbert noted that some classes of PDEs, like the Laplace equation, only admit solutions which are analytic functions, and that these equations share the property of being the Euler–Lagrange equations of a certain class of variational problems of the form $\min \int H(\nabla u)$, where H is convex and smooth. To make the question more precise, convexity and smoothness can be quantified, and boundary data can be imposed (possibly non-analytic). We can formulate the question by assuming that D^2H is both bounded from below and from above by positive multiples of the identity, say $\Lambda_{min}I \leq D^2H \leq \Lambda_{max}I$, and wonder whether all solutions of the variational problem inherit the regularity of H.

We proved in Sect. 2.3 that harmonic functions are analytic, but the arguments used there cannot be translated to more general equations, so we will concentrate on C^∞ regularity. Since the Euler–Lagrange equation of $\min \int H(\nabla u)$ is the non-linear PDE $\nabla \cdot (\nabla H(\nabla u)) = 0$ the question is whether all solutions of this PDE are C^∞ if H is C^∞ and satisfies $\Lambda_{min}I \leq D^2H \leq \Lambda_{max}I$. Differentiating the equation (which is possible, for instance, because of the regularity $u \in H^2$, which can be obtained thanks to the techniques of Sect. 4.4) provides

$$\nabla \cdot (D^2H(\nabla u)\nabla u') = 0$$

for every partial derivative u'. By setting $A(x) = D^2 H(\nabla u(x))$ we saw in Theorem 5.13 that $A \in C^{k,\alpha}$ implies $\nabla u' \in C^{k,\alpha}$. Since $H \in C^\infty$, the regularity of A only depends on ∇u, so that we have $\nabla u \in C^{k,\alpha} \Rightarrow \nabla u' \in C^{k,\alpha}$, i.e. $\nabla u \in C^{k+1,\alpha}$. This allows us to improve the regularity by one degree at each iteration of the argument, and provides $u \in C^\infty$. The techniques due to Morrey and Campanato that we presented are not the only possible ones, but the results are quite classical. The first results proving smoothness (up to analyticity) of the solutions of elliptic second-order PDEs starting from some initial regularity date back, for instance, to Bernstein [28]. Yet, this requires a initialization, i.e. we need at least $u \in C^{1,\alpha}$ in order to have $A \in C^{0,\alpha}$.

This was the state of the art in the first half of the twentieth century (see for instance [169], which improved Bernstein results); then, independently, in the '50s both Ennio De Giorgi [72, 73], and John Forbes Nash [159, 160] filled this gap, by proving the following remarkable theorem: it u solves $\nabla \cdot (A(x)\nabla u) = 0$, without assuming any regularity of A but only the bounds $\Lambda_{min}\mathbf{I} \leq A(x) \leq \Lambda_{max}\mathbf{I}$, then $u \in C^{0,\alpha}$. By applying this to u' with $A = D^2 H(\nabla u)$ one obtains $u' \in C^{0,\alpha}$, i.e. $u \in C^{1,\alpha}$. De Giorgi's proof arrived first, followed by Nash's proof slightly later, but the latter was more general: indeed, Nash proved the same result for solutions of parabolic equations $\partial_t u = \nabla \cdot (A(x)\nabla u)$, which includes the elliptic case when the solution is time-independent. De Giorgi's proof passed through the following statement: all functions u which satisfy an inequality of the form

$$\int_{B(x_0,r)} |\nabla (u-k)_\pm|^2 \, \mathrm{d}x \leq \frac{C}{(R-r)^2} \int_{B(x_0,R)} (u-k)_\pm^2 \, \mathrm{d}x$$

for every k are necessarily locally Hölder continuous. The class of these functions is now usually referred to as the *De Giorgi class*. Subsequently, Jürgen Moser gave an alternate proof of these results in a celebrated short paper [154], and this will be the proof we present in the next section.

As it can be seen in both Moser's proof and in the definition of the De Giorgi class, the order structure of $\mathbb{R}$ plays an important role in the theory, and the same result cannot be obtained for vector-valued functions solutions of elliptic systems, in contrast to the theory of Sect. 5.2, which could be adapted to systems (as is done, for instance, in [95, 97] and in many classical texts about the subject, which we avoided here for simplicity). Later on counter-examples in the vector-valued case appeared, as in De Giorgi [75] and Giusti and Miranda [102].

5.4   Moser's Proof

In this section we consider solutions and subsolutions of $\nabla \cdot (A(x)\nabla u) = 0$ (and later we will add a right-hand side) with a matrix field A without any regularity assumption, but only $\Lambda_{min}\mathbf{I} \leq A(x) \leq \Lambda_{max}\mathbf{I}$ (together with measurability in x, of course). Since we will not be able to apply any approximation argument (differently, for instance, from the case $A = \mathbf{I}$ discussed in Sect. 2.3) it is crucial to clarify the notion of solution that we use:

Definition 5.15 A function u is said to be a solution of $\nabla \cdot (A(x)\nabla u) = 0$ in Ω if it belongs to $H^1_{loc}(\Omega)$ and $\int_\Omega A(x)\nabla u \cdot \nabla\varphi = 0$ for every function $\varphi \in C^1_c(\Omega)$. In this case, the same equality also holds for functions $\varphi \in H^1_0(\Omega')$ for every $\Omega' \subset \Omega$ such that we have $u \in H^1(\Omega')$.

A function u is said to be a subsolution of $\nabla \cdot (A(x)\nabla u) = 0$ (and we write $\nabla \cdot (A(x)\nabla u) \geq 0$) in Ω if it belongs to $H^1_{loc}(\Omega)$ and $\int_\Omega A(x)\nabla u \cdot \nabla\varphi \leq 0$ for every function $\varphi \in C^1_c(\Omega)$ with $\varphi \geq 0$. In this case, the same equality also holds for non-negative functions $\varphi \in H^1_0(\Omega')$ for every $\Omega' \subset \Omega$ such that we have $u \in H^1(\Omega')$.

In the sequel, we will often use the term solution and subsolution without explicitly saying "of $\nabla \cdot (A(x)\nabla u) = 0$". We have the following results

Proposition 5.16

1. *If u is a solution in Ω and $f : \mathbb{R} \to \mathbb{R}$ is convex and f and f' are Lipschitz continuous, then $f(u)$ is a subsolution. The same is true if u is only a subsolution but f is also assumed to be non-decreasing.*
2. *If f is convex and u is a solution such that $f(u) \in L^2_{loc}(\Omega)$, then $f(u)$ is a subsolution. The same is true if f is convex and non-decreasing, and u is a subsolution.*
3. *If u is a solution or a non-negative subsolution, then $|u|^p$ is a subsolution for every $p \leq 2^*/2$.*
4. *If u is a solution or a non-negative subsolution, then $|u|^p$ is a subsolution for every $p \in (1, \infty)$.*

Proof

1. If u is a solution we take $\varphi = f'(u)\psi$ for $\psi \in C^1_c$. Thanks to $u \in H^1_{loc}$ and f' Lipschitz continuous we have $\varphi \in H^1$, and since φ is compactly supported in Ω, it is then an admissible test function, so that we have

$$0 = \int_\Omega A(x)\nabla u \cdot \nabla\varphi \, dx = \int_\Omega A(x)\nabla(f(u)) \cdot \nabla\psi \, dx + \int_\Omega A(x)\nabla u \cdot \nabla u f''(u)\psi \, dx.$$

Assuming $\psi \geq 0$ the last term is non-negative, so that we obtain $\int A(x)\nabla(f(u)) \cdot \nabla\psi \leq 0$ and $f(u)$ (which is H^1_{loc} because f is Lipschitz) is a subsolution. If u is only assumed to be a subsolution then we need to require $f' \geq 0$ in order to use $\varphi = f'(u)\psi$ as a non-negative test function and obtain $0 \geq \int A(x)\nabla u \cdot \nabla\varphi$, from which the result follows.

2. If f is convex but we remove the assumption that f and f' are Lipschitz continuous we approximate f with an increasing sequence f_n of Lipschitz and $C^{1,1}$ convex functions. We then obtain that $f_n(u)$ are subsolutions and the Caccioppoli inequality provides local bounds of their H^1 norm in terms of the L^2 norm on larger domains. If $f(u) \in L^2_{loc}$ then the L^2 norms of $f_n(u)$ are

bounded (we use $f_n(u) \le f(u)$ together with a uniform lower bound with an affine function $f_n(u) \ge au + b$) and this implies a similar bound on their H^1 norms, thus obtaining at the limit $f(u) \in H^1_{loc}$. From this we obtain that $f(u)$ is a subsolution.

3. If u is a solution or subsolution, it is H^1_{loc} and then L^{2p}_{loc} for every $p \le 2^*/2$. We can then apply the previous point to the convex function $s \mapsto |s|^p$ (the positivity is needed for subsolutions since otherwise this function is not monotone).

4. Given an exponent $p \ge 1$ we can always write it as $p = p_1^n$ for large n and $p_1 \le 2^*/2$. The previous argument shows that the power p_1 of a subsolution is always a subsolution, so that we can iterate it and obtain that $|u|^p$ is a subsolution.

$\square$

5.4.1 L^∞ Bounds

We now quantify some of the estimates that we used in the previous proposition and we obtain the following:

Proposition 5.17 *If u is a nonnegative subsolution in Ω, and $B_r := B(x_0, r)$ and $B_R := B(x_0, R)$ are two balls compactly contained in Ω with $r \in (R/2, R)$, then for every $p \in (1, 2^*/2)$ we have*

$$\left(\fint_{B_r} |u|^{2p} \, dx \right)^{1/p} \le C \frac{R^2}{(R-r)^2} \fint_{B_R} |u|^2 \, dx,$$

for a constant C only depending on p, Λ_{min}, Λ_{max}.

Proof We start from the Caccioppoli inequality (Proposition 5.8), which provides

$$\fint_{B_r} |\nabla u|^2 \, dx \le \frac{C}{(R-r)^2} \fint_{B_R} |u|^2 \, dx,$$

where we wrote averages instead of integrals since r and R are assumed to be comparable up to a factor $1/2$. Then we use the Sobolev inequality (see the Box 5.2)

$$\left(\fint_{B_r} |u|^{2p} \, dx \right)^{1/p} \le C \left(r^2 \fint_{B_r} |\nabla u|^2 \, dx + \fint_{B_r} |u|^2 \, dx \right).$$

Combining the two, we obtain

$$\left(\fint_{B_r} |u|^{2p} \, dx \right)^{1/p} \le C \left(\frac{Cr^2}{(R-r)^2} \fint_{B_R} |u|^2 \, dx + \fint_{B_r} |u|^2 \, dx \right)$$

which, together with $\fint_{B_r} |u|^2 \le (R/r)^d \fint_{B_R} |u|^2$ and $r \in (R/2, R)$, gives the claim.

$\square$

Theorem 5.18 *If u is a nonnegative subsolution in Ω and $B_{R_0} := B(x_0, R_0)$ is compactly contained in Ω, then $u \in L^\infty(B_{R_0/2})$ and we have*

$$|u(x)| \le C \left(\fint_{B_{R_0}} |u|^2 \, dx \right)^{1/2} \qquad \text{for a.e. } x \in B_{R_0/2}$$

for a constant C only depending on d, Λ_{min}, Λ_{max}.

Proof We fix a sequence of radii $R_k = R_0/2 + 2^{-(k+1)} R_0$ for $k \ge 0$ and a sequence of exponents $m_k = p^k$ where p is a fixed exponent in $(1, 2^*/2)$. We apply Proposition 5.17 to u^{m_k} with $r = R_{k+1}$ and $R = R_k$. Since $R_k \in (R_0/2, R_0)$ for every k, the assumption $r \in (R/2, R)$ is satisfied. We then have $R^2/(R-r)^2 = R_k^2/(R_k - R_{k+1})^2 \le 4^{k+2}$ hence

$$\left(\fint_{B_{R_{k+1}}} |u|^{2m_{k+1}} \, dx \right)^{1/p} \le C 4^k \fint_{B_{R_k}} |u|^{2m_k} \, dx.$$

We raise both sides to the power $1/(2m_k)$ so that we obtain norms:

$$\left(\fint_{B_{R_{k+1}}} |u|^{2m_{k+1}} \, dx \right)^{\frac{1}{2m_{k+1}}} \le (C4^k)^{\frac{1}{2m_k}} \left(\fint_{B_{R_k}} |u|^{2m_k} \, dx \right)^{\frac{1}{2m_k}}.$$

We set $y_k := \log(\|u\|_{L^{2m_k}(B_{R_k})})$, where the norms are computed in terms of the normalized Lebesgue measure on each ball. We just proved

$$y_{k+1} \le \frac{\log C + k \log 4}{2m_k} + y_k \quad \Rightarrow y_k \le y_0 + \sum_{j=0}^{k-1} \frac{\log C + j \log 4}{2m_j}.$$

Since the series of $\frac{\log C + j \log 4}{2m_j}$ is convergent (remember that m_j grows exponentially) we have $y_k \le C + y_0$ for all k, and then

$$\|u\|_{L^{2m_k}(B(x_0,R_k))} \le C \|u\|_{L^2(B(x_0,R_0))}.$$

We then use

$$\|u\|_{L^{2m_k}(B(x_0,R/2))} \le \left(\frac{R_k}{R/2} \right)^{\frac{d}{2m_k}} \|u\|_{L^{2m_k}(B(x_0,R_k))}$$

and $\|u\|_{L^{2m_k}(B(x_0,R/2))} \to \|u\|_{L^\infty(B(x_0,R/2))}$ in order to obtain the claim. $\square$

5.4.2 Continuity

We have just proven that solutions and non-negative subsolutions are locally L^∞ functions. We need now to study better local L^∞ bounds in order to prove a certain decay of the oscillation of the solutions, if we want to prove that they are Hölder continuous.

We then act as follows.

Lemma 5.19 *If u is a solution in Ω valued in an interval $[a, b]$ and $h : [a, b] \to \mathbb{R}$ is a convex function such that $h'' \geq |h'|^2$ then $v := h(u)$ is a subsolution such that*

$$\int_{B(x_0,R)} |\nabla v|^2 \, dx \leq C R^{d-2}$$

whenever $B(x_0, 2R)$ is compactly contained in Ω, for a constant C only depending on d, Λ_{min}, Λ_{max}.

Proof We come back to the proof that v is a subsolution, i.e. we test the equation against $\varphi = h'(u)\psi$. We take $\psi = \eta^2$ where η is a cut-off function with $\eta = 1$ on $B(x_0, R)$, spt $\eta \subset B(x_0, 2R)$ and $|\nabla \eta| \leq \frac{1}{R}$. We first assume that h is smooth, and the result will be true for general h by approximation: we have

$$0 = \int_{B_{2R}} A(x)\nabla u \cdot \nabla \varphi dx = \int_{B_{2R}} A(x)\nabla u \cdot \nabla u h''(u)\eta^2 dx + 2\int_{B_{2R}} A(x)\nabla u \cdot \nabla \eta h'(u)\eta dx,$$

from which we obtain

$$\int_{B_{2R}} A(x)\nabla v \cdot \nabla v \eta^2 \, dx = \int_{B_{2R}} A(x)\nabla u \cdot \nabla u |h'(u)|^2 \eta^2 \, dx$$

$$\leq \int_{B_{2R}} A(x)\nabla u \cdot \nabla u h''(u)\eta^2 \, dx$$

$$= -2\int_{B_{2R}} A(x)\nabla v \cdot \nabla \eta \eta \, dx.$$

Applying the Hölder inequality to the right-hand side exactly as for the proof of the Caccioppoli inequality we get

$$\Lambda_{min}\int_{B_R} |\nabla v|^2 \, dx \leq \int_{B_{2R}} A(x)\nabla v \cdot \nabla v \eta^2 \, dx \leq 4\int_{B_{2R}} A(x)\nabla \eta \cdot \nabla \eta \, dx$$

$$\leq 4\Lambda_{max}\int_{B_{2R}} |\nabla \eta|^2 \, dx = C R^{d-2},$$

which concludes the proof for h smooth. For general h, we first note that h is bounded on $[a, b]$ so that $v = h(u)$ is bounded, hence L^2, and this guarantees (point

2. in Proposition 5.16) that v is a subsolution. Then, we observe that the condition $h'' \geq |h'|^2$ is equivalent to e^{-h} being concave, so that it is enough to approximate e^{-h} with a sequence of smooth concave functions k_n and take $h_n := -\log(k_n)$; we write the inequality of the claim for h_n and then use the lower semicontinuity of the L^2 norm of the gradient.   □

We define $\mathrm{osc}(u, E)$ as $\mathrm{ess\,sup}\{u(x) : x \in E\} - \mathrm{ess\,inf}\{u(x) : x \in E\}$.

Theorem 5.20 *There exists a constant $c < 1$ only depending on d, Λ_{min}, Λ_{max} such that $\mathrm{osc}(u, B(x_0, R/2)) \leq c\,\mathrm{osc}(u, B(x_0, 2R))$ whenever $B(x_0, 2R)$ is compactly contained in Ω and u is a solution in Ω.*

Proof Fix a ball $B(x_0, 2R)$ compactly contained in Ω. We know that u is locally L^∞, hence $\mathrm{osc}(u, B(x_0, 2R)) < \infty$. Up to adding a constant and multiplying by a constant we can assume $-1 \leq u \leq 1$ a.e. in $B(x_0, 2R)$ and $\mathrm{osc}(u, B(x_0, 2R)) = 2$. The claim is proven if we are able to find a universal constant strictly smaller than 2 which bounds $\mathrm{osc}(u, B(x_0, R/2))$ from above. Looking at u on $B(x_0, R)$ one of the two sets $\{u > 0\} \cap B(x_0, R)$ or $\{u < 0\} \cap B(x_0, R)$ has measure smaller than $|B(x_0, R)|/2$. We assume for simplicity that we have $|\{u > 0\} \cap B(x_0, R)| \leq |B(x_0, R)|/2$. Fix $\varepsilon > 0$ and take $v = h(u)$ with $h : [-1, 1] \to \mathbb{R}$ defined via $h(s) = (-\log(1 + \varepsilon - s))_+$, i.e.

$$h(s) = \begin{cases} 0 & \text{if } s < \varepsilon, \\ -\log(1 + \varepsilon - s) & \text{if } s \geq \varepsilon. \end{cases}$$

The function h satisfies $h''(s) \geq |h'(s)|^2$ and in particular is convex, so that v is a nonnegative subsolution. Then, thanks to Theorem 5.18, we have

$$\|v\|^2_{L^\infty(B(x_0, R/2))} \leq C \fint_{B(x_0, R)} |v|^2 \, dx \leq CR^2 \fint_{B(x_0, R)} |\nabla v|^2 \, dx \leq C,$$

where the second inequality is a Poincaré inequality as in Proposition 2.2 and Box 5.2 justified by the fact that v vanishes on a set which has at least half of the measure of the ball $B(x_0, R)$, and the last inequality comes from Lemma 5.19. Since this last constant C does not depend on ε we can pass to the limit $\varepsilon \to 0$ and obtain

$$-\log(1 - u(x)) \leq C \quad \text{for every } x \in B(x_0, R/2) \quad \Rightarrow u(x) \leq 1 - e^{-C} < 1.$$

This shows that in $B(x_0, R/2)$ the oscillation is at most $2 - e^{-C} < 2$ and proves the result.   □

It is then easy to obtain the following final result, which is the main result of [154] and [72]:

Theorem 5.21 *There exists an exponent α only depending on d, Λ_{min}, Λ_{max} such that all solutions in Ω admit a representative which is locally $C^{0,\alpha}$ in Ω.*

Proof Consider a set $\Omega' \subset \Omega$ on which u is essentially bounded: $|u(x)| \leq M$ for a.e. $x \in \Omega'$. Then, consider a set Ω'' such that $B(x, \delta) \subset \Omega'$ for all $x \in \Omega''$ and for a fixed $\delta > 0$. Take two Lebesgue points x, y of u. Assume $x, y \in \Omega''$. For those points we have $|u(x) - u(y)| \leq \text{osc}(u, B(x, r))$, where $r = |x - y|$, as well as $|u(x)|, |u(y)| \leq M$. Let k be the largest integer such that $4^k r \leq \delta$. We have $\text{osc}(u, B(x, r)) \leq c^k \text{osc}(u, B(x, 4^k r))$ for a constant $c >$ obtained in Theorem 5.20, but we also have $\text{osc}(u, B(x, 4^k r)) \leq 2M$ since $B(x, 4^k r)) \subset \Omega'$. Then $|u(x) - u(y)| \leq 2Mc^k$, where $4^{k+1} r > \delta$, i.e. $k > \frac{\log \delta - \log r}{\log 4} - 1$. Then $c^k < Cr^\alpha$ where $\alpha = \frac{-\log c}{\log 4}$. This proves $|u(x) - u(y)| \leq C|x - y|^\alpha$, i.e. u is $C^{0,\alpha}$ on the Lebesgue points of u in Ω''. This Hölder continuous function can be extended, preserving the modulus of continuity, to the whole Ω'', and this is still a representative of u on Ω''. If we do this construction on a countable and increasing sequence of sets Ω''_n these representatives are each an extension of the previous one, and this provides a representative of u which is locally $C^{0,\alpha}$.   □

5.4.3   The Non-Homogeneous Case

For completeness, we want to study what happens for solutions of $\nabla \cdot (A(x)\nabla u) = \nabla \cdot F$.

We will first start with the case of solutions $u \in H^1_0(\Omega)$ (i.e. with homogeneous Dirichlet boundary conditions) and define the notion of subsolution in this same case.

Definition 5.22 A function u is said to be a subsolution of the Dirichlet problem $\nabla \cdot (A(x)\nabla u) = \nabla \cdot F$ (and we write $\nabla \cdot (A(x)\nabla u) \geq \nabla \cdot F$) in Ω if it belongs to $H^1_0(\Omega)$ and $\int_\Omega A(x)\nabla u \cdot \nabla \varphi \leq \int_\Omega F \cdot \nabla \varphi$ for every function $\varphi \in H^1_0(\Omega)$ with $\varphi \geq 0$.

We have the following result.

Proposition 5.23

1. *If u is a non-negative subsolution and $h : \mathbb{R}_+ \to \mathbb{R}$ is a non-decreasing Lipschitz function such $h(0) = 0$, then we have*

$$\Lambda^2_{min} \int_\Omega h'(u)|\nabla u|^2 \, dx \leq \int_\Omega |F|^2 h'(u) \, dx.$$

2. *If u is a non-negative subsolution and $F \in L^d(\Omega)$ then $u \in L^p(\Omega)$ for all $p < \infty$.*
3. *If u is a non-negative subsolution and $F \in L^d(\Omega)$ then $u^p \in H^1_0(\Omega)$ for all $p \in [1, \infty)$.*

4. *If u is a non-negative subsolution then for every $p \in [0, \infty)$ we have*

$$\Lambda_{min}^2 \int_\Omega u^p |\nabla u|^2 \, dx \le \int_\Omega |F|^2 u^p \, dx. \qquad (5.6)$$

5. *If u is a non-negative subsolution and $F \in L^q(\Omega)$ for some $q > d$, then $u \in L^\infty(\Omega)$ and $\|u\|_{L^\infty}$ is bounded by a constant only depending on $d, q, \Lambda_{min}, \Omega$ and $\|F\|_{L^q}$.*

Proof

1. Let us use $\varphi = h(u)$, and assume that $h(u)$ is in H_0^1. We then obtain

$$\int_\Omega A(x)\nabla u \cdot \nabla u h'(u) \, dx \le \int_\Omega F \cdot \nabla u h'(u) \, dx.$$

This implies

$$\Lambda_{min} \int_\Omega h'(u) |\nabla u|^2 dx \le \int_\Omega |F| |\nabla u| h'(u) dx$$

$$\le \left(\int_\Omega |F|^2 h'(u) dx \right)^{\frac{1}{2}} \left(\int_\Omega |\nabla u|^2 h'(u) dx \right)^{\frac{1}{2}}$$

which provides the claim since in particular this can be applied when $h(u)$ is a non-decreasing Lipschitz function of u which vanishes at 0 (because in this case we do have $h(u) \in H_0^1$).

2. If $d = 2$ then we have $u \in L^P(\Omega)$ for all $p < \infty$ because this is true for every H^1 function. If $d > 2$, then we have $|F|^2 \in L^{d/2}$ and $(\frac{d}{2})' = \frac{d}{d-2} = \frac{2^*}{2}$. Assume now $u \in L^P$ for a certain p and take $h(s) = s^{m+1}$ with $m2^*/2 = p$. This function cannot be used in the estimate that we just proved since it is not globally Lipschitz but can be approximated by a sequence of Lipschitz continuous functions h_n such that $h'_n \le h'$ (take, for instance, $h'_n(s) = \min\{h'(s), n\}$ and define h_n accordingly) for which we have

$$\Lambda_{min}^2 \int_\Omega h'_n(u) |\nabla u|^2 \, dx \le \int_\Omega |F|^2 h'_n(u) \, dx \le \int_\Omega |F|^2 h'(u) \, dx < \infty,$$

where the finiteness of the last integral comes from $|F|^2 \in L^{d/2}$ and $h'(u) = cu^m \in L^{(d/2)'}$. Passing to the limit this implies

$$\int_\Omega u^m |\nabla u|^2 \, dx < \infty \quad \Rightarrow \quad u^{\frac{m}{2}+1} \in H^1 \quad \Rightarrow \quad u^{\frac{m}{2}+1} \in L^{2^*},$$

which means $u \in L^{p+2^*}$. This proves that the integrability of u can be improved by induction and $u \in L^P(\Omega)$ for all $p < \infty$.

3. Now that we know $u \in L^p(\Omega)$ for all $p < \infty$ the same argument as before provides $u^{\frac{m}{2}+1} \in H^1$ for all $m = 2p/2^*$. Note that we can also use $p < 1$ and even $p = 0$ as the only assumption on p is that $h(s) = s^{m+1}$ is non-decreasing, so that $m = 0$ is allowed.

4. Given $p \in [0, \infty[$ we define $h(s) = \frac{s^{p+1}}{p+1}$ so that $h'(s) = s^p$ and we know $h(u) \in H_0^1$. This allows us to obtain the desired estimate as in the first point of the proof.

5. Let us start from the case $|\Omega| = 1$. In this case the L^m norm of u is increasing in m. Set $m_0 := \inf\{m \geq 1 : \|u\|_{L^m} > 1\}$. If $m_0 = +\infty$ then $\|u\|_{L^m} \leq 1$ for all m and $\|u\|_{L^\infty} \leq 1$. In this case there is nothing to prove. If $m_0 > 1$ then $\|u\|_{L^m} \leq 1$ for all $m < m_0$ and, by Fatou's lemma, we also have $\|u\|_{L^{m_0}} \leq 1$. If $m_0 = 1$ we then have $\|u\|_{L^{m_0}} = \|u\|_{L^1} \leq C\|u\|_{L^2} \leq C\|\nabla u\|_{L^2}$ and we know from Lemma 5.9 that u is bounded in H^1 in terms of $\|F\|_{L^2}$, and hence of $\|F\|_{L^q}$. In any case, we obtain that $\|u\|_{L^{m_0}}$ is bounded in terms of $\|F\|_{L^q}$, and for $m > m_0$ we have $\|u\|_{L^m} \geq 1$, an inequality which we will use to simplify the computations.

Using (5.6) we have

$$\frac{4}{(p+2)^2} \int_\Omega |\nabla(u^{\frac{p}{2}+1})|^2 \, dx \leq C \int_\Omega |F|^2 u^p \, dx \leq C \left(\int_\Omega u^{\frac{pq}{q-2}} \, dx \right)^{\frac{q-2}{q}},$$

where the constant C depends on Λ_{min} and $\|F\|_{L^q(\Omega)}$. Let us fix an exponent β such that $\beta < \frac{2^*}{2}$ but $\beta > \frac{q}{q-2}$. This is possible because $q > d$. We then use the Sobolev injection of H_0^1 into $L^{2\beta}$ and obtain

$$\left(\int_\Omega u^{(p+2)\beta} \right)^{\frac{1}{\beta}} \leq C(p+2)^2 \left(\int_\Omega u^{\frac{pq}{q-2}} \right)^{\frac{q-2}{q}}.$$

We now raise to the power $1/(p+2)$ and assume $pq/(q-2) \geq m_0$. We then have

$$\|u\|_{L^{p\beta}} \leq \|u\|_{L^{(p+2)\beta}} \leq C(p+2)^{\frac{2}{p+2}} \|u\|_{L^{\frac{pq}{q-2}}}^{\frac{p}{p+2}} \leq C(p+2)^{\frac{2}{p+2}} \|u\|_{L^{\frac{pq}{q-2}}},$$

where we used $\|u\|_{L^{\frac{pq}{q-2}}} \geq 1$. Setting $r = \beta(q-2)/q > 1$ and $m_k = m_0 r^k$ we then have

$$\|u\|_{L^{m_{k+1}}} \leq C(p_k+2)^{\frac{2}{p_k+2}} \|u\|_{L^{m_k}},$$

where $p_k = m_k \frac{q-2}{q}$. Passing to the logarithms and using the exponential behavior of p_k as in the proof of Theorem 5.18 we then obtain

$$\|u\|_{L^{m_k}} \leq C\|u\|_{L^{m_0}} \leq C,$$

where all constants depend on $d, q, \Lambda_{min}, \Omega$ and $||F||_{L^q}$.

For the case where $|\Omega| \neq 1$, we simply perform a scaling: if u is a subsolution in Ω with the matrix field $A(x)$ and the datum $F(x)$ then we define u_R, A_R and F_R as $u_R(x) = u(Rx)$, $A_R(x) = A(Rx)$ and $F_R(x) = F(Rx)$, and we obtain $\nabla \cdot (A_R \nabla u_R) \geq R \nabla \cdot F_R$ in $\frac{1}{R}\Omega$. Indeed, for every non-negative test function we have $\int_\Omega A(x)\nabla u \cdot \nabla \varphi \leq - \int_\Omega F \cdot \nabla \varphi$ and the change of variable $x = Ry$ gives

$$\int_{\frac{1}{R}\Omega} A_R(y)\nabla u(Ry) \cdot \nabla \varphi(Ry)\, dy \leq - \int_\Omega F_R(y) \cdot \nabla \varphi(Ry)\, dy,$$

which gives the desired inequality using $\nabla u(Ry) = \frac{1}{R}\nabla u_R(y)$ and $\nabla \varphi(Ry) = \frac{1}{R}\nabla \varphi_R(y)$. Choosing R such that $|\frac{1}{R}\Omega| = 1$ provides the desired estimate on $||u_R||_{L^\infty} = ||u||_{L^\infty}$.

$\square$

We can now obtain a precise result on solutions.

Theorem 5.24 *If $u \in H_0^1(\Omega)$ is a solution of $\nabla \cdot (A(x)\nabla u) = \nabla \cdot F$ in Ω with $F \in L^q(\Omega)$ for some $q > d$, then $u \in L^\infty(\Omega)$ and $||u||_{L^\infty} \leq C(d, q, \Lambda_{min}, \Omega)||F||_{L^q}$. Moreover, if $\Omega = B(x_0, R)$ we have $C(d, q, \Lambda_{min}, \Omega) = R^{1-\frac{d}{q}}C(d, q, \Lambda_{min})$.*

Proof First we prove that $||u||_{L^\infty}$ can be bounded in terms of $||F||_{L^q}$, without being precise about the exact dependence of the bound on this norm.

Consider $h : \mathbb{R} \to \mathbb{R}$ given by $h(s) = \sqrt{s^2 + \varepsilon^2} - \varepsilon$. The function h is convex, smooth, and 1-Lipschitz. We set $v = h(u)$. Taking $\varphi = h'(u)\psi$ as a test function, with $\psi \in C_c^1$ and non-negative, we obtain

$$\int_\Omega A(x)\nabla v \cdot \nabla \psi\, dx + \int_\Omega A(x)\nabla u \cdot \nabla u h''(u)\psi\, dx = - \int_\Omega F h'(u) \cdot \nabla \psi\, dx.$$

Since the second term in the left-hand side is nonnegative, we obtain

$$\int_\Omega A(x)\nabla v \cdot \nabla \psi\, dx \leq - \int_\Omega F h'(u) \cdot \nabla \psi\, dx.$$

This is valid for $\psi \in C_c^1$ but, by density, also for $\psi \in H_0^1$ since both $A\nabla v$ and $F h'(u)$ are in L^2. Hence v is a non-negative subsolution with right-hand side $\nabla \cdot (F h'(u))$. We then obtain

$$||\sqrt{u^2 + \varepsilon^2} - \varepsilon||_{L^\infty} \leq C(||F h'(u)||_{L^q}) \leq C(||F||_{L^q})$$

(where we used $|h'| \leq 1$). Sending $\varepsilon \to 0$ we obtain

$$||u|_{L^\infty} \leq C(||F||_{L^q}).$$

We now need to investigate more precisely the dependence of the bound. The map $F \mapsto u$ is well-defined (thanks to Lemma 5.9) from L^2 to H^1 but we proved that it maps L^q into L^∞ and that it is bounded on the unit ball of L^q. It is linear, so we necessarily have $||u||_{L^\infty} \leq C||F||_{L^q}$. The constant C depends here on all the data, including the domain. When $\Omega = B_R := B(x_0, R)$ we perform the same scaling as in the last part of the proof of Proposition 5.23, and we obtain

$$||u||_{L^\infty} = ||u_R||_{L^\infty} \leq C||RF_R||_{L^q(B_1)}$$

$$= CR \left(\int_{B_1} |F(Ry)|^q \, dy \right)^{1/q} = CR^{1-\frac{d}{q}} \left(\int_{B_R} |F(y)|^q \, dy \right)^{1/q}.$$

$\square$

We are now able to consider the full case, where we consider a right-hand side F without imposing Dirichlet boundary conditions:

Theorem 5.25 *If $u \in H^1_{loc}(\Omega)$ is a solution in Ω of $\nabla \cdot (A(x)\nabla u) = \nabla \cdot F$ with $F \in L^q_{loc}(\Omega)$ for some $q > d$, then $u \in C^{0,\alpha}_{loc}(\Omega)$ for an exponent α depending on q, d, Λ_{min} and Λ_{max}.*

Proof Consider a ball $B_R := B(x_0, R)$ and decompose u into $u = \tilde{u} + w$ with $w \in H^1_0(B_R)$ a solution of $\nabla \cdot (A(x)\nabla w) = \nabla \cdot F$. Then $\tilde{u}$ solves $\nabla \cdot (A(x)\nabla \tilde{u}) = 0$. Set $\phi(R) := \mathrm{osc}(u, B_R)$. Theorem 5.20 proves that we have

$$\mathrm{osc}(\tilde{u}, B_R) \leq C \left(\frac{r}{R} \right)^{\alpha_0} \mathrm{osc}(\tilde{u}, B_R) \leq C \left(\frac{r}{R} \right)^{\alpha_0} \left(\mathrm{osc}(u, B_R) + ||w||_{L^\infty(B_R)} \right)$$

for a certain exponent α_0 depending on d, Λ_{min} and Λ_{max}. Then we obtain, using Theorem 5.24,

$$\phi(r) \leq C \left(\frac{r}{R} \right)^{\alpha_0} (\phi(R) + CR^{1-\frac{d}{q}}) + CR^{1-\frac{d}{q}}$$

$$\leq C \left(\frac{r}{R} \right)^{\alpha_0} \phi(R) + CR^{1-\frac{d}{q}} \leq C \left(\frac{r}{R} \right)^{\alpha_0} \phi(R) + CR^{\alpha_1},$$

where α_1 is any exponent strictly smaller than α_0 and smaller than $1 - \frac{d}{q}$. We then obtain, using once more Lemma 5.6, the estimate $\phi(r) \leq Cr^{\alpha_1}$, i.e. $u \in C^{0,\alpha_1}_{loc}$. $\square$

We note that the assumption $q > d$ is essential throughout this analysis, since in the case $A = I$ and $F = \nabla u$ we are just dealing with the Sobolev injection of $W^{1,q}$ in $C^{0,\alpha}$.

5.5 Discussion: L^p Elliptic Regularity and Boundary Data

The results in this chapter mainly concern the regularity of the solutions of elliptic equations in terms of Hölder spaces, in the spirit of a statement of the form "if $\Delta u = f$ with $f \in C^{k,\alpha}$, then $u \in C^{k+2,\alpha}$" (a result which is included in Theorem 5.14). The general idea is that solutions gain two degrees of regularity compared to the data in the right-hand side, provided the coefficients of the equation are compatible with this smoothness. A natural question, different but following the same scheme, would be whether a similar statement is true in L^p spaces. Starting from the easiest case of the Laplacian, the question is whether it is true that $\Delta u = f$ with $f \in L^p$ implies $u \in W^{2,p}$.

Since any solution of $\Delta u = f$ can be written as $u = \tilde{u} + \Gamma * f$, where $\tilde{u}$ is harmonic and Γ is the fundamental solution of the Laplacian (see Sect. 2.3), and harmonic functions are C^∞, we only consider the regularity of $\Gamma * f$. We already know from Proposition 2.12 that $f \in L^p \Rightarrow D^2(\Gamma * f) \in L^p$ is true for $p = 2$. We want to extend this to other exponents p.

The strategy for this amounts to a standard procedure in harmonic analysis, called real interpolation: if a linear operator $T : L^p \cap L^q \to L^p \cap L^q$ is bounded both in the L^p norm and in the L^q norm, then for every $r \in (p, q)$ it is also bounded in L^r (i.e. if there are constants C_p and C_q such that $||Tf||_{L^p} \le C_p||f||_{L^p}$ and $||Tf||_{L^q} \le C_q||f||_{L^q}$, then there exists a constant C_r, which can be explicitly computed in terms of p, q, r, C_p, C_q, such that $||Tf||_{L^r} \le C_r||f||_{L^r}$). Actually, the conditions for this interpolation result are even weaker, since we do not necessarily need the continuity of T in L^p and L^q, but a weaker condition; we need the existence of a constant C such that

$$|\{|Tf| > t\}| \le C_p \frac{||f||_{L^p}^p}{t^p} \quad \text{for all } t > 0,$$

and the same replacing p with q. We usually say that T maps L^p into the Lorentz space $L^{p,\infty}$ (also called $L^{p,w}$ for "L^p weak"; see [106] for the theory of Lorentz spaces) and L^q into $L^{q,\infty}$. This condition is satisfied when $||Tf||_{L^p} \le C_p||f||_{L^p}$ because of the estimate $|\{|g| > t\}|t^p \le ||g||_{L^p}^p$ valid for every function g. This interpolation result is called Marcinkiewicz interpolation, see [143, 202], as well as [98] where the whole content of this section is explained in detail.

In the case of the operator T defined via $Tf := (\Gamma * f)_{ij}$ (the second derivative of the convolution $\Gamma * f$ w.r.t. the variables x_i and x_j) we know that T is continuous in L^2 (see Proposition 2.12). The idea is then to prove the weaker estimate above in the case $p = 1$. Note that if $D^2\Gamma$ was an L^1 function the continuity of T would be trivial in all the L^p spaces; unfortunately, $D^2\Gamma(x) \approx |x|^{-d}$ fails L^1 integrability close to 0. On the other hand, the function $x \mapsto |x|^{-d}$ is a typical example of an $L^{1,\infty}$ function since $\{x : |x|^{-d} > t\} = B(0, t^{-1/d})$ and thus $|\{x : |x|^{-d} > t\}| = c/t$. It is then reasonable to expect that $f \in L^1$ would imply $Tf \in L^{1,\infty}$. If the condition to be in $L^{1,\infty}$ could be expressed as the finiteness of a norm or of a convex functional, it would be clear that the convolution with an L^1 density f would preserve it (it is always true that convex functionals which are invariant under translations decrease by convolution with probability measures), and decomposing f into positive and negative parts and rescaling by their masses one would have $||g * f||_X \leq ||g||_X ||f||_{L^1}$ for every translation-invariant norm $||\cdot||_X$. Unfortunately, the $L^{1,\infty}$ space is a very pathological space on which it is impossible to define a norm with the desired properties and the proof of $(\Gamma * f)_{ij} \in L^{1,\infty}$ is more involved than just a convolution argument. In particular, the proof in [98, Section 9.4] strongly uses the condition $D^3\Gamma(x) \approx |x|^{-1-d}$.

This allows us to prove that T maps L^1 into $L^{1,\infty}$ and L^2 into L^2. By interpolation, it also maps L^r into L^r for all $r \in (1, 2)$. By considering a sequence $f_n \overset{*}{\rightharpoonup} \delta_0$ it is not difficult to understand that T cannot map L^1 into L^1.

It is then possible to consider the case $p \in (2, \infty)$ via a duality argument. Indeed, we can take $f, g \in C_c^\infty(\Omega)$ and write

$$\int_\Omega (Tf)g \, dx = \int_\Omega (\Gamma * f)_{ij} g \, dx = \int_\Omega (\Gamma * f)g_{ij} \, dx$$

$$= \int_\Omega f(Tg) \, dx \leq ||f||_{L^p} ||Tg||_{L^{p'}} \leq C||f||_{L^p} ||g||_{L^{p'}},$$

where we used the fact that both the second-order derivations and the convolution with even kernels are self-adjoint operators in L^2, and the $L^{p'}$ boundedness of T, since $p' < 2$. By arbitrariness of g, the above estimate means $||Tf||_{L^p} \leq C||f||_{L^p}$, which proves the result for $p > 2$.

We then obtain the following result.

Theorem 5.26 *For every open domain $\Omega \subset \mathbb{R}^d$ and any $p \in (1, \infty)$ there exists a constant C such that*

$$||\Gamma * f||_{W^{2,p}(\Omega)} \leq C||f||_{L^p}.$$

In particular, if u solves $\Delta u = f$ in Ω and $\Omega' \subset \Omega$ is a compactly contained subdomain, we have the estimate

$$||u||_{W^{2,p}(\Omega')} \leq C(||f||_{L^p} + ||u||_{L^p(\Omega)}),$$

for a constant C depending on p, d, Ω' and Ω.

Note that an alternative interpolation argument could be based on the fact that T maps L^2 into L^2 and L^∞ into BMO (a space which is not so different from $\mathcal{L}^{2,d}$). Indeed, one can take Theorem 5.12, rewrite it replacing $C^{0,\alpha}$ with $\mathcal{L}^{2,d+2\alpha}$, and adapt the proof to the case $\alpha = 0$ (actually, the assumption $\alpha > 0$ is not needed, only $\alpha < 1$ is important). As a particular case, we find that the solution of $\Delta u = \nabla \cdot F$ is such that $\nabla u \in \mathcal{L}^{2,d}$ whenever $F \in \mathcal{L}^{2,d}$. If one takes $f \in L^\infty \subset \mathcal{L}^{2,d}$ and a solution of $\Delta u = f$, it is possible to differentiate the equation and find $\Delta u' = f'$, view f' as the divergence of a vector field $F = fe$, where the unit vector e is oriented as the direction of the derivative denoted by $'$, and deduce $\nabla u' \in \mathcal{L}^{2,d}$, i.e. $D^2 u \in \mathcal{L}^{2,d}$. It is then possible to apply an interpolation theorem due to Stampacchia ([187]; see also [97, 101] and [11] as some examples of books explaining this approach to elliptic regularity in detail) to prove that T maps L^p into L^p for $p > 2$, and then use the duality argument above to extend to $p < 2$. Note that this approach does not use the fundamental solution of the Laplacian and can be more easily extended to the case of equations with variable coefficients.

Before stating a result on variable coefficients we discuss the role of the data on the boundary. Indeed, all the estimates presented so far (in this section but also in the rest of the chapter) are local, in the sense they are valid in the interior of the domain where the equation is solved, i.e. on subsets $\Omega' \subset \Omega$ with coefficients depending on the distance between Ω' and $\partial\Omega$. The question is whether we can obtain regularity results (in $C^{k,\alpha}$ or in $W^{k,p}$) for Dirichlet problems and/or other boundary data.

We will see shortly a sketch of a technique to prove these regularity results for $\Delta u = f$ with homogeneous boundary conditions $u = 0$ on $\partial\Omega$ when $\partial\Omega$ is flat.

Let $\Omega \subset \mathbb{R}^d$ be a compact domain with a flat part in $\partial\Omega$ and $R : \mathbb{R}^d \to \mathbb{R}^d$ a reflection which fixes this part. For instance, if $\Omega = [0, 1]^d$, we can consider $R(x_1, \ldots, x_d) = (-x_1, \ldots, x_d)$. The map R is linear, self-adjoint, with determinant equal to -1, and $R^2 = id$.

Assume that $u \in H_0^1(\Omega)$ solves in the weak sense $\Delta u = f$. Define $\tilde{\Omega} := \Omega \cup R(\Omega)$ and

$$\tilde{u}(x) = \begin{cases} u(x) & \text{if } x \in \Omega, \\ -u(Rx) & \text{if } x \in R(\Omega). \end{cases}$$

We define in analogous way $\tilde{f}$. Note that the condition $u \in H_0^1$ (i.e. zero on the boundary) makes $\tilde{u}$ well-defined on $\Omega \cap R(\Omega)$ and we have $\tilde{u} \in H_0^1(\tilde{\Omega})$. This is not the case for f if we do not assume any boundary condition on it, but $f \in L^p(\Omega)$ clearly implies $\tilde{f} \in L^p(\tilde{\Omega})$.

We can check that we also have $\Delta\tilde{u} = \tilde{f}$ in $\tilde{\Omega}$. This is left as an exercise (see Exercise 5.12). Using several reflections like the one presented here, we conclude the following.

Proposition 5.27 *Assume that $u \in H_0^1(\Omega)$ is a weak solution of $\Delta u = f$ (i.e. we have $\int \nabla u \cdot \nabla\phi = -\int f\phi$ for every $\phi \in C_c^1(\Omega)$) with $f \in L^p(\Omega)$ and Ω is a cube. Then $u \in W^{2,p}(\Omega)$.*

The main idea in proving the above result consists in reflecting enough times so as to see f and u as restrictions of functions $\tilde{f}$ and $\tilde{u}$ on a larger domain $\tilde{\Omega}$ containing $\overline{\Omega}$ in its interior, and satisfying $\Delta\tilde{u} = \tilde{f}$. Since reflections preserve the L^p behavior, we obtain $\tilde{u} \in W^{2,p}$. Note that this same idea would not work as well for other regularity results, which require better than mere summability for f, since the reflection with the change of sign does not preserve continuity or higher-order regularity. On the other hand, Exercise 5.13 proposes a different reflection formula for Neumann boundary conditions, which preserves first-order regularity.

The same techniques can then be extended to the case of non-zero boundary data, since if we want to prove $u \in W^{2,p}$ necessarily we need to assume that the trace of u on $\partial\Omega$ is the trace of a $W^{2,p}$ function, so we can write the boundary condition as $u - g \in H_0^1(\Omega)$ for some $g \in W^{2,p}$, and then we apply the previous result to $v = u - g$, which solves $\Delta v = f - \Delta g \in L^p$.

It is then possible to extend the result to other domains which are not the cube after a suitable diffeomorphism $S : \Omega \to Q$ where Q is a cube. Yet, $u \circ S^{-1}$ no longer solves on Q a constant-coefficient equation, but a more complicated elliptic equation involving variable coefficients depending on the Jacobian of S. Hence, proving regularity results for the Laplace equation up to the boundary on a domain with non-flat boundaries is equivalent to proving the same regularity results for variable coefficients on the cube or, equivalently, local regularity results (since the reflection tricks transforms the boundary into an interior region). These results usually require regularity of the coefficients, which translate into regularity of S and hence of $\partial\Omega$. This justifies the interest in general equations, and the final result which can be obtained by a non-trivial combination of all the techniques seen so far (+ other technical ingredients) is the following (see, for instance [98, Sections 9.5 and 9.6]).

Theorem 5.28 *Assume that $\Omega \subset \mathbb{R}^d$ has $C^{1,1}$ boundary and let u be a solution of $\sum_{i,j} a^{ij} u_{ij} = f$ in Ω, with $u - g \in H_0^1(\Omega)$. Assume $g \in W^{2,p}(\Omega)$ and $a^{ij} \in C^0(\overline{\Omega})$. Then we have $u \in W^{2,p}(\Omega)$.*

5.6 Exercises

Exercise 5.1 Write and prove a Caccioppoli-type inequality between the $L^p(B_r)$ norm of ∇u and the $L^p(B_R)$ norm of u for solutions of $\nabla \cdot (|\nabla u|^{p-2} \nabla u) = 0$.

Exercise 5.2 Given $u \in W_{loc}^{1,p}(\mathbb{R}^d)$, assume that we have

$$\int_{B(x_0,r)} |\nabla u|^p \, dx \leq C r^d f(r),$$

where $f : \mathbb{R}_+ \to \mathbb{R}_+$ is defined as $f(r) = -\log r$ for $r \leq e^{-1}$ and $f(r) = 1$ for $r \geq e^{-1}$. Prove that u is continuous and that we have

$$|u(x) - u(y)| \leq C |x - y| f(|x - y|)^{1/p}$$

(possibly for a different constant C).

Exercise 5.3 Consider the equation $\Delta u = f(u)$, where $f : \mathbb{R} \to \mathbb{R}$ is a given C^∞ function. Assume that $u \in H^1(\Omega)$ is a weak solution of this equation and f has polynomial growth of order m (i.e. $|f(t)| \le C(1 + |t|)^m$. In dimension $d = 1, 2$, prove $u \in C^\infty$. In higher dimensions, prove it under the restriction $m < 2/(d-2)$.

Exercise 5.4 Let $u \ge 0$ be a continuous function on an open domain Ω which is a weak solution of

$$\begin{cases} \Delta u = \sqrt{u} & \text{in } \Omega, \\ u = 1 & \text{on } \partial\Omega. \end{cases}$$

Prove that we have $u \in C^{2,1/2}_{loc}(\Omega)$ and $u \in C^\infty$ if $u > 0$. Prove that we have $u > 0$ if Ω is contained in a ball of radius R sufficiently small (how much?). For larger domains Ω, is it possible to have $\min u = 0$?

Exercise 5.5 Given a function $H : \mathbb{R}^d \to \mathbb{R}$ which is both elliptic and $C^{1,1}$ (i.e. there are two positive constants c_0, c_1 such that $c_0 \mathbf{I} \le D^2 H \le c_1 \mathbf{I}$), let $u \in H^1_{loc}(\Omega)$ be a weak solution of $\nabla \cdot \left(\nabla H(\nabla u)\right) = 0$ in Ω.

1. Prove that we have $u \in H^2_{loc} \cap C^{1,\alpha}_{loc}$.
2. Prove the same result for solutions of $\nabla \cdot \left(\nabla H(\nabla u)\right) = f$, $f \in L^p_{loc}$ for p large enough.

Exercise 5.6 Let $u \in H^1_{loc}$ be a weak solution of $\nabla \cdot \left((3 + \sin(|\nabla u|^2))\nabla u\right) = 0$. Prove that we have $u \in C^{0,\alpha}_{loc}$. Also prove $u \in C^\infty$ on any open set where $\sup |\nabla u| < 1$.

Exercise 5.7 Let $u \in H^1_{loc} \cap L^\infty_{loc}$ be a weak solution of $\nabla \cdot \left(A(x, u)\nabla u\right) = 0$, where

$$A_{ij}(x, u) = (1 + u^2|x|^2)\delta_{ij} - u^2 x_i x_j.$$

Prove that we have $u \in C^\infty$.

Exercise 5.8 Consider a Lipschitz open domain $\Omega \subset \mathbb{R}^d$, a measurable map $\tau : \Omega \to \Omega$ and a scalar function $u \in H^1_{loc}(\Omega)$ which is a weak solution in Ω of

$$\nabla \cdot \left(\frac{2 + u(\tau(x))^2}{1 + u(\tau(x))^2}\nabla u\right) = 0.$$

1. If τ is the identity, prove that $u + \arctan u$ is a harmonic function.
2. If $\tau \in C^\infty$, prove $u \in C^\infty(\Omega)$.

Exercise 5.9 Let p_n be a sequence of C^1 functions on a smooth open domain $\Omega \subset \mathbb{R}^d$ which is bounded in $L^\infty \cap H^1$ and g a given C^2 function on Ω. Let $u_n \in H^1(\Omega)$

be the unique weak solution of

$$\begin{cases} \Delta u_n = \nabla u \cdot \nabla p_n & \text{in } \Omega, \\ u_n = g & \text{on } \partial\Omega. \end{cases}$$

1. Find a minimization problem solved by u_n.
2. Prove that we have $\|u_n\|_{L^\infty} \leq \|g\|_{L^\infty}$.
3. Prove that u_n is bounded in $H^1(\Omega)$.
4. Prove that u_n is locally Hölder continuous, with a modulus of continuity on each subdomain Ω' compactly contained in Ω which is independent of n (but can depend on Ω').
5. If p_n weakly converges in H^1 to a function p prove that we have $u_n \to u$ (the convergence being strong in H^1 and locally uniform), where u is the unique solution of the same problem where p_n is replaced by p.

Exercise 5.10 Consider the equation $\Delta u = f(|\nabla u|^2)$, where $f : \mathbb{R} \to \mathbb{R}$ is a given C^∞ function. Prove that, if $u \in C^1(\bar{\Omega}) \subset H^1(\Omega)$ is a weak solution of this equation, then we have $u \in C^\infty$. What is the difficulty in removing the assumption $u \in C^1(\bar{\Omega})$? Which assumption should we add to f so as to prove the result for H^1 weak solutions?

Exercise 5.11 Let $Q = (a_1, b_1) \times (a_2, b_2) \times \ldots (a_n, b_n) \subset \mathbb{R}^d$ be a rectangle and $f \in C_c^\infty(Q)$. Let $u \in H_0^1(Q)$ be the solution of $\Delta u = f$ with $u = 0$ on ∂Q. Prove $u \in C^\infty(\bar{Q})$. Can we say $u \in C_c^\infty(Q)$? Do functions f such that $u \in C_c^\infty(Q)$ exist?

Exercise 5.12 Prove the claim $\Delta \tilde{u} = \tilde{f}$ concerning reflections in Sect. 5.5.

Exercise 5.13 If $u \in H^1(\Omega)$ is a weak solution of $\Delta u = f$ in Ω with Neumann boundary conditions (i.e. $\int \nabla u \cdot \nabla \phi = -\int f\phi$ for every $\phi \in C^1(\Omega)$) with $f \in L^p(\Omega)$ and Ω is a cube, prove $u \in W^{2,p}(\Omega)$. If moreover $f \in W^{1,p}(\Omega)$ prove $u \in W^{3,p}(\Omega)$.

Exercise 5.14 Let $T \subset \mathbb{R}^2$ be an equilateral triangle and $f \in L^p(T)$. Let $u \in H_0^1(T)$ be the solution of $\Delta u = f$ with $u = 0$ on ∂T. Prove $u \in W^{2,p}(T)$.

Hints

Hint to Exercise 5.1 Test the equation against $\varphi = u\eta^p$.

Hint to Exercise 5.2 Follow the same proof as for Campanato spaces and Proposition 5.5.

Hint to Exercise 5.3 Differentiate the equation to obtain a right-hand side of the form $\nabla \cdot F$ with $F \in L^p$, $p > d$, and then go on with Hölder regularity.

Hint to Exercise 5.4 Start from u bounded to obtain $u \in C^{0,\alpha}$ and go on with Schauder's estimates. To study $\min u$, use $u \le 1$ to obtain $\Delta u \le 1$ and compare to the solution of $\Delta u = 1$.

Hint to Exercise 5.5 Combine regularity via duality and the strategy for Hilbert's 19th problem.

Hint to Exercise 5.6 Start from Moser's result to obtain $u \in C^{0,\alpha}$. View it as an equation of the form $\nabla \cdot \nabla H(\nabla u)$ with H uniformly convex when $|\nabla u|$ is small.

Hint to Exercise 5.7 Start from Moser's result and go on with Schauder's estimates.

Hint to Exercise 5.8 For the second question, start from Moser's result and go on with Schauder's estimates. For the first, compute the gradient of $u + \arctan u$.

Hint to Exercise 5.9 Write the equation as $\nabla \cdot (e^{-P_n} \nabla u_n) = 0$.

Hint to Exercise 5.10 Start from the fact that the right-hand side is bounded to obtain $u \in W^{2,p} \subset C^{1,\alpha}$ and go on. Without at least $u \in \text{Lip}$ this does not work and the right-hand side could be only L^1.

Hint to Exercise 5.11 Use the reflection method.

Hint to Exercise 5.12 Take a test function φ on $\tilde{\Omega}$ and define a test function ψ on Ω via $\psi(x) := \varphi(x) - \varphi(Rx)$.

Hint to Exercise 5.13 Use reflections without changing the sign. Prove that they preserve the Sobolev regularity of f.

Hint to Exercise 5.14 Use six reflections.

Chapter 6
Variational Problems for Sets

In this chapter we present some classical and less classical optimization problems where the unknown, instead of being a function, is a set (of course, identifying sets with their indicator functions can let us back to the functional setting). These problems, which also belong to the calculus of variations in a broad sense, are sometimes called *shape optimization* problems, in particular when the important feature of the set is its shape (rather than its size and/or its position in the space) and when partial differential equations are involved.

We will collect here a few problems all involving in some sense the notion of perimeter or of length of the set. The first one is the most classical one: minimizing the perimeter for fixed volume. The second one replaces the perimeter with another quantity, called $\lambda_1(\Omega)$, which can be both defined as the minimal value of a variational problem in $H_0^1(\Omega)$ or as an eigenvalue of a differential operator; in this case the perimeter and more generally the use of lower-dimensional measures appears in the proof of the result (the optimality of the ball) that we present, and not in the statement. The last class of problems is less classical and involves optimal networks, i.e. one-dimensional closed and connected sets of finite length. Presenting and studying these problems will require us to introduce the so-called *Hausdorff measures*, i.e. the lower-dimensional measures that we can define on subsets of $\mathbb{R}^d$ (even for non-integer dimensions). So far in the rest of the book we essentially only used the natural surface measure on the boundary of a set, which we denoted by $\mathcal{H}^{d-1}$ but we never rigorously introduced. In Sect. 6.3 we will mainly use, instead, the 1D case $\mathcal{H}^1$: we will present some problems very naturally arising in modeling (choosing the best urban transportation system or the best reinforcement for a given membrane), and then the main cornerstone in the proof of the existence of optimal sets. This key ingredient will be the so-called *Gołąb theorem*, which states the lower semicontinuity of the length (in the sense of the $\mathcal{H}^1$ measure) in the class of connected sets, for the Hausdorff convergence.

The chapter closes with a digression on variants and alternative proofs for the isoperimetric inequality and, as usual, a section of exercises.

© The Author(s), under exclusive license to Springer Nature Switzerland AG 2023   243
F. Santambrogio, *A Course in the Calculus of Variations*, Universitext,
https://doi.org/10.1007/978-3-031-45036-5_6

6.1   The Isoperimetric Problem

This section is devoted to one of the most classical problems in the calculus of variations, consisting in finding the set with maximal volume among those with fixed perimeter. Because of this constraint, it is called the *isoperimetric problem*. A legend connects the first instance of this optimization problem with the foundation of the ancient city of Carthage. When Queen Dido arrived on the Tunisian coasts from the East she wanted to build a new city but was only allowed by the local tribes to use a fixed (and supposedly small) amount of land: she was given a bull's hide, and told she could take as much land for her new city as she could enclose with it. Then, she had two clever ideas. The first was not to cover the land with the hide, but have the hide cut into a very long and thin strip of leather which she would use to enclose a portion of land. The second was to choose the position where to put the leather, and hence the shape of the land portion she would enclose with the strip, in an optimal way (and here is where mathematics comes into play: she chose a semi-circle having the coastline on its diameter).[1]

Mathematically, Dido solved the problem

$$\max\{|A| \ : \ A \subset \Omega, \mathrm{Per}(A; \Omega) = L\},$$

where Ω represented the existing land (and $\Omega^c := \mathbb{R}^2 \setminus \Omega$ the sea), $\mathrm{Per}(\cdot\,; \Omega)$ stands for the perimeter *inside* Ω (not counting the coastline), and L was the length of the strip. In the sequel of this section, after introducing precise mathematical definitions and properties concerning the perimeter $\mathrm{Per}(A)$ of a set A and its variant $\mathrm{Per}(A; \Omega)$, we will mainly concentrate on the case where Ω is the whole space. In this last case it is very easy to see that this problem is equivalent to the following one:

$$\min\{\mathrm{Per}(A) \ : \ |A| = V\}$$

in the sense that if A_0 solves the problem with fixed perimeter, then it must also solve this second one, choosing $V = |A_0|$. Moreover, both these problems are equivalent,

---

[1] One can find the legend concerning the bull's hide in Virgil, [194], Book 1, lines 365–368 but without any mention of Dido's tricks to optimize her city.

Devenere locos, ubi nunc ingentia cernis
moenia surgentemque novae Karthaginis arcem,
mercatique solum, facti de nomine Byrsam,
taurino quantum possent circumdare tergo.

The first part of the trick (cutting the hide into strips) is then described more explicitly by Justinin in [123] but, still, no word about the choice of a circular shape, which seems to be a myth among mathematicians. See also [156] for a discussion about this legend and which parts of it have been added in the mathematical folklore. Being married to a historian with Greek and Phoenician expertise, I can say that the importance given to this part of Dido's legend by mathematicians is not really shared by classicists.

by scaling properties, to the minimization of the ratio

$$A \mapsto \frac{\mathrm{Per}(A)}{|A|^{\frac{d-1}{d}}},$$

where d is the dimension of the ambient space. This last formulation of the problem has the advantage of avoiding constraints on the set A, which allows us more easily to write optimality conditions in the form of a PDE. This is proposed, for instance, in Exercise 6.2, but not developed in detail as we will not need it to prove the main—and very classical—result of the section, i.e. that this problem is solved by the ball. Also the adaptation to the case where Ω is a half-space (a rough approximation of Dido's coastline setting) is proposed as an exercise (Exercise 6.4).

6.1.1 BV Functions and Finite Perimeter Sets

A very useful variant of the space of Sobolev functions is the space of BV functions, which can be defined once we know the space of vector measures on a domain $\Omega \subset \mathbb{R}^d$ (see Box 3.5).

Box 6.1 Memo—*BV Functions of Several Variables*

The space of functions with bounded variation on an open domain Ω, called $BV(\Omega)$ is defined via

$$BV(\Omega) := \{u \in L^1(\Omega) : \nabla u \in \mathcal{M}^d(\Omega)\},$$

where the gradient is to be intended in the sense of distributions in Ω (i.e. $\langle \nabla u, \varphi \rangle := -\int u \nabla \varphi$ for every function $\varphi \in C_c^\infty(\Omega)$). The condition for a function $u \in L^1(\Omega)$ to belong to $BV(\Omega)$ (i.e. for the distributional gradient to be written as a measure) is the following:

$$\sup \left\{ \left| \int_\Omega u \nabla \varphi \, dx \right| : \varphi \in C_c^\infty(\Omega), ||\varphi||_{L^\infty} \le 1 \right\} < +\infty.$$

The norm on the space BV is given by $||u||_{BV} := ||u||_{L^1} + ||\nabla u||_{\mathcal{M}}$ (and $||\nabla u||_{\mathcal{M}}$ coincides with the above sup).

We can see that the space $W^{1,1}$, where the gradients are in L^1, is a subset of BV, since when a gradient is an L^1 function it is also a measure.

$BV(\Omega)$ is a Banach space, which is continuously injected in all L^p spaces for $p \le d/(d-1)$ (for $d = 1$, we already saw in Box 1.8 that BV is injected

(continued)

Box 6.1 (continued)
in L^∞). If Ω is bounded, the injection is compact for every $p < d/(d-1)$ and in particular in L^1.

Thanks to this compact injection and to the Banach–Alaoglu Theorem in the space of measures, given a sequence u_n which is bounded in $BV(\Omega)$, when Ω is bounded it is always possible to extract a subsequence u_{n_k} strongly converging in $L^1(\Omega)$ to a function u, with $\nabla u_{n_k} \overset{*}{\rightharpoonup} \nabla u$.

Some non-trivial indicator functions may belong to the space BV, differently than what happens for Sobolev spaces. For smooth sets A we can indeed see, as a simple consequence of the divergence theorem, that we have

$$\nabla \mathbb{1}_A = -\mathbf{n} \cdot \mathcal{H}^{d-1}_{|\partial A},$$

where $\mathbf{n}$ is the exterior unit normal to A, and $\mathcal{H}^{d-1}$ is the standard $(d-1)$-dimensional surface measure.

Definition 6.1 We say that a set $A \subset \Omega$ is a set of finite perimeter if $\mathbb{1}_A \in BV(\Omega)$, and we define its perimeter $\mathrm{Per}(A; \Omega)$ as $\|\nabla \mathbb{1}_A\|_{\mathcal{M}(\Omega)}$. When $\Omega = \mathbb{R}^d$ we simply write $\mathrm{Per}(A)$ instead of $\mathrm{Per}(A; \mathbb{R}^d)$.

Note that the perimeter of A defined in this way depends on the domain Ω: the two numbers $\mathrm{Per}(A; \Omega)$ and $\mathrm{Per}(A; \Omega')$ can be different even if $A \subset \Omega \cap \Omega'$. More precisely, this perimeter corresponds to the part of the boundary of A which is not in $\partial\Omega$.

We now want to state some equivalent definitions of the perimeter, in terms of approximation via sequences of sets for which the notion of perimeter is clearly well-defined. For simplicity, we state it in the case of the whole space (i.e. $\Omega = \mathbb{R}^d$).

Proposition 6.2 *For every set A of finite perimeter in $\mathbb{R}^d$ we have*

$$\mathrm{Per}(A) := \inf \left\{ \liminf_n \mathcal{H}^{d-1}(\partial A_n) : A_n \in C, \ \mathbb{1}_{A_n} \to \mathbb{1}_A \text{ in } L^1(\mathbb{R}^d) \right\},$$

where the class C can be chosen as the class of all smooth and bounded sets or as that of all bounded polyhedra.

Proof From the semicontinuity of the norm in the space of measures it is clear that the inf in the right-hand side is larger than $\mathrm{Per}(A)$. Indeed, for any set $A \in C$ (for both choices of C) we have $\mathrm{Per}(A) = \|\nabla \mathbb{1}_A\|_{\mathcal{M}} = \mathcal{H}^{d-1}(\partial A)$ and for any sequence $A_n \in C$ with $\mathbb{1}_{A_n} \to \mathbb{1}_A$ in L^1 and $\liminf_n \mathcal{H}^{d-1}(\partial A_n) < \infty$ we can see that the sequence $u_n := \mathbb{1}_{A_n}$ is bounded in BV and, up to subsequences, we have $\nabla \mathbb{1}_{A_n} \overset{*}{\rightharpoonup} v$, where the limit v can only be $\nabla \mathbb{1}_A$ because of the convergence in the sense of distributions of $\mathbb{1}_{A_n}$ to $\mathbb{1}_A$. Then, we obtain $\mathrm{Per}(A) = \|\nabla \mathbb{1}_A\|_{\mathcal{M}} \le \liminf_n \|\nabla \mathbb{1}_{A_n}\|_{\mathcal{M}} = \liminf_n \mathrm{Per}(A_n)$.

We then only need to show the opposite inequality, i.e. to find a suitable sequence of sets $A_n \in C$. We start from the case where C is the class of smooth compact sets. We fix a set A of finite perimeter and set $u = \mathbb{1}_A$. Take a smooth and compactly supported convolution kernel $\eta_n \rightharpoonup \delta_0$ defined by rescaling a fixed kernel η_1, and define $v_n := \eta_n * u$. The function v_n is smooth and, for each n, uniformly continuous (since it has the same modulus of continuity as η_n up to a multiplicative factor $\|u\|_{L^1}$). In particular, since it is L^1 and uniformly continuous, it tends to 0 at infinity (see Exercise 6.1). In particular, its level sets $\{v_n \ge r\}$ are all bounded for every $r > 0$.

We have $\int |\nabla v_n| \le \|\nabla u\|_{\mathcal{M}}$ since the norm of the gradient is a convex functional invariant under translations, hence it decreases under convolution. We use the coarea formula to write

$$\int_0^1 \mathcal{H}^{d-1}(\{v_n = r\}) \, dr = \int_{\mathbb{R}^d} |\nabla v_n| \le \mathrm{Per}(A).$$

If we fix $\delta > 0$ we can then choose a number $r_n \in [\delta, 1 - \delta]$ such that

$$\mathrm{Per}(\{v_n \ge r_n\}) \le \mathcal{H}^{d-1}(\{v_n = r_n\}) \le \frac{\mathrm{Per}\, A}{1 - 2\delta}$$

and r_n is not a critical value for v_n (since, by Sard's lemma, a.e. value is not critical).

Box 6.2 Important Notion—*Co-area Formula*

Consider a Lipschitz function $f : \mathbb{R}^d \to \mathbb{R}^m$, where $n \ge m$. Define, at any differentiability point of f, the Jacobian factor $Jf(x) := \sqrt{\sum_\alpha |\det(M_\alpha)|^2}$ where the matrices M_α are all the $m \times m$ minors of Df. We then have the following change-of-variables formula (see [84, Section 3.43])

$$\int_{\mathbb{R}^d} g(x) \, Jf(x) \, dx = \int_{\mathbb{R}^m} \left(\int_{f^{-1}(\{y\})} g \, d\mathcal{H}^{n-m} \right) dy$$

which is valid for any measurable function $g : \mathbb{R}^m \to \mathbb{R}$ for which the integrals make sense (for instance $g \ge 0$ or $g \in L^1(\mathbb{R}^d)$). Note that, when $m = 1$, we have $Jf = |\nabla f|$, and when $m = n$ we have $Jf = |\det(Df)|$. The latter case gives the standard change-of-variable formula.

See [84, Chapter 3] for proofs and details.

248 6 Variational Problems for Sets

Sard's lemma states that, given a C^k function $f : \mathbb{R}^d \to \mathbb{R}^m$, with $k \geq \max\{n - m + 1, 1\}$, the set of critical values is Lebesgue-negligible in $\mathbb{R}^m$. In the case $m = 1$ this requires C^n functions.

In particular, the set $A_n := \{v_n \geq r_n\}$ is a smooth set (as a consequence of the implicit function theorem, and we use here $\nabla v_n \neq 0$), and we then take $u_n = \mathbb{1}_{A_n}$. As we already pointed out as a consequence of $\lim_{|x|\to\infty} v_n(x) = 0$, such a level set is compact. By construction the perimeter of A_n is bounded by $\frac{\text{Per } A}{1-2\delta}$ and u_n is bounded in BV. We want to prove that we have $u_n \to u$ in L^1. Up to subsequences, we have $u_n \to w$ strongly in L^1 and a.e. Since $u_n \in \{0, 1\}$, we also have $w \in \{0, 1\}$ a.e. Consider a point x which is a Lebesgue point (see Box 5.1) for u.

In particular, this implies $v_n(x) \to u(x)$ as $n \to \infty$. Consider the case $w(x) = 1$. Then $u_n(x) = 1$ for large n, i.e. $v_n(x) \geq r_n \geq \delta$. Then $u(x) \geq \delta$, which means $u(x) = 1$. Analogously, $w(x) = 0$ implies $u(x) = 0$. Finally we have $w = u$ and $u_n \to u$ strongly in L^1.

This means that we found a sequence of sets A_n such that $\mathbb{1}_{A_n} \to \mathbb{1}_A$ and their perimeter is bounded by $(1 - 2\delta)^{-1} \text{Per}(A)$, which shows

$$\frac{\text{Per}(A)}{1 - 2\delta} \geq \inf\left\{\liminf_n \mathcal{H}^{d-1}(\partial A_n) : A_n \in C, \mathbb{1}_{A_n} \to \mathbb{1}_A \text{ in } L^1(\mathbb{R}^d)\right\},$$

and the claim follows since $\delta > 0$ is arbitrary.

Replacing the class of smooth domains with that of polyhedra is easy, using a suitable approximation argument. Given a smooth compact domain A and a number $\delta > 0$ it is indeed possible to find a polyhedron $\tilde{A}$ with $\|\mathbb{1}_A - \mathbb{1}_{\tilde{A}}\| < \delta$ and $\text{Per}(\tilde{A}) < \text{Per}(A)(1 + \delta)$ by a simple triangulation procedure (it is enough to cover ∂A with local charts and approximate the smooth function defining the boundary with piecewise affine functions). This allows us to find a sequence $\tilde{A}_n$ of polyhedra showing the desired result. $\qquad\square$

6.1.2 The Isoperimetric Property of the Ball

In this section we will prove the following inequality: given a set A and a ball $B = B(0, R)$ with the same volume as A (i.e. $|A| = \omega_d R^d$, where ω_d stands for the volume of the unit ball in $\mathbb{R}^d$), we have $\text{Per}(A) \geq \text{Per}(B)$. This means that the ball minimizes the perimeter among sets with fixed volume, i.e. it solves the isoperimetric problem. The idea to prove this fact is

- to prove that a minimizer for the perimeter under volume constraints exists, using the theory of finite perimeter sets and the properties of the BV space,
- to prove that any set which is not a ball can be modified via a suitable transformation which preserves the volume but reduces the perimeter.

Actually, the very first step is delicate, since we are considering arbitrary sets in $\mathbb{R}^d$, and the injection of $BV(\mathbb{R}^d)$ into $L^1(\Omega)$ is not compact. It would indeed be possible for a minimizing sequence A_n to escape at infinity, so that $\mathbb{1}_{A_n}$, despite being bounded in BV, does not converge in L^1, since its pointwise limit would be 0, but this does not preserve the L^1 norm.

Two possible strategies to overcome this difficulty in dimension 2 are proposed in the exercises (see Exercises 6.6, 6.15). Since the same strategies do not work in higher dimensions, for the general proof we first add an additional constraint, imposing the sets A to be contained in a given ball $B(0, R_0)$, and minimizing among them, and later prove that the minimizer is the same ball whatever R_0 is. We follow the approach of [74], as described, for instance, in [142, Chapter 14].

We start from an easy existence result.

Proposition 6.3 *Given a number $V > 0$ and a radius R_0 such that $|B(0, R_0)| \geq V$, the optimization problem*

$$\min\{\operatorname{Per}(A) \ : \ A \subset B(0, R_0), |A| = V\}$$

admits at least one solution.

Proof We consider a minimizing sequence A_n and define $u_n := \mathbb{1}_{A_n}$. Since $\operatorname{Per}(A_n)$ is bounded and $|A_n|$ is fixed, the sequence u_n is bounded in $BV(\Omega)$ where $\Omega = B(0, R_0 + 1)$ (we add 1 to the radius so that $\partial B(0, R_0)$ is now included in the interior of the domain $\Omega = B(0, R_0 + 1)$ and there is no ambiguity between the perimeter inside Ω or in the whole space). We can then extract a subsequence u_{n_k} which converges strongly in $L^1(\Omega)$ and a.e. to a function $u \in BV(\Omega)$ and $\nabla u_{n_k} \overset{*}{\rightharpoonup} \nabla u$. The a.e. convergence shows that u is also an indicator function $u = \mathbb{1}_A$ and $A \subset B(0, R_0)$. We then have, because of the strong L^1 convergence, $|A| = \int u = \lim_k \int u_{n_k} = \lim_k |A_{n_k}| = V$ and, because of the weak-$*$ convergence of the gradients, $\operatorname{Per}(A) = ||\nabla u||_{\mathcal{M}} \leq \liminf_k ||\nabla u_{n_k}||_{\mathcal{M}} = \liminf_k \operatorname{Per}(A_{n_k})$. This shows that A solves the problem. $\qquad \square$

We now want to prove that, for each R_0 and $V \leq |B(0, R_0)|$, the ball $B := B(0, r)$ with $\omega_d r^d = V$ is a minimizer of the above problem. This will then be used to prove that, for every set A with $|A| = V$, we have $\operatorname{Per}(A) \geq \operatorname{Per}(B)$. This strategy, consisting in first proving the existence of a minimizer under an additional constraint depending on a parameter R_0 (here, being included in $B(0, R_0)$) and then proving that this minimizer is the same for every R_0, is dangerous: indeed, for the problem without constraints we could make use of simpler optimality conditions (such as those of Exercise 6.2), but this could be impossible in the constrained case, as they are based on perturbations of the minimizer which could violate the constraints; on the other hand, for the problem without constraints the existence itself of a minimizer would be more delicate and may be impossible to prove via the direct method of the calculus of variations. In our case, we will hence need to prove that the minimizer is a ball (we will prove that any ball of radius r contained in $B(0, R_0)$ is a minimizer) by using necessary optimality conditions which compare it to other competitors which are guaranteed to be contained in $B(0, R_0)$. This will be done thanks to the notion of Steiner symmetrization described in Sect. 6.1.3.

Proposition 6.4 *Given a number $V > 0$ and a radius R_0 such that $|B(0, R_0)| \geq V$, any solution of the optimization problem*

$$\min \{\mathrm{Per}(A) \, : \, A \subset B(0, R_0), |A| = V\}$$

is necessarily a ball of radius r with $\omega_d r^d = V$.

Proof We take an optimal set A and apply the symmetrization described in Sect. 6.1.3 with respect to an arbitrary $e \in \mathbb{S}^{d-1}$. First, we note that $A \subset B(0, R_0)$ implies $S_e(A) \subset B(0, R_0)$, hence we should have $\mathrm{Per}(S_e(A)) \geq \mathrm{Per}(A)$. This means that we have equality and the set A' of Lebesgue points of A is convex in the direction e (Theorem 6.8). Since this is valid for any e, we deduce that A' is convex. Now we can apply Proposition 6.9, which implies that A and $S_e(A)$ are translations of one another. This implies that A is a ball, thanks to the next lemma. □

Lemma 6.5 *A convex set A such that for every e the set $S_e(A)$ is a translation of A is necessarily a ball.*

Proof Without loss of generality, we can assume that the origin is the barycenter of A. This does not change the shape of A or $S_e(A)$, or the conclusion, but it allows us to obtain $A = S_e(A)$ for every e. In particular, A is symmetric w.r.t. every hyperplane passing through its barycenter. Now, take two points $x, x' \in \partial A$. By choosing e oriented as $x - x'$ we obtain that x and x' are at the same distance (by symmetry) from the hyperplane passing through 0 and orthogonal to e. Yet, they also have the same projection onto this hyperplane, and hence $|x| = |x'|$. Since all points of ∂A are at the same distance from the origin, A is a ball. □

We can then conclude:

Proposition 6.6 *Given a number $V > 0$, let B be a ball with radius R such that $\omega_d R^d = V$. Then, for any measurable set $A \subset \mathbb{R}^d$ with $|A| = V = |B|$ we have $\mathrm{Per}(A) \geq \mathrm{Per}(B)$.*

Proof We can assume that A is a finite perimeter set, otherwise there is nothing to prove. Then, using Proposition 6.2, we consider a sequence of smooth and compact sets A_n such that $\mathrm{Per}(A_n) \to \mathrm{Per}(A)$ and $|A_n| \to |A|$. Each set A_n is bounded, hence contained in a ball of radius $R_0(n)$, depending on n. We can assume $R_0(n)$

large enough so that $V \leq \omega_d R_0(n)^d$. We then have

$$\text{Per}(A_n) \geq \min \left\{ \text{Per}(\tilde{A}) \; : \; \tilde{A} \subset B(0, R_0(n)), |\tilde{A}| = V \right\} = \text{Per}(B_n) = d\omega_d R_n^{d-1},$$

where R_n is a radius such that $\omega_d R_n^d = |A_n|$. We then have $R_n \to R$. This shows $\text{Per}(A) \geq \text{Per}(B)$ and proves the isoperimetric inequality. $\qquad\qquad\square$

6.1.3 Steiner Symmetrization of Sets With Respect to a Hyperplane

In this section we consider the following construction: given a set $A \subset \mathbb{R}^d$ with $|A| < \infty$ and a vector $e \in \mathbb{S}^{d-1}$, we construct a set $S_e(A)$ which has the same measure as A but is symmetric w.r.t. the hyperplane $\{x \; : \; x \cdot e = 0\}$. For simplicity of notation we describe this construction when $e = e_d$ is the last element of the canonical basis, but it can of course be adapted to any unit vector e.

In this case, let us define

$$\ell(x_1, \ldots, x_{d-1}) := \mathcal{L}^1(\{x_d \; : \; (x_1, \ldots, x_{d-1}, x_d) \in A\}),$$

where $\mathcal{L}^1$ is the one-dimensional Lebesgue measure. For a.e. $(x_1, \ldots, x_{d-1}) \in \mathbb{R}^{d-1}$ we have $\ell(x_1, \ldots, x_{d-1}) < +\infty$. We then define

$$S_e(A) := \left\{ (x_1, \ldots, x_{d-1}, x_d) \; : \; |x_d| < \frac{\ell(x_1, \ldots, x_{d-1})}{2} \right\}.$$

We can easily see that $S_e(A)$ is symmetric w.r.t. $\{x_d = 0\}$ and has the same measure as A. We now want to prove that $S_e(A)$ has a smaller perimeter than A, and in many cases strictly smaller.

We start from the easiest case.

Proposition 6.7 *Assume that A is a polyhedron and that none of its faces contains segments directed as $e = e_d$. Then we have*

$$\text{Per}(S_e(A)) + \frac{6\mathcal{L}^{d-1}(B)^2}{\text{Per}(A)} \leq \text{Per}(A)$$

where B is the set of all points $(x_1, \ldots, x_{d-1}) \in \mathbb{R}^{d-1}$ such that the set $\{x_d \; : \; (x_1, \ldots, x_{d-1}, x_d) \in A\}$ is disconnected (i.e. it is composed of more than one segment).

Proof Let π be the projection onto the $(d-1)$-dimensional plane orthogonal to e. The set $\pi(A)$ is a polyhedron which can be decomposed into a finite number of polyhedra F_i such that, for each i, there is a number $N_i \geq 1$ and some affine functions $f_{i,j}^{\pm}$ for $j = 1, \ldots, N_i$ with

$$A \cap \pi^{-1}(F_i) = \bigcup_{j=1}^{N_i} \{x \in \mathbb{R}^d : x_d \in [f_{i,j}^-(x_1, \ldots, x_{d-1}), f_{i,j}^+(x_1, \ldots, x_{d-1})]\}.$$

Note that we have $B = \bigcup_{i \,:\, N_i > 1} F_i$. Moreover, we have

$$\ell(x_1, \ldots, x_{d-1}) = \sum_{j=1}^{N_i} f_{i,j}^+(x_1, \ldots, x_{d-1}) - f_{i,j}^-(x_1, \ldots, x_{d-1})$$

and

$$\text{Per}(A) = \sum_i \sum_{j=1}^{N_i} \int_{F_i} \left(\sqrt{1 + |\nabla f_{i,j}^+|^2} + \sqrt{1 + |\nabla f_{i,j}^-|^2} \right) \, dx,$$

while

$$\text{Per}(S_e(A)) = 2 \sum_i \int_{F_i} \sqrt{1 + \left| \nabla \frac{\ell}{2} \right|^2} \, dx = \sum_i \int_{F_i} \sqrt{4 + |\nabla \ell|^2} \, dx,$$

where all these integrals on F_i are actually integrals of constant functions. We then use the convexity of the function $v \mapsto \sqrt{1 + |v|^2}$ to obtain

$$\frac{1}{2N_i} \sum_{j=1}^{N_i} \int_{F_i} \left(\sqrt{1 + |\nabla f_{i,j}^+|^2} + \sqrt{1 + |\nabla f_{i,j}^-|^2} \right) \, dx$$

$$\geq \int_{F_i} \sqrt{1 + \left| \nabla \frac{1}{2N_i} \sum_{j=1}^{N_i} f_{i,j}^+(x_1, \ldots, x_{d-1}) - f_{i,j}^-(x_1, \ldots, x_{d-1}) \right|^2} \, dx.$$

We deduce

$$\text{Per}(A) \geq \sum_i \sum_{j=1}^{N_i} \int_{F_i} \sqrt{4N_i^2 + |\nabla \ell|^2} \, dx.$$

From $N_i \geq 1$ we immediately obtain $\mathrm{Per}(A) \geq \mathrm{Per}(S_e(A))$ but we can do more. Indeed, we have

$$\mathrm{Per}(A) - \mathrm{Per}(S_e(A)) \geq \sum_{i\,:\,N_i>1} \int_{F_i} \left(\sqrt{16 + |\nabla \ell|^2} - \sqrt{4 + |\nabla \ell|^2} \right) \, dx.$$

We then use

$$\sqrt{16 + |\nabla \ell|^2} - \sqrt{4 + |\nabla \ell|^2} = \frac{12}{\sqrt{16 + |\nabla \ell|^2} + \sqrt{4 + |\nabla \ell|^2}} \geq \frac{6}{\sqrt{16 + |\nabla \ell|^2}}$$

and, thanks to the Hölder inequality, we have

$$\sum_{i:N_i>1} \int_{F_i} \frac{1}{\sqrt{16 + |\nabla \ell|^2}} dx \sum_{i:N_i>1} \int_{F_i} \sqrt{16 + |\nabla \ell|^2} dx$$

$$\geq \left(\sum_{i:N_i>1} \mathcal{L}^{d-1}(F_i) \right)^2 = \mathcal{L}^{d-1}(B)^2.$$

Together with

$$\sum_{i\,:\,N_i>1} \int_{F_i} \sqrt{16 + |\nabla \ell|^2} \, dx \leq \sum_{i} \int_{F_i} \sqrt{4N_i^2 + |\nabla \ell|^2} \, dx \leq \mathrm{Per}(A)$$

we obtain the desired inequality. □

By approximation, we can obtain the following result:

Theorem 6.8 *Let A be a finite perimeter set in $\mathbb{R}^d$ and e a unit vector. Then we have $\mathrm{Per}(S_e(A)) \leq \mathrm{Per}(A)$. Moreover, if $\mathrm{Per}(S_e(A)) = \mathrm{Per}(A)$ then the set of Lebesgue points of A is convex in the direction e: whenever it contains two points x, x' with $x - x'$ oriented as e, it also contains the segment $[x, x']$.*

Proof Given A there exists a sequence of polyhedral sets A_n such that $\mathbb{1}_{A_n} \to \mathbb{1}_A$ in L^1 and $\mathrm{Per}(A_n) \to \mathrm{Per}(A)$. Without loss of generality, we can assume that all A_n are polyhedra with no face containing vectors parallel to e (otherwise we just need to tilt them a little bit). We then have

$$\mathrm{Per}(S_e(A_n)) \leq \mathrm{Per}(S_e(A_n)) + \frac{6\mathcal{L}^{d-1}(B_n)^2}{\mathrm{Per}(A_n)} \leq \mathrm{Per}(A_n).$$

The L^1 convergence of $\mathbb{1}_{A_n}$ to $\mathbb{1}_A$ can be seen as L^1 convergence of functions defined on $\mathbb{R}^{d-1}$, valued in $L^1(\mathbb{R})$. This implies in particular the convergence of the functions ℓ_n associated with the sets A_n to ℓ (associated with A), and hence the sets $S_e(A_n)$ converge (in the sense that their indicator functions converge in L^1 to the indicator function of the limit) to $S_e(A)$. By semicontinuity of the perimeter we obtain

$$\text{Per}(S_e(A)) \leq \liminf_n \text{Per}(S_e(A_n)) \leq \liminf \text{Per}(A_n) = \text{Per}(A),$$

which proves the first part of the claim. Moreover, in the case of equality $\text{Per}(S_e(A)) = \text{Per}(A)$ we also obtain $\lim_n \frac{6\mathcal{L}^{d-1}(B_n)^2}{\text{Per}(A_n)} = 0$, i.e. $\mathcal{L}^{d-1}(B_n) \to 0$. This means that for a.e. point of $\mathbb{R}^{d-1}$ the set $\{x_d : (x_1, \ldots, x_{d-1}, x_d) \in A_n\}$ is a segment for large n. Hence, for a.e. point of $\mathbb{R}^{d-1}$ the set $\{x_d : (x_1, \ldots, x_{d-1}, x_d) \in A\}$ is also a segment. If we denote by A' the set of Lebesgue points of A, we then deduce that for every point of $\mathbb{R}^{d-1}$ the set $\{x_d : (x_1, \ldots, x_{d-1}, x_d) \in A'\}$ is also a segment.[2] If this was not the case, there would be two points $x_a := (x_1, \ldots, x_{d-1}, a)$, $x_b := (x_1, \ldots, x_{d-1}, b) \in A'$ and a number $c \in (a, b)$ such that $x_c := (x_1, \ldots, x_{d-1}, c) \notin A'$. Yet, if the points x_a and x_b are of density 1 for A, knowing that almost any vertical section of A is convex, we also deduce that x_c is of density 1 (more precisely, the proportion of points in $B(x_c, r)$ not belonging to A is at most the sum of the proportion of points in $B(x_a, r)$ not belonging to A and of the proportion of points in $B(x_b, r)$ not belonging to A, these two proportions tending to 0 by the fact that x_a and x_b are of density 1). This proves the claim. $\square$

We also have the following result:

Proposition 6.9 *Let A be a bounded convex set in $\mathbb{R}^d$ and e a unit vector. If $\text{Per}(S_e(A)) = \text{Per}(A)$ then A is a translation of $S_e(A)$ in a direction parallel to the vector e.*

Proof If A is convex we can write

$$A = \left\{ x = (x_1, \ldots, x_{d-1}, x_d) : \begin{array}{l} (x_1, \ldots, x_{d-1}) \in \pi(A), \\ f^-(x_1, \ldots, x_{d-1}) \leq x_d \leq f^+(x_1, \ldots, x_{d-1}) \end{array} \right\},$$

where f^+ is a suitable concave function and f^- a suitable convex function. We can then perform globally a computation similar to the one we performed in Proposition 6.7 and obtain

$$\text{Per}(A) = \mathcal{H}^{d-1}(\partial_e A) + \int_{\pi(A)} \left(\sqrt{1 + |\nabla f^+|^2} + \sqrt{1 + |\nabla f^-|^2} \right) dx,$$

---

[2] This simple and clarifying fact was not explicitly mentioned in the original paper by De Giorgi, [74], but is an observation contained in [91] and due to G. Alberti and N. Fusco.

where $\partial_e A$ is the vertical part of the boundary[3] of A, situated in $\pi^{-1}(\partial(\pi(A)))$. On the other hand we have

$$\text{Per}(S_e(A)) = \mathcal{H}^{d-1}(\partial_e(S_e(A)) + \int_{\pi(A)} 2\sqrt{1 + |\nabla \ell|^2}\, dx,$$

where $\ell = f^+ - f^-$. Moreover, by construction, $\mathcal{H}^{d-1}(\partial_e(S_e(A)) = \mathcal{H}^{d-1}(\partial_e A)$ since above each point of $\partial(\pi(A)) = \partial(\pi(S_e(A)))$ the two sets contain a segment of the same length. Hence, in the case of equality $\text{Per}(S_e(A)) = \text{Per}(A)$, we can conclude that we have a.e. the equality $\sqrt{1 + |\nabla f^+|^2} + \sqrt{1 + |\nabla f^-|^2} = 2\sqrt{1 + |\nabla \ell|^2}$ (since there would be equality of the integrals, but we have a pointwise inequality because of the same convexity argument as in Proposition 6.7). This implies, by strict convexity and using $\ell = f^+ - f^-$, that we have $\nabla f^+ = -\nabla f^- = \nabla \ell/2$. We then find $f^+ = \ell/2 + c^+$ and $f^- = -\ell/2 + c^-$ for two real constants $c^\pm$. By subtracting these two equalities we obtain $c^+ = c^-$, which shows that A is the translation of $S_e(A)$ by a vector $c^+ e$.   □

6.2   Optimization of the First Dirichlet Eigenvalue

The starting point of this section is the following variational problem in a fixed open domain Ω:

$$\min\left\{ R(u) := \frac{\int_\Omega |\nabla u|^2\, dx}{\int_\Omega u^2\, dx} \; : \; u \in H_0^1(\Omega) \setminus \{0\} \right\}.$$

The ratio $R(u)$ is called the Rayleigh quotient. Finding a lower bound on it is equivalent to finding a constant in the Poincaré inequality for functions in $H_0^1(\Omega)$. We first study this problem for fixed Ω:

Proposition 6.10 *For every bounded domain Ω the problem*

$$\min\{R(u) \; : \; u \in H_0^1(\Omega) \setminus \{0\}\}$$

admits a solution. Such a solution can be chosen non-negative, is C^∞ in the interior of Ω and satisfies $-\Delta u = \lambda_1(\Omega)u$ where $\lambda_1(\Omega) > 0$ is the smallest positive number

---

[3] Another option to get rid of this vertical part is to observe that the proofs of Propositions 6.7 and 6.8 could be adapted to show that the Steiner symmetrization decreases not only the total perimeter $\text{Per}(A)$ but also the relative perimeters $\text{Per}(A; \Omega)$ for every open set Ω which is—after choosing a coordinate frame where $e = e_d$—a cylinder of the form $\{x \in \mathbb{R}^d \; : \; (x_1, \ldots, x_{d-1}) \in \Omega'\}$ for an open set $\Omega' \in \mathbb{R}^{d-1}$, and that the equality $\text{Per}(A) = \text{Per}(S_e(A))$ implies the equality $\text{Per}(A; \Omega) = \text{Per}(S_e(A); \Omega)$ in any such open set. This allows us to restrict to open sets compactly contained in $\pi(A)$.

λ *such that a non-zero solution to* $-\Delta u = \lambda(\Omega)u$ *exists. Moreover,* $\lambda_1(\Omega)$ *is also the value of the problem.*

Proof Let us start from the existence of a solution. We take a minimizing sequence $u_n \in H_0^1(\Omega)$ and it is not restrictive, by multiplying it by a constant, to assume $||u_n||_{L^2} = 1$. Then, the bound on $R(u_n)$ becomes a bound on $||\nabla u_n||_{L^2}$ and u_n is bounded in H^1. We can then extract a subsequence which converges weakly in H^1 to a function $u \in H_0^1(\Omega)$. The weak convergence $\nabla u_n \rightharpoonup \nabla u$ in L^2 implies $||\nabla u||_{L^2} \leq \liminf_n ||\nabla u_n||_{L^2}$, while the strong convergence $u_n \to u$ in L^2 (due to the compact embedding of H^1 into L^2) proves $||u||_{L^2} = 1$. Hence $R(u) \leq \liminf_n R(u_n)$ and u is a minimizer. Of course any multiple of u is also a minimizer, but the same is true for $|u|$. Hence, we can assume that u is non-negative.

We now look at the optimality conditions. We take a function $\phi \in H_0^1(\Omega)$ and look at $R(u + \varepsilon\phi)$. We have

$$R(u + \varepsilon\phi) = \frac{\int_\Omega |\nabla u|^2 \, dx + 2\varepsilon \int_\Omega \nabla u \cdot \nabla \phi \, dx + o(\varepsilon)}{1 + 2\varepsilon \int_\Omega u\phi \, dx + o(\varepsilon)}$$

$$= \left(\int_\Omega |\nabla u|^2 \, dx + 2\varepsilon \int_\Omega \nabla u \cdot \nabla \phi \, dx + o(\varepsilon) \right)$$

$$\times \left(1 - 2\varepsilon \int_\Omega u\phi \, dx + o(\varepsilon) \right)$$

$$= \int_\Omega |\nabla u|^2 \, dx + 2\varepsilon \left(\int_\Omega \nabla u \cdot \nabla \phi \, dx - M \int_\Omega u\phi \, dx \right) + o(\varepsilon),$$

where $M = \int_\Omega |\nabla u|^2 = R(u)$ equals the value of the minimum. By optimality, we then obtain

$$\int_\Omega \nabla u \cdot \nabla \phi \, dx - M \int_\Omega u\phi \, dx = 0$$

for every ϕ, which means $-\Delta u = Mu$ in the weak sense. From $u \in L^2$ we obtain $u \in H_{loc}^2$ and, differentiating the equation, we then obtain $u \in H_{loc}^4$ and so on, up to $u \in C^\infty$.

This also proves that u is an eigenfunction of the (negative) Laplacian operator, with eigenvalue equal to the minimal value of the problem. Moreover, for any eigenvalue λ, if v is a corresponding eigenfunction, i.e. a solution of $-\Delta v = \lambda v$, we easily see (by testing against v and integrating by parts) that we have $R(v) = \lambda$. This shows that among all the eigenvalues the one corresponding to the minimizer is the minimal one, so we write $M = \lambda_1(\Omega)$. $\qquad\square$

Remark 6.11 By proving that solutions of $-\Delta u = \lambda u$ are analytic (which we did not do in Sect. 2.3, where we only proved it for $\lambda = 0$) or applying a strong maximum principle for elliptic PDEs (see for instance [98, Theorem 3.5]), it is also possible to prove that, when Ω is connected, there is a unique (up to multiplicative constants) minimizer u, but we will not do this in this book.

When studying the problem in a fixed domain Ω, we see that a geometric quantity called $\lambda_1(\Omega)$ appears. This quantity is also the inverse of the optimal constant in the Poincaré inequality on $H_0^1(\Omega)$. Since when $\Omega' \supset \Omega$ all the functions in $H_0^1(\Omega)$ can be extended to 0 and such an extension belongs to $H_0^1(\Omega')$, we easily see that we have $\lambda_1(\Omega') \leq \lambda_1(\Omega)$.

A natural question in shape optimization (the branch of optimization theory where the unknown is a set Ω and we optimize quantities related to the solution of some PDEs in Ω) is to find the domain Ω which minimizes $\lambda_1(\Omega)$ under volume constraints.[4] The answer is, again, that the optimal set is the ball, as we state in the next theorem, also known as the *Faber–Krahn inequality* (see [85, 129, 130]).

Theorem 6.12 *Let $\Omega \subset \mathbb{R}^d$ be an arbitrary bounded domain and B the ball with the same measure as Ω. Then we have $\lambda_1(\Omega) \geq \lambda_1(B)$.*

We note that in the proof of this fact the definition of λ_1 as an eigenvalue is not important, and the result would stay true even if Ω was unbounded, but with finite measure, if defining $\lambda_1(\Omega) := \inf\{R(u) : u \in C_c^\infty(\Omega), u \neq 0\}$.

Proof The proof consists in taking an arbitrary competitor in the minimization problem defining $\lambda_1(\Omega)$ and associating with it a function in $H_0^1(B)$ with a smaller Rayleigh quotient. The construction will be made by a radially decreasing rearrangement, which makes this new competitor also radial, but this is not important. For this, it is sufficient to take $u \geq 0$ in $C_c^\infty(\Omega)$ (we can restrict to $u \geq 0$ thanks to Proposition 6.10 and then to $u \in C_c^\infty$ by density) and consider u°, its radially decreasing rearrangement (see Definition 6.13), which is clearly supported on B. The *Pólya–Szegő inequality* (see Remark 6.15 below) proves that we have $\|\nabla u^\circ\|_{L^2} \geq \|\nabla u\|_{L^2}$ and the definition itself guarantees $\|u^\circ\|_{L^2} = \|u\|_{L^2}$, so that $R(u^\circ) \leq R(u)$. Hence, we have $\lambda_1(B) \leq R(u^\circ) \leq R(u)$ and, u being arbitrary, we obtain $\lambda_1(B) \leq \lambda_1(\Omega)$. $\square$

Definition 6.13 Given a nonnegative function $u \in L^1(\mathbb{R}^d)$ we define its radially decreasing rearrangement as the unique function u° such that its level sets $\{u^\circ > t\}$ are balls with the same measure of the corresponding level set $\{u > t\}$. The function u° can be defined as

$$u^\circ(x) = \int_0^\infty \mathbb{1}_{B(0,R(t))} \, dt$$

where $R(t) := (|\{u > t\}|/\omega_d)^{1/d}$ is the radius of a ball with the same measure as the level set $\{u > t\}$.

---

[4] This recalls the isoperimetric problem, where λ_1 was replaced by the perimeter. Yet, we note that the perimeter is no longer involved, which makes it very inappropriate that some mathematicians still call these kinds of optimization problems "generalized isoperimetric inequalities".

Theorem 6.14 *For every* $u \in C_c^\infty(\mathbb{R}^d)$, $u \geq 0$, *and every increasing function* φ : $\mathbb{R}_+ \to \mathbb{R}$ *with* $\varphi(0) = 0$ *we have* $\int \varphi(u) = \int \varphi(u^\circ)$. *For every function* $f : \mathbb{R}_+ \times \mathbb{R}_+ \to \mathbb{R}$ *with* $f(0,s) = f(t,0) = 0$ *for every* $s, t \in \mathbb{R}_+$ *which is convex in the second variable we have*

$$\int_{\mathbb{R}^d} f(u, |\nabla u|)\, dx \geq \int_{\mathbb{R}^d} f(u^\circ, |\nabla u^\circ|)\, dx.$$

Remark 6.15 As a particular case of the above theorem, we have $\|\nabla u\|_{L^p} \geq \|\nabla u^\circ\|_{L^p}$, a fact which is known as the *Pólya–Szegő inequality* (see [170]).

Proof The layer-cake representation of the integral of a nonnegative function $\int u = \int_0^\infty |\{u > t\}|\, dt$ shows that the integral of a function only depends on the measure of its level sets, which easily shows that we have $\int \varphi(u) = \int \varphi(u^\circ)$ for every φ (the condition $\varphi(0) = 0$ is required for the finiteness of the integral when integrating on the whole space, while the monotone increasing behavior of φ is not strictly needed, even if it easily shows the correspondence between the level sets).

For the integrals depending on the gradient, we first observe that every convex function f vanishing at 0 can be written as $f(s) = sg(1/s)$ for a convex and nonincreasing function g. Indeed, we take $g(1/s) = f(s)/s$, which is clearly a nondecreasing function of s because of the convexity of f and $f(0) = 0$; this proves that g is nonincreasing. Then, we write

$$g(s) = sf(1/s) = sf^{**}(1/s) = s \sup_t \frac{t}{s} - f^*(t) = \sup_t t - sf^*(t),$$

which proves that g is also convex as a sup of affine functions of s. We should then consider the quantity

$$\int_{\mathbb{R}^d} f(u, |\nabla u|)dx = \int_{\mathbb{R}^d} |\nabla u| g\left(u, \frac{1}{|\nabla u|}\right)$$

$$dx = \int_0^\infty dt \int_{\{u=t\}} g\left(t, \frac{1}{|\nabla u|}\right) d\mathcal{H}^{d-1},$$

where the correspondence between f and g is done in the second variable, and the last equality is due to the co-area formula (Box 6.2). We then use the fact that for a.e. t the gradient ∇u does not vanish on $\{u = t\}$ (Sard's lemma) and $\{u = t\} = \partial\{u > t\}$. By the isoperimetric inequality, since $\{u > t\}$ and $\{u^\circ > t\}$ have the same measure and the latter is a ball, we have $\mathcal{H}^{d-1}(\{u = t\}) \geq \mathcal{H}^{d-1}(\{u^\circ = t\})$. Moreover, we have, again by the coarea formula

$$|\{u > t_0\}| = \int_{t_0}^\infty dt \int_{\{u=t\}} \frac{1}{|\nabla u|} d\mathcal{H}^{d-1},$$

and a corresponding formula holds for u°. This implies that for a.e. t we have

$$\int_{\{u=t\}} \frac{1}{|\nabla u|} \, d\mathcal{H}^{d-1} = \int_{\{u^\circ=t\}} \frac{1}{|\nabla u^\circ|} \, d\mathcal{H}^{d-1}.$$

We then use the convexity of g to write, for a.e. t

$$\int_{\{u=t\}} g\left(t, \frac{1}{|\nabla u|}\right) d\mathcal{H}^{d-1} = \mathcal{H}^{d-1}(\{u=t\}) \fint_{\{u=t\}} g\left(t, \frac{1}{|\nabla u|}\right) d\mathcal{H}^{d-1}$$

$$\geq \mathcal{H}^{d-1}(\{u=t\})g\left(t, \fint_{\{u=t\}} \frac{1}{|\nabla u|} \, d\mathcal{H}^{d-1}\right)$$

$$= \mathcal{H}^{d-1}(\{u=t\})g\left(t, \frac{\int_{\{u=t\}} \frac{1}{|\nabla u|} \, d\mathcal{H}^{d-1}}{\mathcal{H}^{d-1}(\{u=t\})}\right),$$

and the fact that g is nonincreasing allows us to obtain

$$\int_{\{u=t\}} g\left(t, \frac{1}{|\nabla u|}\right) d\mathcal{H}^{d-1} \geq \mathcal{H}^{d-1}(\{u^\circ=t\}) \, g\left(t, \frac{\int_{\{u^\circ=t\}} \frac{1}{|\nabla u^\circ|} \, d\mathcal{H}^{d-1}}{\mathcal{H}^{d-1}(\{u^\circ=t\})}\right).$$

If one replaces from the beginning u with u° all these inequalities are equalities, in particular because $|\nabla u^\circ|$ is constant on $\{u^\circ = t\}$ since u° is radially symmetric. This shows

$$\int_{\mathbb{R}^d} |\nabla u| g\left(u, \frac{1}{|\nabla u|}\right) dx \geq \int_{\mathbb{R}^d} |\nabla u^\circ| g\left(u^\circ, \frac{1}{|\nabla u^\circ|}\right) dx,$$

and the claim.   □

6.3   Optimal 1D Networks and the Gołąb Theorem

In this section we briefly present some variational problems which all share the following features: the unknown is a one-dimensional connected set whose length is prescribed or minimized, and the optimization problem involves, besides its length, a penalization or a constraint which forces it to be spread as much as possible in the domain (as close as possible to some points of the domain, for instance) or, more generally, to be as large as possible. These 1D connected sets could be called "networks". The important point compared to many other problems in applied mathematics usually classified as network optimization is that here the network itself is the main unknown (differently from what happens, for instance, in the problems related to Wardrop equilibria described in Sect. 4.6, where the network is given and we optimize the traffic on it).

We will discuss some examples of such problems together with their interpretation, and provide the tools to prove the existence of a solution. The qualitative description of the minimizers and their regularity will only be informally described with references, as they are outside the scope of an elementary book such as this one.

6.3.1   Steiner Trees, Irrigation and Transportation Networks, and Membrane Reinforcement

This section begins with a very classical problem: the Steiner problem of the optimal connection of a given number of points.

Steiner Trees Many optimization problems are known in mathematics and computer science under the name of Steiner tree problems, and they all involve the connection of a given set of points under some constraints. The most widely known is probably the Steiner tree problem on graphs. Given a non-oriented graph with a cost on each edge, and a subset of vertices (called terminals), we look for a tree of minimum total cost (equal to the sum of the costs of the edges composing the tree) that contains all such vertices (but may include additional vertices). This generalizes, for instance, two famous combinatorial optimization problems: the search for a shortest path in a graph and the problem of finding a minimum spanning tree. Indeed, if we only prescribe two terminals then the Steiner tree problem reduces to finding the shortest path connecting them. If all the vertices of the graph are terminals, then the Steiner tree problem is equivalent to finding a minimum spanning tree (i.e. a connected set of edges with minimal total cost arriving at each vertex of the graph). However, while these two precise problems are solvable in polynomial time, this is not the case for the Steiner tree problem, which is an NP-hard problem. More precisely, its decision variant is NP-complete: given the graph, the costs and the terminals, we wonder whether there exists a tree of cost less than some given threshold connecting all the terminals; this problem was among Karp's original list of 21 NP-complete problems, [125].

However, the Steiner tree problem on graphs can be classified as discrete optimization, while the variant which is closer to the spirit of the calculus of variations is the Euclidean Steiner tree problem. In this variant we do not look at a graph, a set of terminals is prescribed as a finite subset of the Euclidean space $\mathbb{R}^d$, and any possible tree is admitted. For this, we first need to give a precise definition of length for an arbitrary measurable subset of $\mathbb{R}^d$ (or, more generally, of a metric space). We will choose as a definition of the length the value of the one-dimensional Hausdorff measure $\mathcal{H}^1$ (see Box 6.4).

The problem then consists in solving

$$\min\left\{\mathcal{H}^1(\Sigma) \quad : \quad \Sigma \subset \mathbb{R}^d \text{ compact, connected, and containing } D\right\}, \quad (6.1)$$

where $D \subset \mathbb{R}^d$ is a given set of points, usually finite. In this case, it can be proven that the optimal set Σ is a network composed of finitely many segments whose endpoints either belong to D or are extra points, called *Steiner points* which are of order 3 (i.e. three segments stem out of these points) and the three segments at each Steiner point lie on the same plane and meet at 120°. Moreover, the network is a tree, i.e. it has no cycles, and the maximum number of Steiner points that a Steiner tree can have is #$D-2$. The case where #$D = 3$ is at the basis of the whole analysis, and shows two possible behaviors: if the angles of the triangle formed by the three given points are all less than 120 degrees, the solution is given by a Steiner point located at the so-called *Fermat point*; otherwise the solution is given by the two sides of the triangle which form an angle of 120 or more degrees. We refer to [100] for a general presentation of the problem, and to [166] for modern mathematical results in a more general framework, including the case where D is not finite (see also [167] for an example where D is fractal), or the case where the ambient space is a general metric space.

Box 6.4 Important Notion—*Hausdorff Measures*

Hausdorff measures are a notion of k-dimensional "surface" measures for sets contained in a higher-dimensional Euclidean space or even in a metric space (X, d).

With each $k \geq 0$ we associate a constant ω_k whose general definition is given by $\omega_k = \pi^k \left(\int_0^\infty t^{k/2} e^{-t}\, dt\right)^{-1}$ (the denominator being the so-called Euler's Γ function) but whose main property is that, when k is an integer, it coincides with the volume of the unit ball in $\mathbb{R}^k$. Given $k \geq 0$, we consider a subset $A \subset X$, a positive number $\delta > 0$, and we define

$$\mathcal{H}^k_\delta(A) := \frac{\omega_k}{2^k} \inf\left\{\sum_i (\operatorname{diam}(A_i))^k \ : \ A \subset \bigcup_i A_i, \ \operatorname{diam}(A_i) < \delta\right\},$$

where the infimum is taken among countable coverings of A. This means that we try to cover A with small sets and consider the sum of the power k of their diameters. We then define

$$\mathcal{H}^k(A) := \sup_{\delta > 0} \mathcal{H}^k_\delta(A) = \lim_{\delta \to 0} \mathcal{H}^k_\delta(A).$$

(continued)

Box 6.4 (continued)
It is possible to prove the following facts:

- The quantity $\mathcal{H}^k_\delta$ is additive on pairs of sets whose distance is at least 2δ; this property is inherited by $\mathcal{H}^k$, which is additive on Borel sets and is hence a nonnegative measure. On the other hand, it is not necessarily a finite measure.
- If $X = \mathbb{R}^d$ and $A \subset \mathbb{R}^k \times \{0\} \subset \mathbb{R}^d$ then $\mathcal{H}^k(A) = \mathcal{L}^k(A)$.
- When $k = 0$ we have $\mathcal{H}^0(A) = \#A$, i.e. $\mathcal{H}^0$ is the counting measure.
- If $A = \gamma([0, 1])$ is the image of an injective curve, then $\mathcal{H}^1(A) =$ Length(γ).
- If for a set A we have $\mathcal{H}^{k_0}(A) \in (0, \infty)$ for some k_0, then we have $\mathcal{H}^k(A) = +\infty$ for every $k < k_0$ and $\mathcal{H}^k(A) = 0$ for every $k > k_0$. It is possible to define the Hausdorff dimension of a set A as $\inf\{k \,:\, \mathcal{H}^k(A) = 0\}$.
- Whenever $\varphi = X \to Y$ is an L-Lipschitz function we have $\mathcal{H}^k(\varphi(A)) \leq L^k \mathcal{H}^k(A)$.

One can also consider non-integer k and sets with non-integer dimension are often called *fractals*. As an example, we can consider the standard triadic Cantor set C in $[0, 1]$, which has $\mathcal{H}^k(C) = 2^{-k}\omega_k$ for $k = \log 2/\log 3$.

We refer to [15, Chapter 2] for proofs and details.

Existence Results: Hausdorff Convergence and the Gołąb Theorem In the case of the Euclidean Steiner tree problem with finitely many terminals we can reduce the study to a finite-dimensional issue, after proving that we only need to choose at most $\#D - 2$ extra points and the topology of the network (which is the most difficult part of the optimization: once the topology is fixed, the choice of the position of the points amount to minimizing a sum of length, i.e. a convex function). Yet, this is no longer possible in the most general variants of the problem (and will not be possible for the other problems that we will present in the rest of the section). Hence, we need an abstract theorem which guarantees the existence of an optimal set. This requires a notion of convergence on sets to be chosen which is relevant for these problems. Note that the choice which was made in Sect. 6.1 (i.e. identifying sets with their indicator functions and using the space BV) is no longer reasonable, since here we face sets which are often Lebesgue-negligible (indeed, if $\mathcal{H}^1(A) < +\infty$ and $A \subset \mathbb{R}^d$ for $d > 1$ we necessarily have $\mathcal{L}^d(A) = 0$), which would then be identified with the 0 function... Moreover, the choice of the space BV was adapted to deal with perimeters, while here, even when $d = 2$ and hence perimeters are also given by lengths, we consider the length of sets which are not necessarily boundaries of other sets.

Box 6.5 Important Notion—*Hausdorff Convergence*

Definition On a bounded metric space (X, d) we define the Hausdorff distance d_H between two closed subsets $A, B \subset X$ via

$$d_H(A, B) := \max\left\{ \sup\{d(x, A) : x \in B\}, \sup\{d(x, B) : x \in A\} \right\}$$
$$= \inf\{\delta > 0 : A \subset B_\delta, B \subset A_\delta\}.$$

We can prove the following facts

- d_H is a distance on the set $C(X)$ of closed subsets of X;
- a sequence A_n converges for d_H (we say that it Hausdorff converges) to A if and only if the functions $x \mapsto d(x, A_n)$ uniformly converge to $x \mapsto d(x, A)$;
- if (X, d) is complete then $(C(X), d_H)$ is also complete;
- if (X, d) is compact then $(C(X), d_H)$ is also compact.

The notion of Hausdorff convergence is very closely related to that of Kuratowski convergence (see Definition 7.7). When the ambient space X is compact, the two notions coincide.

We refer to [15, Chapter 4] for proofs and details.

With the notion of Hausdorff convergence it is then possible to prove existence results using the direct method of the calculus of variations, since from any sequence of competitors Σ_n, as soon as they are contained in some compact set, it is possible to extract a subsequence which converges in the Hausdorff sense. The other ingredient consists in the semicontinuity of the functional to be minimized, and in the case of the length functional this is exactly the content of the so-called Gołąb theorem that we state below and prove in the next subsection.

Theorem 6.16 *Let X be a compact subset of the Euclidean space $\mathbb{R}^d$ and $C_n \subset X$ a sequence of closed connected subsets Hausdorff-converging to a closed set $C \subset X$. Then C is connected and we have*

$$\mathcal{H}^1(C) \leq \liminf_n \mathcal{H}^1(C_n).$$

The proof will be the object of Sect. 6.3.2.

Remark 6.17 We note that the assumptions that the sets are connected cannot be removed in the Gołąb theorem: otherwise just take a sequence of sets A_n defined by numbering the rational points of $[0, 1]^2$ and adding a segment of length $2^{-j}/n$

around the j-th rational point $q_j \in [0, 1]^2$ for $j \leq n$. Since $A_n \supset \{q_1, q_2, \ldots, q_n\}$ and the set $\{q_j : j \in \mathbb{N}\}$ is dense, we can see that we have $d_H(A_n, [0, 1]^2) \to 0$ but $\mathcal{H}^1(A_n) \leq 1/n$ while $\mathcal{H}^1([0, 1]^2) = +\infty$. However, this does not mean that connectedness is necessary, nor that it is the only reasonable assumption to require for the semicontinuity to hold. For instance, in [70], the authors introduce what they call the *concentration property* (a set $A \subset \Omega$ is said to satisfy it if, roughly speaking, inside every ball $B(x, r) \subset \Omega$ centered at a point $x \in A$ we can find a smaller sub-ball of comparable radius which contains enough length of A) and prove the same lower semicontinuity result whenever C_n is a sequence of sets satisfying the concentration property in a uniform way. Anyway, as the reader may guess from the fact that we did not provide a precise definition for this property, connectedness is a much easier notion to deal with.

Remark 6.18 We also note that, even under connectedness assumptions, the result of the Gołąb theorem could not hold in the case of $\mathcal{H}^k$ for $k > 1$. For instance, we can consider the same rational points q_j as before and define $A_n \subset [0, 1]^3$ via $A_n = [0, 1]^2 \times \{1\} \cup \bigcup_{j=1}^{n} \partial B(q_j, 2^{-j}/n) \times [0, 1]$. We then have a sequence of connected sets A_n composed of a horizontal square and many vertical cylinders. We also have $d_H(A_n, [0, 1]^3) \to 0$ but $\mathcal{H}^2(A_n) \leq 1 + 1/n$ while $\mathcal{H}^2([0, 1]^3) = +\infty$.

Even if the reader will have already guessed the procedure, we describe in detail how to prove the existence of an optimal set for Problem (6.1).

Proposition 6.19 *If $D \subset \mathbb{R}^d$ is compact and contained in a compact connected set with finite $\mathcal{H}^1$ measure, then the problem*

$$\min\left\{\mathcal{H}^1(\Sigma) : \Sigma \subset \mathbb{R}^d \text{ compact, connected, and containing } D\right\}$$

admits at least one solution.

Proof We take a minimizing sequence Σ_n. Let π be the projection onto the convex hull of D: such a map is 1-Lipschitz and it is the identity on D. It reduces the Hausdorff measures, so that $\tilde{\Sigma}_n := \pi(\Sigma_n)$ is again a minimizing sequence. The sets $\tilde{\Sigma}_n$ are contained in the same compact set (the convex hull of D, which is contained in a ball), so that we can extract a Hausdorff convergent subsequence. We call Σ its limit. The set Σ contains D: indeed, for every $x \in D$ we had $d(x, \tilde{\Sigma}_n) = 0$ and by the uniform convergence of the distance functions we find $d(x, \Sigma) = 0$ for all $x \in D$, i.e. $D \subset \Sigma$. Moreover Σ is connected, thanks to Theorem 6.16. Hence, Σ is an admissible competitor. Again, Theorem 6.16 gives the semicontinuity of the $\mathcal{H}^1$ measure and proves that Σ is a minimizer. $\square$

Irrigation and Transportation Networks: The Average Distance Problem We present here a very natural problem proposed in [54]. Given a compact domain $\Omega \subset \mathbb{R}^d$, a nonnegative density $f \in L^1(\Omega)$, and a number $L > 0$, solve

$$\min\left\{ \int_\Omega d(x, \Sigma) f(x) \, dx : \Sigma \subset \Omega \text{ compact, connected, and such that } \mathcal{H}^1(\Sigma) \leq L \right\}.$$

This can be interpreted in many ways. The name *irrigation*, which was first used to describe these problems, comes from considering Σ as a network of irrigation channels, and f as the density of the need for water. There is a budget constraint preventing the use of networks which are too long, and among the admissible ones we try to be as close as possible (on average, but weighted with f) to all points of Ω. Another natural interpretation views Σ as a transportation network: we want to build a subway system, connected, arriving as close as possible on average (weighted with the population density f) to all points of the city Ω, but the budget constraint bounds the length of the network. If we forget the possible modeling interpretation, a reasonable name for this problem, which has been used, is the *average distance problem*, even if, unfortunately, it does not convey the information that the unknown is a network.

Of course many variants are easy to consider, replacing the length constraint with a penalization and/or replacing the distance in the integral with other increasing functions of it. In general we can consider problems of the form

$$\min\left\{ \int_\Omega \psi(d(x, \Sigma)) f(x) \, dx + g(\mathcal{H}^1(\Sigma)) \quad : \quad \Sigma \subset \Omega \text{ compact, connected} \right\}$$

for some non-decreasing and l.s.c. functions $\psi, g : \mathbb{R}_+ \to [0, +\infty]$ (this includes the previous case when $\psi = id$ and $g = I_{[0,L]}$). For suitable choices of ψ and g, the above problem also has an interpretation in terms of computer graphics or image processing: if f stands for an object whose shape can be roughly simulated by that of a 1D network but which is observed through a blurred density, we could be interested in replacing it with its *skeleton*, i.e. a truly 1D object which passes through the middle of the density, as close as possible to the important parts of the density, but without using too much length.

This leads to a huge branch of generalizations (and this was the original scope of [54]) where, instead of prescribing a unique density f, two densities $f^\pm$ are prescribed, and we compute the value of an optimal transport problem (see Sect. 4.6) where the cost depends on Σ. A possibility is

$$c_\Sigma(x, y) := \min\{|x - y|, d(x, \Sigma) + d(y, \Sigma)\},$$

which consists in giving two options: walking from x to y or walking to the subway system Σ, move on it to a new point, and then walk towards the destination y, and the cost is the walking time. More generally, one could consider a cost depending on $\mathcal{H}^1(\gamma \cap \Sigma)$ and $\mathcal{H}^1(\gamma \setminus \Sigma)$, to be minimized among all curves γ connecting x to y, where the part outside Σ is more expensive than the one on Σ.

Similar arguments to those used for the Euclidean Steiner problem can prove the existence of a minimizer. Note that in the case where the cost involves $\mathcal{H}^1(\gamma \cap \Sigma)$ and $\mathcal{H}^1(\gamma \setminus \Sigma)$ a variant of the Gołąb theorem is needed, where we consider the semicontinuity of $(C, K) \mapsto \mathcal{H}^1(C \setminus K)$ among pairs of compact sets, the first one being connected, w.r.t. the Hausdorff convergence of both of them (see [42]).

Then, [54] raised a certain number of interesting questions, that we briefly list here:

1. (Regularity). Study the regularity properties of the solutions, which are expected to be composed of a finite number of smooth curves connected through a finite number of singular points.
2. (Absence of loops). Study the topological properties of the solutions, which are expected to be loop-free. When $d = 2$ this can be expressed by saying that $\mathbb{R}^2 \setminus \Sigma$ is connected.
3. (Triple points). Study the nature of the singular points, which are expected to be triple junctions with angles of 120 degrees.
4. (Distance from the boundary). Study the cases when the optimal solutions do not touch the boundary $\partial \Omega$, which is expected to occur at least for convex Ω.
5. (Behavior for small lengths). Study the asymptotic behavior of the optimal Σ_L as $L \to 0$. It is expected that Σ is a smooth curve without singular points located close to the point x_0 which minimizes $\int_\Omega |x - x_0| f(x) \, dx$.
6. (Behavior for large lengths). Study the asymptotic behavior of the minimal value and of the optimizers as $L \to \infty$. For the optimizers, find the weak limits as $L \to \infty$ of $\frac{1}{L} \mathcal{H}^1_{|\Sigma_L}$.

The answers to most of these questions have been given in subsequent papers, most of which are restricted to the case where $d = 2$ and f is bounded from below and above by positive constants. For instance in [57] the topological question is clarified, thus giving a full answer to question 2 and partial answers to questions 1 and 3: it is proven that Σ is composed of a finite number of curves connected through a finite number of singular points which are all triple junctions, but the regularity of these curves and hence the angle condition had not been settled yet (note that one needs to define tangent vectors, and hence discuss regularity, in order to fully answer question 3). The regularity is widely discussed in [179], where the blow-up of the optimal Σ (i.e. the limits in the Hausdorff convergence of the sets $\frac{1}{r}(\Sigma \cap \overline{B(x_0, r)} - x_0)$ as $r \to 0$) are studied. More precisely, [179] proved that Σ has a blow-up limit (a full limit, not up to subsequences) $K \subset \overline{B(0, 1)}$ at every point $x_0 \in \Sigma$, and characterized the possible limits K according to the nature of the point x_0: K is a radius of the unit ball if x_0 is an endpoint, it is the union of two radii if x_0 is an ordinary point (the two radii forming a diameter or an angle larger than 120°), or is the union of three radii forming angles of 120° if x_0 is a triple junction. Moreover, a partial regularity result was proven, namely that every branch of Σ is a $C^{1,1}$ curve in a neighborhood of every point x_0 such that diam($\{x \in \Omega : d(x, \Sigma) = |x - x_0|\}$) is sufficiently small. Since this assumption is always satisfied if x_0 is a triple junction, the three branches starting at a triple junction are indeed $C^{1,1}$ near the singularity, and their tangents form angles of 120 degrees at x_0, which fully answers question 3. The general regularity of the optimal Σ is discussed and summarized in [136], and the fact that $C^{1,1}$ is the best that we can expect, even for $f \in C^\infty$, is presented in [190]. In all these papers, as well in the survey [137], the possibility that Σ contained corners far from the triple junction was left open, but considered as a technical difficulty to overcome since it was expected that in the end these curves would be smooth. Yet, [183] provided clever counterexamples which showed that the presence of corners was indeed possible. We will not discuss the other questions listed above, but only mention that the answer to Question 6 has been given in [153] with a Γ-convergence technique very similar to the one that we will explain in Sect. 7.3. By the way, the techniques developed in [153] can be used to improve the results presented in Sect. 7.3 or in the original paper [38].

Optimal Membrane Reinforcement We consider now a different problem inspired by mechanics: the domain $\Omega \subset \mathbb{R}^2$ represents a membrane fixed on its boundary, and the function f represents a force exerted on it and pushing it down, so that the shape of the membrane will be represented by a function u vanishing on $\partial\Omega$ and solving $\Delta u = f$; we can then reinforce the membrane on a set $\Sigma \subset \Omega$ where we are able to impose $u = 0$, but the length of Σ is either penalized or constrained. More precisely, given $\Omega \subset \mathbb{R}^2$, $\Sigma \subset \overline{\Omega}$, and $f \in L^2(\Omega)$, we consider the energy

$$E(u) := \frac{1}{2} \int_{\Omega \setminus \Sigma} |\nabla u|^2 \, dx + \int_\Omega fu \, dx,$$

defined for $u \in H_0^1(\Omega \setminus \Sigma)$. We denote by u_Σ the unique minimizer of E over $H_0^1(\Omega \setminus \Sigma)$, which is also the unique solution of

$$\Delta u_\Sigma = f \tag{6.2}$$

in the weak sense in $\Omega \setminus \Sigma$. The profile u_Σ stands for the displacement of the membrane when it is subject to the force field f but the boundary is fixed and the membrane is also "glued" along Σ. The compliance of the membrane is then defined as

$$C(\Sigma) := -E(u_\Sigma) = \frac{1}{2} \int_{\Omega \setminus \Sigma} |\nabla u_\Sigma|^2 \, dx = -\frac{1}{2} \int_\Omega fu_\Sigma \, dx,$$

where the value of the optimal energy is computed using (6.2). Using the last equivalent expression, this value can be interpreted as a measure of how much the membrane resists the force f, and is proportional to the work of such a force.

We then consider the problem

$$\min\left\{C(\Sigma)+\lambda\mathcal{H}^1(\Sigma)=-E(u_\Sigma)+\lambda\mathcal{H}^1(\Sigma):\Sigma\subset\overline{\Omega}\text{ compact and connected}\right\},$$

where the parameter $\lambda>0$ is given, and the length penalization could also be replaced by a length constraint.

We observe, which is useful for lower semicontinuity and then existence (see Exercise 6.17), that we have

$$C(\Sigma)=\sup\left\{-\int_\Omega fu\,dx-\frac{1}{2}\int_{\Omega\setminus\Sigma}|\nabla u|^2\,dx\ :\ u\in C_c^\infty(\Omega\setminus\Sigma)\right\}.$$

We note anyway that this problem only makes sense in $\mathbb{R}^2$. Indeed, in higher dimensions, one-dimensional sets do not play a role in the definition of Sobolev functions, and in particular the set $C_c^\infty(\Omega\setminus\Sigma)$ is dense in $H_0^1(\Omega)$, for every closed subset Σ of finite length. This is due to the fact that 1D sets in dimension $d>2$ have zero 2-capacity.

Box 6.6 Important Notion—*p-Capacity of a Set*

Definition Given a number $p\le d$ we define the p-capacity of a set $A\subset\mathbb{R}^d$ as

$$\inf\left\{\int_{\mathbb{R}^d}|\nabla u|^p\,dx\ :\ u\in C_c^\infty(\mathbb{R}^d),u=1\text{ on a neighborhood of }A\right\}.$$

For instance, the p-capacity of a point is 0 (and the condition $p\le d$ is crucial), which exactly fits the condition for Sobolev functions to be well-defined on points. If A is a smooth surface of positive and finite $\mathcal{H}^k$ measure, the condition to have zero capacity becomes $p\le d-k$.

We refer to [84, Chapter 4] for proofs and details.

We could then generalize the problem to arbitrary dimension and, replacing the exponent 2 with p, provided we assume $p>d-1$, we consider

$$E_p(u):=\frac{1}{p}\int_{\Omega\setminus\Sigma}|\nabla u|^p\,dx+\int_\Omega fu\,dx,$$

and then

$$C_p(\Sigma) := -E_p(u_\Sigma) = \frac{1}{p'} \int_{\Omega \setminus \Sigma} |\nabla u_\Sigma|^p \, dx = -\frac{1}{p'} \int_\Omega f u_\Sigma \, dx,$$

where p' is the dual exponent of p. We then solve

$$\min \left\{ C_p(\Sigma) + \lambda \mathcal{H}^1(\Sigma) = -E(u_\Sigma) + \lambda \mathcal{H}^1(\Sigma) \ : \ \Sigma \subset \overline{\Omega} \text{ compact, connected} \right\}.$$

It is not possible, on the other hand, to modify the dimension of Σ without imposing additional constraints, since there is no Gołąb theorem for $\mathcal{H}^k$, $k > 1$. Among other generalizations, one could remove the condition $u = 0$ on $\partial\Omega$, and define u_Σ as a minimizer of E_p among functions $u \in W^{1,p}(\Omega)$ which vanish on Σ, i.e. on the closure in $W^{1,p}(\Omega)$ of $C_c^\infty(\overline{\Omega} \setminus \Sigma)$. In this case u_Σ would solve $\Delta_p u = f$ with Neumann boundary conditions on $\partial\Omega$.

It is interesting to observe (as pointed out in [55]) that, when we have $f \geq 0$, the average distance problem can be recovered at the limit $p \to \infty$ of the Neumann case. Indeed, using Γ-convergence techniques (see Sects. 7.1 and 7.2) one can see that as $p \to \infty$ the penalization $\frac{1}{p} \int_{\Omega \setminus \Sigma} |\nabla u|^p$ becomes a constraint $\|\nabla u\| \leq 1$, and the function which maximizes $\int f u$ among those functions $u \in \mathrm{Lip}_1$ satisfying $u = 0$ on Σ is exactly $d(\cdot, \Sigma)$.

For all these problems, besides existence results, the regularity of the optimal Σ is very interesting and much more delicate than in the average distance case, as any small perturbation of Σ globally affects u_Σ: it has been addressed in [65] and [51]. Moreover, we stress that these are typical examples of shape optimization problems (see [50, 113, 114]), where the unknown is a set and the functional depends on the solution of a PDE on such a set. This was already the case for the optimization of the Dirichlet eigenvalue, and we point out that the eigenvalue case with length constraints on Σ has also been studied, in [191].

6.3.2   Proof of the Gołąb Semicontinuity Theorem

The semicontinuity result which is the object of this section was first proven in [103], but the proofs in this section are essentially taken from [15] up to some minor modifications.

Lemma 6.20 *Assume that $A \subset X$ is a connected subset of a metric space X and $x_0 \in A$. Then, if $r > 0$ is such that $A \setminus B(x, r) \neq \emptyset$, we have $\mathcal{H}^1(A \cap B(x_0, r)) \geq r$.*

Proof Consider the function $\varphi : A \to \mathbb{R}$ defined by $\varphi(x) = \min\{r, d(x, x_0)\}$. This function is continuous, so that it sends connected sets into connected sets. Moreover, it is 1-Lipschitz, so that it reduces the Hausdorff measures. Furthermore, we have $\varphi(A) = [0, r]$ because $x_0 \in A$ and A contains points at distance larger than r from x_0. We then obtain

$$r = \mathcal{H}^1(\varphi(A)) \leq \mathcal{H}^1(\varphi(A \setminus B(x_0, r))) + \mathcal{H}^1(\varphi(A \cap B(x_0, r)))$$
$$= \mathcal{H}^1(\varphi(A \cap B(x_0, r))) \leq \mathcal{H}^1(A \cap B(x_0, r)),$$

where we used $\varphi(A \setminus B(x_0, r)) = \{r\}$ and hence $\mathcal{H}^1(\varphi(A \setminus B(x_0, r))) = 0$. □

Proposition 6.21 *Let X be a compact connected metric space with $\mathcal{H}^1(X) < +\infty$. Then there exist a Lipschitz continuous curve $\gamma : [0, 1] \to X$ parameterized with constant speed, with $|\gamma'| \leq 2\mathcal{H}^1(X)$, and such that $\gamma([0, 1]) = X$.*

Proof We first observe that the space X satisfies the assumptions of Theorem 1.21, i.e. it is thin. Indeed, using Lemma 6.20, for every $r < \mathrm{diam}(X)/2$ and every point $x \in X$ we have $\mathcal{H}^1(B(x, r)) \geq r$, which shows that the maximal number of disjoint balls of radius r that one can find in X is at most C/r.

This implies that every pair of points in X can be connected by an AC curve, and hence by reparameterization by a Lipschitz curve. We then take two points $x_0, y_0 \in X$ and a Lipschitz curve $\omega_0 : [0, \ell_0] \to X$ connecting them. Such a curve can be chosen to be a geodesic parameterized with unit speed (so that $\ell_0 = \mathrm{Length}(\omega_0)$). Then, we iteratively construct a sequence of curves in the following way: we set $\Gamma_k := \bigcup_{i=0}^{k} \omega_i([0, 1])$ and, if $\Gamma_k \neq X$, we consider the function $X \ni x \mapsto d(x, \Gamma_k)$. We call d_k the maximal value of this function, which is strictly positive (otherwise $\Gamma_k = X$, since Γ_k is closed). We then consider a point x_{k+1} which maximizes this function and $y_{k+1} \in \Gamma_k$ such that $d(x_{k+1}, y_{k+1}) = d_k$. We consider a Lipschitz curve ω, again a geodesic parameterized by constant speed, with $\omega(0) = x_{k+1}$ and $\omega(1) = y_{k+1}$ and set $t_0 := \inf\{t > 0 : \omega(t) \in \Gamma_k\}$. We then set $z_{k+1} = \omega(t_0)$ and reparameterize ω so as to obtain a Lipschitz curve $\omega_{k+1} : [0, \ell_{k+1}] \to X$ which is parameterized by unit speed and such that $\omega_{k+1}(0) = z_{k+1}, \omega_{k+1}(\ell_{k+1}) = x_{k+1}$ and $\omega_{k+1}(t) \notin \Gamma_k$ for $t > 0$. We have $\mathrm{Length}(\omega_{k+1}) = \ell_{k+1} \geq d(x_{k+1}, z_{k+1}) \geq d_k$. In this way we have $\mathcal{H}^1(\omega_{k+1}([0, \ell_{k+1}])) \geq d_k$ and $\sum_k \mathcal{H}^1(\omega_k([0, \ell_k])) \leq \mathcal{H}^1(X)$ (since the images of these curves are essentially disjoint). We deduce $d_k \to 0$ and since we have $X \subset (\bigcup_{i=1}^{k} \Gamma_i)_{d_k}$ we deduce that $\Gamma_\infty := \bigcup_{i=1}^{\infty} \Gamma_i$ is dense.

We now build a sequence of curves γ_k in the following way. Each γ_k will be defined on $[0, 2L_k]$ where $L_k := \mathcal{H}^1(\Gamma_k) = \sum_{i=0}^{k} \ell_i$. The first curve $\gamma_0 : [0, 2L_0] \to X$ is defined by

$$\gamma_0(t) = \begin{cases} \omega_0(t) & \text{if } t \leq \ell_0, \\ \omega_0(2\ell_0 - t) & \text{if } t \geq \ell_0. \end{cases}$$

Once γ_k is defined, define γ_{k+1} as follows: let t_k be an instant (it will not be unique) such that $\gamma_k(t_k) = z_{k+1}$, then take

$$\gamma_{k+1}(t) = \begin{cases} \gamma_k(t) & \text{if } t \leq t_k, \\ \omega_{k+1}(t - t_k) & \text{if } t_k \leq t \leq t_k + \ell_{k+1}, \\ \omega_{k+1}(t_k + 2\ell_{k+1} - t) & \text{if } t_k + \ell_{k+1} \leq t \leq t_k + 2\ell_{k+1}, \\ \gamma_k(t - (t_k + 2\ell_{k+1})) & \text{if } t \geq t_k + 2\ell_{k+1}. \end{cases}$$

This means that we transform γ_k into γ_{k+1} by "inserting" the curve ω_{k+1} inside γ_k first in the forward direction, and then backward. All curves γ_k are parameterized by unit speed and the image of γ_k equals the set Γ_k (and most points of Γ_k are visited twice by γ_k). Moreover, they are all closed curves with $\gamma_k(0) = \gamma_k(2L_k) = x_0 \in X$. We can then reparameterize all these curves into curves $\tilde{\gamma}_k$ with constant speed on the same interval, say $[0, 1]$. In this way we have $\mathrm{Lip}(\tilde{\gamma}_k) = 2L_k \leq 2\mathcal{H}^1(X)$. Hence, we can apply to this sequence of curves, which are equi-continuous and valued in a compact set, the Ascoli–Arzelà theorem and obtain, up to extracting a subsequence, a uniform limit γ.

If we take an arbitrary point $x \in \Gamma_k$ for every $j \geq k$ we have $x \in \tilde{\gamma}_j([0, 1])$, hence $x = \tilde{\gamma}_j(s_j)$ for some points $t_j \in [0, 1]$ so that, by extracting a convergent subsequence $t_j \to t$, we find $x = \gamma(s)$. This proves $\Gamma_\infty \subset \gamma([0, 1])$. Since $\gamma([0, 1])$ is compact and hence closed, and Γ_∞ is dense, we deduce $\gamma([0, 1]) = X$.

This proves that there exists a Lipschitz curve with Lipschitz constant at most $2\mathcal{H}^1(X)$ which covers the set X. The proof is not yet complete since we do not know if this curve is parameterized with constant speed.

In order to obtain a curve parameterized with constant speed it is enough to solve

$$\min\{\mathrm{Lip}(\omega) : \omega : [0, 1] \to X \text{ surjective}\}.$$

By applying once more the Ascoli–Arzelà theorem it is clear that a solution ω exists. Suppose that the optimizer ω is constant on a subinterval $(t_0, t_1) \subset [0, 1]$; then we could remove it and reparametrize the curve sending in a piecewise linear way $[0, t_0] \cup [t_1, 1]$ into $[0, 1]$, which would multiply the Lipschitz constant by $1 - (t_1 - t_0) < 1$ and contradict the optimality. So, ω is not constant on any interval, allowing us to apply Proposition 1.13 and reparametrize ω as a constant speed curve without increasing its Lipschitz constant (see also Remark 1.14). □

The above proposition, or simple variants of it, are crucial in all the proofs of Gołąb's theorem. Observe that in our construction we defined a sequence of curves covering the set X twice, and such that their length was exactly equal to twice the $\mathcal{H}^1$ measure of their image. We did not try to preserve this fact in the conclusion (i.e. taking the limit and then extracting a curve parameterized by constant speed), as it is not important for us in the sequel. Yet, it would have been possible to impose this on the curve γ and the original proof by Gołąb [103] passes through this fact: each set C_n is covered by a curve γ_n such that $\mathrm{Lip}(\gamma_n) = \mathrm{Length}(\gamma_n) = 2\mathcal{H}^1(C_n)$, then at the limit we have a curve γ with $\mathrm{Length}(\gamma) \leq \liminf_n 2\mathcal{H}^1(C_n)$ covering the set C. The conclusion comes after proving $\mathcal{H}^1(C) \leq \frac{1}{2}\mathrm{Length}(\gamma)$, which more or less corresponds to saying that γ also covers each point of C at least twice. This was also done in a more recent proof of the same theorem, see [5], in a more general setting (in metric spaces instead of the Euclidean setting studied by Gołąb). However, the last part of the proof (proving that γ covers each point twice) is delicate, so we will use a different strategy, the one presented in [15].[5]

Proof *(of Theorem 6.16)* Let us first prove that C is connected. If not, there are two subsets A_1, A_2 which are closed, non-empty and disjoint such that $A_1 \cup A_2 = C$. Set $\varepsilon := d(A_1, A_2) = \min\{d(x_1, x_2) : x_i \in A_i\} > 0$. The Hausdorff convergence of C_n to C implies that, for n large, we have $C_n \subset (A_1)_{\varepsilon/3} \cup (A_2)_{\varepsilon/3}$. Moreover, both $C_n \cap \overline{(A_1)_{\varepsilon/3}}$ and $C_n \cap \overline{(A_2)_{\varepsilon/3}}$ are non-empty since every point of $C = A_1 \cup A_2$ is a limit of points of C_n. But this is a contradiction to the connectedness of C_n, as we would have $C_n = \left(C_n \cap \overline{(A_1)_{\varepsilon/3}}\right) \cup \left(C_n \cap \overline{(A_2)_{\varepsilon/3}}\right)$, which is a disjoint union of two non-empty closed sets.

We then move to the semicontinuity of the length. Up to extracting a subsequence, we assume $\mathcal{H}^1(C_n) \to L$, and we want to prove $\mathcal{H}^1(C) \leq L$.

First we prove $\mathcal{H}^1(C) < +\infty$. To do this, we define a sequence of measures μ_n via $\mu_n(A) := \mathcal{H}^1(A \cap C_n)$ and, up to subsequences, we have $\mu_n \overset{*}{\rightharpoonup} \mu$ for a nonnegative measure μ with $\mu(X) = L$. Take $x \in X$ and $r < \mathrm{diam}(X)/2$ (note that if $\mathrm{diam}(X) = 0$ there is nothing to prove). Since x is a limit of points in C_n we can fix $r' < r$ and find $x_n \in C_n$ with $x_n \to x$ and $B(x_n, r') \subset B(x, r)$. From Lemma 6.20 we have $\mu_n(B(x_n, r')) \geq r'$, so that we have $\mu_n(\overline{B(x, r)}) \geq r'$. Taking first the limit $n \to \infty$ and then $r' \to r$ we deduce $\mu(\overline{B(x, r)}) \geq r$, which shows $\overline{\Theta}_1(\mu, x)1/2$ for all $x \in C$ (see Box 6.7 below). This shows

$$L = \mu(X) \geq \mu(C) \geq \frac{1}{2}\mathcal{H}^1(C),$$

(where we also used the fact that the mass passes to the limit via weak-* convergence in compact spaces: it is enough to test the weak-* convergence $\mu_n \overset{*}{\rightharpoonup} \mu$ against the function 1 in order to obtain $\mu(X) = L$). This implies that C has finite $\mathcal{H}^1$ measure and almost proves the theorem, up to a factor 2.

---

[5] Note that [15] exploited a variant of Proposition 6.21, as, after producing the curves ω_k and observing that Γ_∞ is dense, instead of using this to cover the whole set with a single curve, the authors proved $\mathcal{H}^1(X \setminus \Gamma_\infty) = 0$, which is very interesting and of course enough for the proof of Gołąb's theorem, but less elementary than what we did. From this point of view, in this part of the proof we have followed [71, Section 30] instead of [15].

Box 6.7 Important Notion—*Differentiation of Measures*

Given a positive measure μ on a metric space X and a real number $k > 0$ we define for every point the upper and lower k-dimensional densities of μ as follows:

$$\overline{\Theta}_k(\mu, x) := \limsup_{r \to 0} \frac{\mu(B(x,r))}{\omega_k r^k}, \qquad \underline{\Theta}_k(\mu, x) := \liminf_{r \to 0} \frac{\mu(B(x,r))}{\omega_k r^k},$$

where the numbers ω_k are some dimensional constants which coincide with the unit volume of the k-dimensional unit ball in $\mathbb{R}^k$ when k is an integer.

The following theorem holds, and its proof is based on suitable covering arguments:

Theorem *Assume that μ is a locally finite measure on X and $B \subset X$ is a Borel set. Then for every $c > 0$ we have the following implications:*

$$\overline{\Theta}_k(\mu, x) \geq c \text{ for all } x \in B \Rightarrow \mu(B) \geq c\mathcal{H}^k(B),$$

$$\overline{\Theta}_k(\mu, x) \leq c \text{ for all } x \in B \Rightarrow \mu(B) \leq 2^k c\mathcal{H}^k(B).$$

We refer to [15, Chapter 2] or [14, Chapter 2] for proofs and details.

We now use Proposition 6.21 to say that $\mathcal{H}^1$-a.e. point $x \in C$ can be expressed as $x = \gamma(t)$ for a Lipschitz curve γ such that $\gamma'(t)$ exists and is not zero. Indeed, for a.e. t the curve γ is differentiable (since it is Lipschitz continuous) and satisfies $|\gamma'|(t) = c > 0$; if we remove the instants t such that this does not hold we are only removing a set of points $x = \gamma(t)$ which is $\mathcal{H}^1$-negligible (as a Lipschitz image of a $\mathcal{H}^1$-negligible set). We now take such a point x and, up to a change of coordinates, we can say $x = 0$ and $\gamma'(t) = e_1$. We fix a small value of $r > 0$ and consider the points $x_k = \gamma(\frac{k}{N}r) \in C$. Using the differentiability of γ, we have $x_k \in B(\frac{k}{N}re_1, r\varepsilon(r))$ where $\lim_{r \to 0} \varepsilon(r) = 0$. For large n, the sets C_n intersect each of the balls $B(\frac{k}{N}re_1, r2\varepsilon(r))$. Applying Lemma 6.22 below[6] we obtain $\mathcal{H}^1(C_n \cap B(x,r)) \geq 2r(1 - \frac{1}{N} - 2\varepsilon(r))$. Taking the limit $n \to \infty$, this gives $\mu(\overline{B(x,r)}) \geq 2r(1 - \frac{1}{N} - 2\varepsilon(r))$. Taking the liminf for $r \to 0$, we obtain that the density of μ at

---

[6] This argument was not present in [15], but fixes a small issue in the proof therein, where the estimate giving density 1 has to be clarified; this issue was also fixed in [166] with a very similar argument. The construction of our Lemma 6.22 allows us anyway to exclude one of the cases considered in Lemma 3.2 of [166] and is slightly shorter. On the other hand, it exploits the Euclidean setting in which we work, while [166] deals with general metric spaces.

x is at least $1 - \frac{2}{N}$ for any N, hence it is 1. This provides

$$L = \mu(X) \geq \mu(C) \geq \mathcal{H}^1(C)$$

and proves the claim.   □

Lemma 6.22 *If $A \subset \mathbb{R}^d$ is a compact connected set in $\mathbb{R}^d$ such that $A \cap B(\frac{k}{N}re_1, \varepsilon) \neq \emptyset$ for all $k = -(N-1), \ldots, N-1$, then we have $\mathcal{H}^1(A \cap B(0, r)) \geq 2r - \frac{2r}{N} - 4\varepsilon$.*

Proof The set A is connected but this is not necessarily the case for $A \cap B(0, r)$. We distinguish two cases: either we have $A \subset B(0, r)$, in which case $A \cap B(0, r)$ is indeed connected, or $A \setminus B(0, r) \neq \emptyset$, in which case every connected component of $A \cap B(0, r)$ must touch $\partial B(0, r)$.

Let us choose some points $x_k \in A \cap B(\frac{k}{N}re_1, \varepsilon)$. If $A \cap B(0, r)$ is connected then considering the 1-Lipschitz function $\varphi(x) = x \cdot e_1$, we see that we have

$$\mathcal{H}^1(A \cap B(0, r)) \geq \mathcal{H}^1(\varphi(A \cap B(0, r))) \geq (x_{N-1} - x_{-(N-1)}) \cdot e_1 \geq 2\frac{N-1}{N}r - 2\varepsilon$$

and the claim is proven.

We consider now the case where the connected components of $A \cap B(0, r)$ all touch $\partial B(0, r)$ and consider in particular the connected components which contain each x_k. For one such connected component $B \subset A \cap B(0, r)$, we have for sure

$$\mathcal{H}^1(B) \geq |i - j|\frac{r}{N} - 2\varepsilon$$

whenever two points x_i, x_j belong to B (via the same argument using the function φ as before), but also

$$\mathcal{H}^1(B) \geq (N - |i|)\frac{r}{N} - \varepsilon$$

whenever a single point x_i belongs to B (to prove this, we use instead $\varphi(x) = r - |x|$). Hence, the set of indices $k = -(N-1), \ldots, N-1$ is partitioned into some classes, each corresponding to one connected component, and each of them contributes to $\mathcal{H}^1(B)$ the maximum between $|i - j|r - 2\varepsilon$, where i and j are the extreme indices of the class (its "diameter"), and $(N - |i|)r - \varepsilon$, where i is an index which maximizes $N - |i|$ in the class.

We need to look at the partition into classes which minimizes the sum of these contributions. It is clear that, in order to minimize the sum of the contributions, the classes have to be connected.[7] Indeed, if a class contains i and j but not h with $i < h < j$, then merging the class of h with that of i and j reduces the total contribution (indeed, the diameter of the union would be—in this very particular case—smaller than the sum of the diameters, and the maximal value of $N - |i|$ on the union is the maximum between the two maximal values).

We then consider the class including $i = 0$. If it includes all the indices we then obtain a contribution of $2(N - 1)\frac{r}{N} - 2\varepsilon$. If not, let us call k_1 the largest negative index not included in this class and k_2 the smallest positive one. We then have three contributions, equal to $[(k-2-1)-(k_1+1)]\frac{r}{N}-2\varepsilon$, $(N+k_1)\frac{r}{N}-\varepsilon$ and $(N-k_2)\frac{r}{N}-\varepsilon$, so that the sum is $2r - \frac{2r}{N} - 4\varepsilon$. The case where the class of 0 includes all negative indices but not all positive ones or vice versa is treated similarly and gives a total contribution of $2r - \frac{2r}{N} - 3\varepsilon$. In all cases, we obtain $\mathcal{H}^1(A \cap B(0, r)) \geq 2r - \frac{2r}{N} - 4\varepsilon$.

$\square$

6.4 Discussion: Direct Proofs, Quantitative Versions, and Variants for the Isoperimetric Inequality

In this section we come back to the isoperimetric problem and provide some sketches of some direct proofs, i.e. proofs where it is directly proven that we have $\text{Per}(A) \geq \text{Per}(B)$ (when B is a ball with the same volume as A) without using variational arguments (i.e. proving that a minimizer of the perimeter exists and that it must be the ball).

A Proof Via Fourier Series We start with a proof which is really two-dimensional. It is due to Hurwitz [118] but can be found, for instance, in [88] together with a variant for the higher dimensional case. It applies when A is a set whose boundary is a Jordan curve, and in particular this requires A to be connected and simply connected. It is however not difficult to reduce to this case in dimension 2 (indeed, it is easy to see that we can restrict to convex sets, as taking the convex hull of a set in dimension 2 increases the volume but reduces the perimeter, differently than what happens in higher dimensions, see Exercise 6.5).

Consider a set $A \subset \mathbb{R}^2$ whose boundary is given by a closed curve γ that we can assume to be parameterized with constant speed on the interval $[0, 2\pi]$ and be

---

[7] This must not be confused with the connectedness in $\mathbb{R}^d$: a class of indices including 0 and 2 but not 1 is disconnected, but this just means that the connected component of $A \cap (B(0, r))$ containing x_0 and x_2 does not pass through x_1.

2π-periodic. If we set $L = \mathrm{Per}(A)$ we have $|\gamma'| = \frac{L}{2\pi}$. We then observe that the area of A can be written as

$$|A| = \frac{1}{2} \int_0^{2\pi} \gamma(t) \wedge \gamma'(t) \, dt.$$

This can be seen, for instance, by approximating γ with polygonal curves and computing the area of each triangle. Here the vector product $v \wedge w$ in $\mathbb{R}^2$ is given by the determinant of the 2×2 matrix whose columns are v and w or, equivalently, to the product of their Euclidean norms times the sine of the (oriented) angle between them. We identify $\mathbb{R}^2$ with $\mathbb{C}$ and see that for two vectors $v, w \in \mathbb{C}$ we have $v \wedge w = \mathrm{Re}(i v \bar{w})$. We then have

$$|A| = \frac{1}{2} \mathrm{Re} \left(\int_0^{2\pi} i \gamma(t) \overline{\gamma'(t)} \, dt \right).$$

We write γ as a Fourier series, i.e. $\gamma(t) = \sum_{n \in \mathbb{Z}} a_n e^{int}$. Then we have $\gamma'(t) = \sum_{n \in \mathbb{Z}} a_n i n e^{int}$ and $\overline{\gamma'(t)} = -i \sum_{n \in \mathbb{Z}} \bar{a}_n n e^{-int}$. Hence, we obtain,

$$|A| = \pi \sum_{n \in \mathbb{Z}} n |a_n|^2,$$

where we used the fact that the functions e^{int} are orthogonal in $L^2([0, 2\pi])$ and each has squared L^2 norm equal to 2π. On the other hand, from the value of $|\gamma'|$, we also have

$$\frac{L^2}{2\pi} = \int_0^{2\pi} |\gamma'(t)|^2 \, dt = 2\pi \sum_{n \in \mathbb{Z}} |n|^2 |a_n|^2.$$

We thus obtain

$$L^2 = 4\pi^2 \sum_{n \in \mathbb{Z}} |n|^2 |a_n|^2 \geq 4\pi^2 \sum_{n \in \mathbb{Z}} n |a_n|^2 = 4\pi |A|.$$

Note that this coincides exactly with the isoperimetric inequality that we wanted to prove, since if write $|A| = \pi r^2$ our goal was to prove $L \geq 2\pi r$, i.e. $L^2 \geq 4\pi^2 r^2 = 4\pi |A|$.

The key point in this proof was the inequality $|n|^2 \geq n$, which is satisfied for every $n \in \mathbb{Z}$. Moreover, note the cases of equality: the only possible way to obtain an equality in the isoperimetric inequality is if $a_n = 0$ for every n such that $|n|^2 > n$, i.e. for $n \neq 0, 1$. This clearly shows that only curves of the form $\gamma(t) = a_0 + a_1 e^{it}$ realize the equality, and they are circles.

It is even possible to use this approach to prove a stability result, i.e. to make the inequality quantitative. This means proving inequalities of the form $\mathrm{Per}(A) \geq$

$\text{Per}(B) + \varepsilon(A)$, where $\varepsilon(A)$ is an error term which quantifies how different A is from a ball. In our case we can define $\omega(t) := \gamma(t) - (a_0 + a_1 e^{it})$, i.e. ω is the difference between γ and the corresponding circle. The Fourier coefficients of ω are the same as those of γ except for the coefficients of the modes $n = 0, 1$, which vanish. We then use the inequality

$$|n|^2 \geq n + \frac{2}{5}(1 + |n|^2),$$

which is valid for $n \in \mathbb{Z} \setminus \{0, 1\}$. To check this, we just need to prove that the difference $g(n) := n^2 - n - \frac{2}{5}(1 + |n|^2) = \frac{3}{5}n^2 - n - \frac{2}{5}$ is non-negative on $[2, \infty) \cup (-\infty, -1]$. The function g is a convex function whose minimum is attained as $n = \frac{5}{6}$, hence it is increasing on $[2, \infty)$, where it is thus non-negative since $g(2) = 0$, and decreasing on $(-\infty, -1]$, where it is thus non-negative since $g(-1) > 0$.

We then obtain

$$L^2 = 4\pi^2 \sum_{n \in \mathbb{Z}} |n|^2 |a_n|^2 \geq 4\pi^2 \sum_{n \in \mathbb{Z}} n|a_n|^2 + \frac{8\pi^2}{5} \sum_{n \neq 0,1} (1 + |n|^2)|a_n|^2$$

$$= 4\pi|A| + \frac{8\pi^2}{5}||\omega||_{H^1}^2.$$

From this inequality it is possible to obtain an inequality of the form

$$\text{Per}(A) \geq \text{Per}(B) + c||\omega||_{H^1}^2 \geq \text{Per}(B) + c'||\omega||_{L^\infty}^2,$$

where the constant c in the first inequality is given by $\frac{4}{5}\pi^2 \text{Per}(B)^{-1}$. As for the second inequality, we use the Sobolev injection $H^1 \subset L^\infty$. This shows that any set whose perimeter is almost optimal in the isoperimetric inequality is, in terms of L^∞ distance of the boundary, close to a ball. This is not the most classical or general way of expressing this quantitative fact (which we will come back to later) but is a very clear statement, presented in [88] and compared to the previous stability results by Bonnesen [31]. Note that similar techniques can also be used to solve isoperimetric problems in higher codimension, i.e. comparing the length of a closed curve (image of $\mathbb{S}^1$) in $\mathbb{R}^3$ with the area of a minimal surface, parameterized by the unit disc, having the curve as a boundary datum, see for instance [8].

A Proof Via Transport Maps We provide here another direct proof of the isoperimetric inequality, this time in arbitrary dimension. The methods are based on optimal transport (see Sect. 4.6) but can also be adapted to use other non-optimal transport maps.

The technique is the following; we take a smooth domain $A \subset \mathbb{R}^d$ and we choose R such that $|A| = |B(0, R)|$. We set $B := B(0, R)$. We then consider the uniform measures on A and B, and consider a transport map T sending one into the other. Assume that the Jacobian matrix of T has only non-negative eigenvalues. It then satisfies the condition

$$\det(DT) = \frac{|B|}{|A|} = 1$$

(with no absolute value at the determinant). Hence we can write

$$\omega_d R^d = |A| = \int_A (\det(DT))^{\frac{1}{d}} dx \le \frac{1}{d} \int_A \nabla \cdot T dx = \frac{1}{d} \int_{\partial A} T d\mathcal{H}^{d-1} \cdot \mathbf{n} \le \frac{R}{d} \mathrm{Per}(A),$$

where the inequalities are obtained thanks to the arithmetic-geometric inequality $\frac{1}{d} \sum_i \lambda_i \ge (\Pi_i \lambda_i)^{\frac{1}{d}}$ applied to the eigenvalues of DT (which gives an inequality between the determinant and the trace, i.e. the divergence of T), and to the fact that T is valued in B, whence $T \cdot \mathbf{n} \le R$. This shows $\mathrm{Per}(A) \ge d\omega_d R^{d-1} = \mathrm{Per}(B)$.

Note that we only need to use a map T such that the Jacobian matrix DT has non-negative eigenvalues and that it transports one measure onto the other. These properties are satisfied by the Brenier map (see Box 4.5). The regularity of such a map can depend on the smoothness of A but it is possible to make the above argument rigorous even when T is not smooth. Indeed, the Monge–Ampère equation will be solved a.e. by $T = \nabla u$ replacing $DT = D^2 u$ with its absolutely continuous part (since u is convex $D^2 u$ is a positive measure), while the divergence theorem requires the use of the full distributional derivatives, but the positivity of the Hessian implies a sign on the singular part. The first appearance of this proof of the isoperimetric inequality in the literature is due to Knothe [128], and we follow here the presentation given by Gromov in the appendix of [145]. Both Knothe and Gromov worked on this before the discovering of the Brenier's map and then used an alternative map, which is now called the Knothe transport map. We do not give the details of the construction of this map here, for which we refer to [128] and to [177, Section 2.3]; roughly speaking, it is a map obtained in the following way: its first component is the monotone map which sends the projection of one measure on the first coordinate axis onto the projection of the other; we then disintegrate both measures according to the value of x_1 and iterate the construction projecting the disintegration on the second axis, and so on. The result is a map of the form $T_K(x_1, \ldots, x_d) = (T_1(x_1), T_2(x_1, x_2), \ldots, T_D(x_1, \ldots, x_d))$, where each map T_i is increasing in the variable x_i. Hence DT_K is a triangular matrix with non-negative entries on the diagonal.

This proof can also be used to prove the uniqueness of the optimal sets.

For this, the use of the Brenier map strongly simplifies the study of the equality cases. Indeed, assume we have equality in the above inequalities. This implies the equality $(\det(DT))^{\frac{1}{d}} = \frac{1}{d}(\nabla \cdot T)$ a.e. and, from the cases of equality in the arithmetic-geometric inequality, we get that the eigenvalues of DT should all be equal. Thanks to the condition on the determinant, they are equal to 1. If the matrix is known to be symmetric (which is the case for Brenier's map, but not for Knothe's), then we get $DT = \mathbf{I}$ and $T(x) = x + x_0$. Since T would be a translation, we conclude $A = B$, up to translations.

It is slightly harder to do this for the Knothe transport, since DT is not symmetric. Yet, if all the eigenvalues are 1, one deduces that T_1 (the first component) is a translation. Let us call x^A and x^B the barycenters of A and B respectively; then, the barycenters of the two projected measures are $(x^A)_1$ and $(x^B)_1$, respectively. If the transport map is a translation, then the barycenters also are obtained by translations from one another, so that we have $(x^B)_1 = T_1((x^A)_1)$. If we assume from the very beginning that A and B are centered at the origin, then the map T_1 is the identity, since it is a translation which fixes the origin. This means $\pi_1(A) = \pi_1(B)$. This is not yet enough to get $A = B$. Yet, we can exploit the fact that the directions of the axes used to define T_K are arbitrary. Hence, if A is a set giving equality in the isoperimetric inequality, then its projections onto any direction coincide with those of the ball. This implies $A \subset B$ and, by equality of the volumes, $A = B$.

Extensions: Anisotropic Isoperimetric Inequalities and Quantitative Variants
Given a bounded and convex body $K \subset \mathbb{R}^d$, containing 0 in its interior, we can associate with it a norm $|| \cdot ||_K$ on $\mathbb{R}^d$ defined via $||x||_K := \sup\{x \cdot y, \, y \in K\}$ (this norm can be non-symmetric if K is not symmetric). We also consider its dual norm $||x||_K^*$, characterized by $\{x \, : \, ||x||_K^* \leq 1\} = K$. Since 0 belongs to the interior of K, which is bounded, there exist two balls centered at 0, one contained in K and one containing K, so that we can find two constants $m_K \leq M_K$ with $m_K ||x|| \leq ||x||_K \leq M_K ||x||$. We can then define an anisotropic perimeter, or K-perimeter, by taking

$$\mathrm{Per}_K(A) := \int_{\partial A} ||\mathbf{n}_A||_K \, d\mathcal{H}^{d-1},$$

where $\mathbf{n}_A$ is the exterior normal to A. This definition works for smooth sets E. For sets of finite perimeter we can just consider $|| - \nabla \mathbb{1}_A ||$ but measuring the norms of the vector measure $\nabla \mathbb{1}_A$ with the norm $|| \cdot ||_K$ instead of using the Euclidean norm. Note that it is important to use here the negative sign on the gradient, since the normal is opposite to the gradient, and K could be non-symmetric.

We then have a natural question about the isoperimetric inequality for this notion of perimeter. The answer was suggested by Wulff [199] at the beginning of the twentieth century for applications in crystallography, and then proven by Gromov in [145]: the optimal set is K itself, so that we have

$$\mathrm{Per}_K(A) \geq \left(\frac{|A|}{|K|}\right)^{1-\frac{1}{d}} \mathrm{Per}_K(K) = d|A|^{1-\frac{1}{d}} |K|^{1/d}. \tag{6.3}$$

The last equality comes from the fact that we always have $|K| = \frac{1}{d} \mathrm{Per}_K(K)$, as can be seen by approximating K with convex polyhedra and computing the volume of each pyramid based on each face and arriving at the origin.

Now that we have addressed the question of how to find the optimal sets and discussed the equality cases (note that for (6.3) it is also true that the only equality case is given by K up to translations and dilations), it is natural to ask about the stability of the optimizers: if a set A is almost optimal in the isoperimetric inequality, is it almost a ball, and in which sense? Can we say the same for the case of Per_K?

We already saw an answer in the 2D case for the isotropic perimeter, but as we already mentioned, it is not possible in general to quantify this proximity between almost optimal sets and optimal sets in an L^∞ sense. The usual approach is the following: first we introduce the isoperimetric deficit $\delta(A)$ of A as follows

$$\delta(A) := \frac{\text{Per}_K(A)}{d|A|^{1-\frac{1}{d}}|K|^{\frac{1}{d}}} - 1.$$

Then, we define the asymmetry index $\mathcal{A}_K(A)$ of a set A via

$$\mathcal{A}_K(A) := \inf\left\{\frac{||\mathbb{1}_A - \mathbb{1}_{x_0+rK}||_{L^1}}{|A|} \ : \ x_0 \in \mathbb{R}^d, \ r^d|K| = |A|\right\},$$

which is the proportion of the volume of the symmetric difference of A and any copy of K, dilated and translated so as to fit the same volume as A. The question is then to estimate $\delta(E)$ from below in terms of $\mathcal{A}_K(A)$.

The best answer to this question has been provided in a paper by Figalli, Maggi and Pratelli [87], where they find an inequality with an optimal exponent:

$$\mathcal{A}_K(A) \leq C(d)\sqrt{\delta(A)}. \tag{6.4}$$

The constant $C(d)$ that they find is not necessarily optimal, but the exponent $\frac{1}{2}$ is optimal, and the constant $C(d)$ they find is independent of K and grows polynomially w.r.t. d.

Before their study, many works had been devoted to the Euclidean case (i.e. $K = B$), first obtaining a proof in dimension 2, [29, 31], then among convex competitors, [89], in arbitrary dimension, then without the convexity constraint but with a non-optimal exponent $\frac{1}{4}$ [109]. The Euclidean question had been closed by Fusco, Maggi and Pratelli in [92], which was the starting point for [87]. Yet, in all these works the main technique is based on the Steiner symmetrization which we already saw in Sect. 6.1.3 or on some of its variants, but unfortunately this tool is less flexible in the anisotropic case. Indeed, a symmetrization procedure exists in order to prove $\text{Per}_K(E) \geq \text{Per}_{S_e(K)}(S_e(E))$ (see [126] and [193]), and this can be iterated in order to obtain the desired isoperimetric inequality, but it is harder to obtain sharp quantitative estimates. In the case of a more general convex body K, a first stability result for the isoperimetric inequality was presented in [82], but with a non-optimal exponent. The strategy of [87], instead, is based on the proof by transport maps (which can be easily adapted to the case of the anisotropic perimeter, using the estimate

$$\int_{\partial A} T \cdot \mathbf{n} \, d\mathcal{H}^{d-1} \leq \int_{\partial A} ||T||_K^* ||\mathbf{n}||_K \, d\mathcal{H}^{d-1} = R \, \text{Per}_K(A),$$

when T sends the uniform measure on A onto the uniform measure on RK, with R chosen such that $R^d|K| = |A|$). Of course, the difficult part of the proof consists in finding which of the inequalities which are used can be made quantitative. Besides [87], we also refer to [175] for a short review on these results.

6.5 Exercises

Exercise 6.1 Prove that if $f : \mathbb{R}^d \to \mathbb{R}$ is uniformly continuous and L^1, then we have $\lim_{|x| \to \infty} f(x) = 0$.

Exercise 6.2 Let A be a solution of

$$\min_A \frac{\mathrm{Per}(A)}{|A|^{\frac{d-1}{d}}},$$

among all finite perimeter sets $A \subset \mathbb{R}^d$. Assume that in a certain open set Ω we can write A as an epigraph of a smooth function: there exists an $f : \mathbb{R}^{d-1} \to \mathbb{R}$ such that for $x = (x_1, \ldots, x_{d-1}, x_d) \in \Omega$ we have $x \in A$ if and only if $x_d > f(x_1, \ldots, x_{d-1})$. Prove that f solves

$$\nabla \cdot \left(\frac{\nabla f}{\sqrt{1 + |\nabla f|^2}} \right) = c$$

for some constant $c \in \mathbb{R}$, and find this constant in terms of A.

Exercise 6.3 Given two smooth and strictly positive functions $a, b : \mathbb{R}^d \to \mathbb{R}$, consider the isoperimetric problem where both the volume and the perimeter are weighted by variable coefficients:

$$\min_A \frac{\int_{\partial A} a \, d\mathcal{H}^{d-1}}{\left(\int_A b \, dx \right)^{\frac{d-1}{d}}},$$

among all smooth domains $A \subset \mathbb{R}^d$. Assume that a solution exists and that it is a convex domain A. Prove that A has a C^∞ boundary.

Exercise 6.4 Using the isoperimetric property of the ball in the whole space, prove that when $\Omega = \mathbb{R}^d_+ := \{x \in \mathbb{R}^d : x_1 > 0\}$ the half ball $B(0, R) \cap \mathbb{R}^d_+$ solves the problem

$$\max\{|A| : A \subset \Omega, \mathrm{Per}(A; \Omega) = L\}.$$

Exercise 6.5 Show that in dimension 3 there are compact sets with perimeter smaller than that of their convex hull, i.e. sets A such that $\text{Per}(co(A)) > \text{Per}(A)$, where $co(A)$ is the smallest convex set containing A.

Exercise 6.6 Prove the following fact: if A_n is a sequence of subsets of $\mathbb{R}^2$ such that $|A_n| + \text{Per}(A_n) \le C$ and we define A_n' as $S_{e_1}(S_{e_2}(A_n))$, where e_1, e_2 is the canonical basis of $\mathbb{R}^2$ the sequence of sets A_n' is contained in a compact domain of $\mathbb{R}^d$. Use this to prove the existence of an optimal set in the isoperimetric problem in the whole space without first restricting to compact sets. What fails for $d > 2$?

Exercise 6.7 Prove the well-known inequality $||u||_{L^{d/(d-1)}} \le C||\nabla u||_{L^1}$ (with optimal constant) using the isoperimetric inequality, and deduce the Sobolev injection $W^{1,p} \subset L^{p^*}$ for $p > 1$.

Exercise 6.8 Let Ω be an open, connected, bounded, and smooth subset of $\mathbb{R}^d$ and $p \in (1, \infty)$. Prove that the following minimization problem admits a solution

$$\min\left\{ \frac{\int_\Omega |\nabla u|^p \, dx}{\int_\Omega |u|^p \, dx} : u \in W_0^{1,p}(\Omega) \setminus \{0\} \right\},$$

that this solution may be taken non-negative, and that it satisfies

$$\Delta_p u = cu^{p-1}$$

for a suitable constant $c < 0$. Assume $p > 2$ and consider $G := |\nabla u|^{p-2}\nabla u$: prove $G \in H^1_{loc}(\Omega)$.

Exercise 6.9 Let Ω be an open, connected, bounded, and smooth subset of $\mathbb{R}^d$. Prove that the following minimization problem admits a solution

$$\min\left\{ \frac{\int_\Omega |\nabla u|^2 \, dx}{\left|\int_\Omega u \, dx\right|^2} : u \in H_0^1(\Omega), \int_\Omega u \, dx \ne 0 \right\},$$

that this solution may be taken to be non-negative, and that any solution is C^∞. Prove that the solution is unique up to multiplicative constants and strictly positive inside Ω.

Exercise 6.10 Given two functions $u, v \ge 0$ in $L^2(\mathbb{R}^d)$ prove that we have $\int u^\circ v^\circ \ge \int uv$ and deduce that the radially decreasing rearrangement does not increase the L^2 distance.

Exercise 6.11 Let $u \ge 0$ be a Lipschitz continuous function with compact support in $\mathbb{R}^d$, and u^* its symmetrization. Prove that u° is also Lipschitz, and $\text{Lip}(u^\circ) \le \text{Lip}(u)$.

Exercise 6.12 Let Ω be an open and bounded subset of $\mathbb{R}^d$. Consider the following minimization problem

$$\min\left\{\int_\Omega \left(\frac{1}{2}|\nabla u(x)|^2 + \cos(u(x))\right) dx \; : \; u \in H_0^1(\Omega)\right\}.$$

1. Prove that it admits a solution, which can be taken to be positive
2. Write the PDE satisfied by the solutions.
3. Prove that the solutions are C^∞ in the interior of the domain Ω.
4. Prove that if $\lambda_1(\Omega) \geq 1$ then the only solution is $u = 0$.
5. Prove that if $\lambda_1(\Omega) < 1$ then $u = 0$ is not a solution.
6. Prove that, if Ω is a ball, there exists a radial solution.

Exercise 6.13 Given an open bounded set $\Omega \subset \mathbb{R}^d$ with volume $|\Omega| > 2\omega_d$ (ω_d denoting the volume of the unit ball in $\mathbb{R}^d$) and a function $f \in L^1(\Omega)$, prove that the following problem has a solution

$$\min\left\{\text{Per}(A) + \text{Per}(B) + \int_{A\cap B} f(x)\,dx \; : \; A, B \subset \Omega, |A| = |B| = \omega_d\right\}.$$

Also compute the above minimal value when Ω is the smallest convex domain containing $B(e, 1)$ and $B(-e, 1)$, where e is an arbitrary unit vector $e \in \mathbb{R}^d$, and $f = 1$ on Ω.

Exercise 6.14 Let $\Omega \subset \mathbb{R}^d$ be the unit cube $\Omega = (0, 1)^d$. Prove that for any measurable set $A \subset \Omega$ there exists a unique solution to the problem

$$\min\left\{\frac{1}{2}\int_\Omega |\nabla u(x)|^2\,dx + \int_A u(x)\,dx - \int_{\Omega\setminus A} u(x)\,dx \; : \; u \in H_0^1(\Omega)\right\}.$$

Let us call such a solution u_A. Write the Euler–Lagrange equation satisfied by u_A. Prove that if $A_n \to A$ (in the sense $\mathbb{1}_{A_n} \to \mathbb{1}_A$ in L^1) then $u_{A_n} \to u_A$ in the strong H_0^1 sense.

Prove that the problem

$$\min\left\{\int_\Omega |u_A(x)|^2\,dx + \text{Per}(A) \; : \; A \subset \Omega\right\}$$

admits a solution. Removing the perimeter penalization, prove that we have

$$\inf\left\{\int_\Omega |u_A(x)|^2\,dx \; : \; A \subset \Omega\right\} = 0$$

and that the infimum is not attained.

Exercise 6.15 Let $A \subset \mathbb{R}^2$ be a bounded measurable set. Let $\pi(A) \subset \mathbb{R}$ be defined via $\pi(A) = \{x \in \mathbb{R} : \mathcal{L}^1(\{y \in \mathbb{R} : (x, y) \in A\} > 0\}$. Prove that we have $\text{Per}(A) \geq 2\mathcal{L}^1(\pi(A))$. Apply this result to the isoperimetric problem in $\mathbb{R}^2$, proving the existence of a solution to

$$\min\left\{ \text{Per}(A) : A \subset \mathbb{R}^2, A \text{ bounded}, |A| = 1 \right\}.$$

Exercise 6.16 Let Γ be the image of an injective Lipschitz curve with endpoints $x, y \in \mathbb{R}^d$ and $[x, y]$ be the segment from x to y. Prove $\mathcal{H}^1(\Gamma) \geq \sqrt{4d_H(\Gamma, [x, y])^2 + |x - y|^2}$. Is the same estimate true if we only assume that Γ is compact and connected?

Exercise 6.17 Prove the existence of a minimizer for the problem

$$\min\left\{ C(\Sigma) + \lambda \mathcal{H}^1(\Sigma) : \Sigma \subset \overline{\Omega} \text{ compact and connected} \right\},$$

where $\lambda > 0$ is given, and we set

$$C(\Sigma) = \sup\left\{ \int_\Omega fu\,dx - \frac{1}{2}\int_{\Omega\setminus\Sigma} |\nabla u|^2\,dx : u \in C_c^\infty(\Omega \setminus \Sigma) \right\}.$$

Exercise 6.18 Prove that the functional $A \mapsto |A|$ is u.s.c. for the Hausdorff convergence but not continuous.

Hints

Hint to Exercise 6.1 If not there are infinitely many disjoint balls of fixed radius where $f \geq \varepsilon > 0$.

Hint to Exercise 6.2 Perturb the function f into $f + \varepsilon\varphi$ and compute the first-order development of the ratio. The value of the constant is $c = -\frac{(d-1)\,\text{Per}(A)}{d|A|}$.

Hint to Exercise 6.3 Since A is locally the epigraph of a convex, and hence Lipschitz, function f, perturb f into $f + \varepsilon\varphi$ and use the PDE to prove its regularity, in the spirit of Chap. 5.

Hint to Exercise 6.4 For any competitor A in this problem consider $\tilde{A}$ obtained by taking the union of A and its symmetric set as a competitor in the standard isoperimetric problem.

Hint to Exercise 6.5 Use for instance $A = B(0, 1) \times ([-L, -L+\varepsilon] \cup [L-\varepsilon, L])$.

Hint to Exercise 6.6 After two Steiner symmetrizations, the set A_n' is such that if it contains a point (x_1, x_2) it also contains the rectangle $[-|x_1|, |x_1|] \times [-|x_2|, |x_2|]$

and the perimeter condition yields the result. On the other hand, even if combined with the volume condition, this does not work for $d > 2$.

Hint to Exercise 6.7 For $u \geq 0$, write $\int |\nabla u|$ as $\int_0^\infty \mathrm{Per}(\{u > t\})\, dt$ and bound this from below in terms of $|\{u > t\}|$. Also write $\int |u|^m = \int_0^\infty m t^{m-1} |\{u > t\}|\, dt$. For the last question, apply the case $p = 1$ to $|u|^\alpha$.

Hint to Exercise 6.8 Follow the same steps as in the proof of Proposition 6.10. For the regularity, use Sect. 4.4.

Hint to Exercise 6.9 Follow again similar steps as in the proof of Proposition 6.10. Then use the equation $\Delta u = c$.

Hint to Exercise 6.10 Write $\int uv\, dx = \int dx \int_0^\infty dt\, ds\, \mathbb{1}_{u>t}(x)\mathbb{1}_{v>s}(x)$ and compare the intersection of these level sets for u, v and u°, v°. For the L^2 distances, expand the squares.

Hint to Exercise 6.11 Use Remark 6.15 and send $p \to \infty$.

Hint to Exercise 6.12 When $\lambda_1(\Omega) \geq 1$ the functional is convex, if not one can Taylor-expand the cosine to find a better competitor than 0. For the existence of a radial solution, replace u with u°.

Hint to Exercise 6.13 Take a minimizing sequence and use the bound on the perimeters to obtain strong L^1 and hence a.e. convergence of the indicator functions. For the specific example of Ω containing $B(e, 1)$ and $B(-e, 1)$, take $A = B(e, 1)$ and $B = B(-e, 1)$.

Hint to Exercise 6.14 The function u_A is the solution of $\Delta u = f$, for $f = 2\mathbb{1}_A - 1$. Strong L^2 convergence of the right-hand side implies strong convergence in H_0^1 of the solutions. When penalizing the perimeter, we obtain strong L^1 convergence of $\mathbb{1}_{A_n}$ on a minimizing sequence, which is also strong L^2 because these functions are bounded in L^∞. Without this penalization we could have $2\mathbb{1}_{A_n} - 1 \rightharpoonup 0$.

Hint to Exercise 6.15 The perimeter decreases by Steiner symmetrization and the perimeter of the symmetrized set is at least twice the measure of its projection. If we replace A with $\varphi(A)$, where $\varphi(x, y) = (\psi(x), y)$, and ψ is a Lipschitz function with $\psi' = 1$ on $\pi(A)$ and $\psi' = 0$ elsewhere, the perimeter again decreases. This allows us to modify a minimizing sequence so that it is compact in x. The same construction can then be performed in y.

Hint to Exercise 6.16 Use $\mathcal{H}^1(\Gamma) \geq |x - z| + |z - y|$ for a point z such that $d(z, [x, y]) = d_H(\Gamma, [x, y])$ and optimize over possible z. If Γ is not a curve this estimate could be false, think of $x = (-1, 0)$, $y = (1, 0)$ and $\Gamma = [x, y] \cup [a, b]$ with $a = (0, 0)$ and $b = (0, L)$, $L \gg 1$.

Hint to Exercise 6.17 Take a minimizing sequence Σ_n which converges to Σ in the Hausdorff sense. Prove the lower semicontinuity of the functional by fixing a function $u \in C_c^\infty(\Omega \setminus \Sigma)$ and proving that for large n we have $u \in C_c^\infty(\Omega \setminus \Sigma)$. Combine with the Gołąb theorem.

Hint to Exercise 6.18 Use $|\Omega| - |A| = \sup\{\int \varphi\,dx \ : \ \varphi \le 1, \varphi \in C_c^\infty(\Omega \setminus A)\}$. For the failure of lower semicontinuity, use a sequence of small sets which become denser and denser.

Chapter 7
Γ-Convergence: Theory and Examples

This chapter is devoted to a very natural question in optimization, which consists in investigating the limit of a sequence of optimization problems. Which is the correct notion of convergence which guarantees the convergence of the minimal values and/or of the minimizers?

Although the question is very natural in optimization in general, its leading approach has been particularly popularized in the framework of the calculus of variations. The reason for this can be found in the connections with the convergence of solutions of some PDEs, which are the Euler–Lagrange equations of variational problems. This also explains the name which has been chosen to denote this convergence: according to the terminology in [76] it is called Γ-convergence and it is to be considered as a variational counterpart of a previously existing notion of convergence for differential operators (hence, PDEs) based on the convergence of the corresponding graphs, called G-convergence (see [185]); Γ was then chosen as a variant of G. As we will see, Γ-convergence is equivalent to a suitable notion of convergence of the epigraphs of the functions to be minimized, and is consequently also called *epiconvergence* but it is much less frequently referred to by this name than Γ-convergence.

In this chapter the first section will be devoted to the abstract theory, for simplicity in a metric setting (as a reference for the topological case we refer to [69], while for a simpler introduction to the topic we refer to [40]); we will then move in the second section to some easy examples in the framework of the calculus of variations where the Γ-limit is in some sense not surprising, or at least shares a similar form with the approximating functionals. This includes a very simple 1D example of homogenization. Then, Sects. 7.3 and 7.4 will be devoted to two, very different, examples where the limit optimization problem is quite different from the approximating ones. One of these examples, related to the perimeter functional and useful for the numerical approximation of perimeter optimization, is a very classical result in the calculus of variations and Γ-convergence and is known (at least to a part of the variational community, in particular the Italian community) as the Modica–

Mortola approximation of the perimeter. The other is an optimal location problem appearing both in logistic and practical optimization issues and in the quantization of measures in signal theory. Its presentation in textbooks about Γ-convergence is less classical. We then finish the chapter with a discussion section about variants of the Modica–Mortola approach for other problems involving lower-dimensional energies, and with an exercise section.

7.1   General Metric Theory and Main Properties

Let us fix a metric space (X, d). Given a sequence of functionals $F_{n'} : X \to \mathbb{R} \cup \{+\infty\}$ we define their Γ-liminf and their Γ-limsup as two other functionals on X, in the following way.

$$(\Gamma\text{-}\liminf F_n)(x) := \inf\{\liminf_n F_n(x_n) : x_n \to x\},$$

$$(\Gamma\text{-}\limsup F_n)(x) := \inf\{\limsup_n F_n(x_n) : x_n \to x\}.$$

We say that F_n Γ-converges to F, and we write $F_n \overset{\Gamma}{\to} F$, if

$$(\Gamma\text{-}\liminf F_n) = (\Gamma\text{-}\limsup F_n) = F.$$

Very often we will write F^- instead of $\Gamma\text{-}\liminf F_n$ and F^+ instead of $\Gamma\text{-}\limsup F_n$.

In practice, proving $F_n \overset{\Gamma}{\to} F$ requires us to prove two facts:

- (*Γ-liminf inequality*): we need to prove $F^- \geq F$, i.e. we need to prove the inequality $\liminf_n F_n(x_n) \geq F(x)$ for any approximating sequence $x_n \to x$; of course it is sufficient to prove it when $F_n(x_n)$ is bounded, at least after passing to a subsequence;
- (*Γ-limsup inequality*): we need to prove $F^+ \leq F$, i.e. we need to find, for every x, a sequence $x_n \to x$ such that $\limsup_n F_n(x_n) \leq F(x)$ (such a sequence is called a *recovery sequence*) or at least, for every $\varepsilon > 0$, find a sequence with $\limsup_n F_n(x_n) \leq F(x) + \varepsilon$ (in which case we call this sequence an *ε-recovery sequence*). We will see later that this inequality can be restricted to suitable dense subsets. In any case, it can obviously be restricted to the set $\{F < +\infty\}$, otherwise the inequality is trivial.

We will see that the notion of Γ-convergence is exactly what we need in order to consider limits of optimization problems, since we have the following main properties (that will be proven in this section):

- if there exists a compact set $K \subset X$ such that $\inf_X F_n = \inf_K F_n$ for any n, then F attains its minimum and $\inf F_n \to \min F$;

- if $(x_n)_n$ is a sequence of minimizers for F_n admitting a subsequence converging to x, then x minimizes F;
- if F_n is a sequence Γ-converging to F, then $F_n + G$ will Γ-converge to $F + G$ for any continuous function $G : X \to \mathbb{R} \cup \{+\infty\}$.

The first two properties above explain the relation between Γ-convergence and the convergence of the minimization problems, i.e. they provide, under suitable conditions, the convergence of the minimal values and of the minimizers. The third property explains why we have developed all this theory instead of just looking for ad-hoc tricks to study the limits of minima and minimizers of a given sequence of problems: because the theory of Γ-convergence allows us to avoid re-doing all the work when similar functionals (for instance, if adding continuous perturbations) are considered.

Note that the theory developed here deals with minimization problems, and indeed the definitions of Γ-liminf, Γ-limsup and Γ-limit are not symmetric. If one wanted to study the limit of a sequence of maximization problems it would be necessary either to change the sign of the functionals, or to re-do the entire theory by changing all the inequalities, following a mirror theory which is sometimes called Γ^+-convergence.

In order to prove the three properties above we will prove a series of technical facts.

Proposition 7.1 *If $F_n \overset{\Gamma}{\to} F$ then for every subsequence F_{n_k} we still have $F_{n_k} \overset{\Gamma}{\to} F$.*

Proof Let us take a sequence $x_{n_k} \to x$ corresponding to the indices of this subsequence and let us complete it to a full sequence using $x_n = x$ for all the other indices. We have

$$\liminf_k F_{n_k}(x_{n_k}) \geq \liminf_n F_n(x_n) \geq (\Gamma\text{-}\liminf F_n)(x),$$

which proves that the Γ-liminf of the subsequence is larger than that of the full sequence.

For the Γ-limsup, let us take an ε-recovery sequence $x_n \to x$ for the full sequence. We then have

$$\limsup_k F_{n_k}(x_{n_k}) \leq \limsup_n F_n(x_n) \leq (\Gamma\text{-}\limsup F_n)(x) + \varepsilon,$$

thus proving $(\Gamma\text{-}\limsup F_{n_k}) \leq (\Gamma\text{-}\limsup F_n) + \varepsilon$.

This proves that passing to a subsequence increases the Γ-liminf and reduces the Γ-limsup, but when they coincide they cannot change, thus proving the claim. □

Proposition 7.2 *Both F^- and F^+ are l.s.c. functionals.*

Proof Let us take $x \in X$ and a sequence $x_k \to x$. We want to prove the semicontinuity of F^-. It is not restrictive to assume $F^-(x_k) < +\infty$ for every k. Fix an arbitrary sequence of strictly positive numbers $\varepsilon_k \to 0$. For each k there is a sequence $x_{n,k}$ such that $\lim_n x_{n,k} = x_k$ and $\liminf_n F_n(x_{n,k}) < F^-(x_k) + \varepsilon_k$. We can hence choose $n = n(k)$ arbitrarily large such that $d(x_{n,k}, x_k) < \varepsilon_k$ and $F_n(x_{n,k}) < F^-(x_k) + \varepsilon_k$. We can assume $n(k+1) > n(k)$ which means that the sets of indices n that we use defines a subsequence. Consider a sequence $\tilde{x}_n$ defined as

$$\tilde{x}_n = \begin{cases} x_{n,k} & \text{if } n = n(k), \\ x & \text{if } n \neq n(k) \text{ for all } k. \end{cases}$$

We have $\lim_n \tilde{x}_n = x$ and

$$F^-(x) \leq \liminf_n F_n(\tilde{x}_n) \leq \liminf_k F_{n(k)}(x_{n(k),k}) \leq \liminf F^-(x_k),$$

which proves that F^- is l.s.c.

We now want to prove the semicontinuity of F^+. We can assume, up to subsequences, that $F^+(x_k)$ is finite for every k and that $\lim F^+(x_k)$ exists and is finite as well. For each k there is a sequence $x_{n,k}$ such that $\lim_n x_{n,k} = x_k$ and $\limsup_n F_n(x_{n,k}) < F^+(x_k) + \varepsilon_k$. This means that we can hence choose $n = n(k)$ such that for every $n \geq n(k)$ we have $d(x_{n,k}, x_k) < \varepsilon_k$ and $F_n(x_{n,k}) < F^+(x_k) + \varepsilon_k$. We can choose the sequence $n(k)$ such that $n(k+1) > n(k)$. We now define a sequence $\tilde{x}_n$ defined as

$$\tilde{x}_n = x_{n,k} \text{ if } n(k) \leq n < n(k+1).$$

We have $\lim_n \tilde{x}_n = x$ and

$$F^+(x) \leq \limsup_n F_n(\tilde{x}_n) \leq \lim_k F^+(x_k),$$

which proves that F^+ is l.s.c.   □

Proposition 7.3 *Assume $F^+ \leq F$ on a set D with the following property: for every $x \in X$ there is a sequence $x_n \to x$ contained in D and such that $F(x_n) \to F(x)$. Then we have $F^+ \leq F$ on X.*

Proof Let us take $x \in X$ and a sequence $x_n \to x$ as in the statement. We then write $F^+(x_n) \leq F(x_n)$ and pass to the limit, using the semicontinuity of F^+ and the assumption $F(x_n) \to F(x)$. This provides $F^+(x) \leq F(x)$, as required.   □

The sets D with the property described in the assumption of Proposition 7.3 are said to be *dense in energy*. This means that they are dense but the approximating sequence that can be found in D for every limit point in X also satisfies an extra property: it makes the functional F converge. The name "energy" comes from the fact that often the functionals are called energies. Note that only F, and not F_n, is involved in this definition.

Proposition 7.4 *If $F_n \overset{\Gamma}{\to} F$ and there exists a compact set $K \subset X$ such that $\inf_X F_n = \inf_K F_n$ for any n, then F attains its minimum and $\inf F_n \to \min F$.*

Proof Let us first prove $\liminf_n (\inf F_n) \geq \inf F$. Up to extracting a subsequence, we can assume that this liminf is a limit, and Γ-convergence is preserved. Now, take $x_n \in K$ such that $F_n(x_n) \leq \inf F_n + \frac{1}{n}$. Extract a subsequence $x_{n_k} \to x_0$ and look at the sequence of functionals F_{n_k}, which also Γ-converge to F. We then have $\inf F \leq F(x_0) \leq \liminf F_{n_k}(x_{n_k}) = \liminf(\inf F_n)$.

Take now a point x, and $\varepsilon > 0$. There exists $x_n \to x$ such that $\limsup F_n(x_n) \leq F(x) + \varepsilon$. In particular, $\limsup_n (\inf F_n) \leq F(x) + \varepsilon$ and, $\varepsilon > 0$ and $x \in X$ being arbitrary, we get $\limsup_n (\inf F_n) \leq \inf F$.

We then have

$$\limsup_n (\inf F_n) \leq \inf F \leq F(x_0) \leq \liminf_n (\inf F_n) \leq \limsup_n (\inf F_n).$$

Hence, all inequalities are equalities, which proves at the same time $\inf F_n \to \inf F$ and $\inf F = F(x_0)$, i.e. the min is attained by x_0. □

Proposition 7.5 *If $F_n \overset{\Gamma}{\to} F$, $x_n \in \operatorname{argmin} F_n$, and $x_n \to x$, then $x \in \operatorname{argmin} F$.*

Proof First notice that $\bigcup_n \{x_n\} \cup \{x\}$ is a compact set on which all the functionals F_n attain their minimum, so we can apply the previous proposition and get $\inf F_n \to \inf F$. We then apply the Γ-liminf property to this sequence, thus obtaining

$$F(x) \leq \liminf F_n(x_n) = \liminf_n (\inf F_n) = \inf F,$$

which proves at the same time $x \in \operatorname{argmin} F$ and $\operatorname{argmin} F \neq \emptyset$. □

Proposition 7.6 *If $F_n \overset{\Gamma}{\to} F$, then for any continuous function $G : X \to \mathbb{R} \cup \{+\infty\}$ we have $F_n + G \overset{\Gamma}{\to} F + G$.*

Proof To prove the Γ-liminf inequality, take $x_n \to x$: we have $\liminf(F_n(x_n) + G(x_n)) = (\liminf F_n(x_n)) + G(x) \geq F(x) + G(x)$. To prove the Γ-limsup inequality, the same ε-recovery sequence $x_n \to x$ can be used: indeed, $\limsup(F_n(x_n) + G(x_n)) = (\limsup F_n(x_n)) + G(x) \leq F(x) + \varepsilon + G(x)$. □

We finish this abstract section by showing the relation between Γ-convergence and the convergence of the epigraphs of the functions F_n.

We recall the following definition (as usual, adapted to metric spaces).

Definition 7.7 In a metric space X a sequence of sets $A_n \subset X$ is said to be *Kuratowski-converging* to A if

- for every $x \in A$ there exists a sequence x_n with $x_n \in A_n$ and $x_n \to x$;
- for every sequence x_n with $x_n \in A_n$ and every x which is a limit of a subsequence x_{n_k} of x_n we have $x \in A$.

As we underlined in Box 6.5, in compact metric spaces this notion of convergence is equivalent to that of convergence in the Hausdorff distance.

We can then state the following result.

Proposition 7.8 *Given a sequence* $F_n : X \to \mathbb{R} \cup \{+\infty\}$ *of functionals on a metric space* X *and another functional* $F : X \to \mathbb{R} \cup \{+\infty\}$ *the following are equivalent*

- $F_n \overset{\Gamma}{\to} F$
- Epi(F_n) *Kuratowski converges to* Epi(F) *in* $X \times \mathbb{R}$.

Proof We will show that the first condition in the Kuratowski convergence of the epigraphs is equivalent to the Γ-limsup inequality and the second to the Γ-liminf inequality.

Indeed, let us assume that we have the Γ-limsup inequality and let us take $(x_0, t_0) \in$ Epi(F). We know that there exists a sequence $x_n \to x_0$ with $\limsup_n F_n(x_n) \leq F(x_0) \leq t_0$. We can assume $F_n(x_n) < +\infty$, at least for n large enough. If we define $t_n := \max\{t_0, F_n(x_n)\}$ we have $(x_n, t_n) \in$ Epi(F_n) and $(x_n, t_n) \to (x_0, t_0)$, which shows the first condition in the Kuratowski convergence. Conversely, if we assume that this condition holds and take a point $x_0 \in X$ we can consider the point $(x_0, F(x_0)) \in$ Epi(F) and we obtain the existence of a sequence $(x_n, t_n) \in$ Epi(F_n) with $(x_n, t_n) \to (x_0, F(x_0))$, which means $t_n \geq F_n(x_n)$ and $t_n \to F(x_0)$. In particular, $\limsup_n F_n(x_n) \leq \limsup_n t_n = F(x_0)$ and the Γ-limsup inequality is satisfied.

We now assume that we have the Γ-liminf inequality and take a sequence $(x_n, t_n) \in$ Epi(F_n) as well as a point (x_0, t_0) which is a limit of a subsequence of such a sequence. Up to replacing the x_n which do not belong to this subsequence with x_0, we can assume that the whole sequence x_n converges to x_0, while we only have $t_{n_k} \to t_0$ on a subsequence. On this subsequence we have $t_0 = \lim_k t_{n_k} \geq \liminf_k F_{n_k}(x_{n_k}) \geq \liminf_n F_n(x_n) \geq F(x_0)$ because of the Γ-liminf inequality. We then obtain $t_0 \geq F(x_0)$, i.e. $(x_0, t_0) \in$ Epi(F), which shows the second condition of the Kuratowski convergence. Conversely, we assume that this condition is satisfied and take a sequence $x_n \to x_0$. There exists a subsequence such that $\lim_k F(x_{n_k}) = \ell := \liminf_n F_n(x_n)$ and we can assume $\ell < +\infty$. Then we have $(x_{n_k}, F(x_{n_k})) \to (x_0, \ell)$ and $(x_{n_k}, F(x_{n_k})) \in$ Epi(F_{n_k}). Up to completing this subsequence with some points $(x_n, t_n) \in$ Epi(F_n) we can apply the second condition of Kuratowski convergence in order to obtain $(x_0, \ell) \in$ Epi(F), i.e. $\ell \geq F(x_0)$, which is the Γ-liminf condition that we wanted to prove.   □

A consequence of the above equivalence, considering that the Hausdorff and the Kuratowski convergence on compact sets are equivalent and compact, is that from every sequence of functions $F_n : X \to \mathbb{R}_+$ one can extract a Γ-converging subsequence (see Exercise 7.3; the limit can of course be infinite-valued, and possibly the constant $+\infty$). This can be used in order to obtain a compact notion of convergence on cost functions, and has been used for instance in [94] for price optimization.

7.2 Γ-Convergence of Classical Integral Functionals

In this section we discuss some first, easy, examples of Γ-convergence in the framework of the calculus of variations, i.e. we consider integral functionals defined on functional spaces.

We start from the most trivial situation, where we consider an integral functional depending on a parameter, one of the involved functions converges in a suitable sense to another function of the same form, and the limit functional is just obtained by using this limit function. We describe this via an easy example.

Example 7.9 Consider $X = L^2(\Omega)$ for a smooth open domain $\Omega \subset \mathbb{R}^d$, and some functions $f_n \in L^2(\Omega)$. Consider

$$F_n(u) := \begin{cases} \int_\Omega \left(\frac{1}{p} |\nabla u|^p + f_n u \right) dx & \text{if } u \in W_0^{1,p}(\Omega), \\ +\infty & \text{if not.} \end{cases}$$

Assume $f_n \rightharpoonup f$. Then we have $F_n \xrightarrow{\Gamma} F$ where F is defined by replacing f_n with f.

The statement of the above example can be proven in the following way. For the Γ-liminf inequality, assume $u_n \to u$ in L^2 and $F_n(u_n) \le C$. Using the L^2 bounds on u_n and f_n we see that the $W^{1,p}$ norm of u_n is bounded, so that we can assume $\nabla u_n \rightharpoonup \nabla u$ in L^p and, by semicontinuity $\int \frac{1}{p} |\nabla u|^p \le \liminf \int \frac{1}{p} |\nabla u_n|^p$. In what concerns the second part of the integral, we clearly have $\int f_n u_n \to \int f u$ since we have weak convergence of f_n and strong of u_n. We then obtain $\liminf F_n(u_n) \ge F(u)$. For the Γ-limsup inequality it is enough to choose u with $F(u) < +\infty$ and take $u_n = u$.

This Γ-convergence result implies the convergence of the minimizers as soon as they are compact for the L^2 strong convergence. This is true for $p \ge \frac{2d}{d+2}$ so that $W^{1,p}$ compactly embeds into L^2. In this case, we can for instance deduce the L^2 convergence of the solutions u_n of $\Delta_p u_n = f_n$ to the solution of $\Delta_p u = f$. Note that the linear case $p = 2$ could be easily studied by bounding the norm $||u_n - u||_{H^1}$ in terms of $||f_n - f||_{H^{-1}}$ (which tends to 0 since L^2 compactly embeds into H^{-1}) but the situation is more complicated in the non-linear case and the Γ-convergence is a useful tool.

As a last remark, we observe that for $p \ge \frac{2d}{d+2}$ the choice of the strong L^2 convergence to establish the Γ-convergence is irrelevant. We could have for instance chosen the weak L^2 convergence and deduced the strong L^2 convergence in the Γ-liminf inequality from the $W^{1,p}$ bound. We could have chosen other norms, but in this case it would have been necessary to first obtain a bound on $||u_n||_{W^{1,p}}$.

The Γ-Limit of the L^p Norm when $p \to \infty$ The next examples that we consider involve the behavior of the L^p norms when $p \to \infty$ and, without any surprise, the limit involves the L^∞ norm.

We first consider the following situation:

Example 7.10 Consider $X = L^1(\Omega)$ endowed with the topology of weak convergence in L^1, where Ω is a measurable space with finite measure. For $p \in [1, +\infty]$, define $F_p : X \to \mathbb{R} \cup \{+\infty\}$ via $F_p(u) = ||u||_{L^p}$ (a functional which takes the value $+\infty$ if $u \in L^1 \setminus L^p$). Then we have, as $p \to \infty$, the Γ-convergence $F_p \xrightarrow{\Gamma} F_\infty$.

The reader can check that, as soon as the measure of Ω is 1, then the statement of the above example is a particular case of the statement of Exercise 7.1, since $F_p \le F_q$ as soon as $p \le q$. When the measure of Ω is finite but not equal to 1 we have $F_p = |\Omega|^{1/p} \tilde{F}_p$, where the functionals $\tilde{F}_p$ are increasing in p, and $\lim_{p\to\infty} |\Omega|^{1/p} = 1$, which gives the same result.

It is often more useful to deal with the following variant, that we fully prove:

Proposition 7.11 *Consider* $X = L^1(\Omega)$ *endowed with the topology of weak convergence in* L^1, *where* Ω *is a measurable space with finite measure. For* $p \in [1, \infty)$, *define* $F_p : X \to \mathbb{R} \cup \{+\infty\}$ *via* $F_p(u) = c(p) \int_\Omega |u|^p \, dx$ *(a functional which takes the value* $+\infty$ *if* $u \in L^1 \setminus L^p$). *Assume that the coefficients* $c(p)$ *satisfy* $\lim_{p\to\infty} c(p)^{1/p} = 1$. *Then we have, as* $p \to \infty$, *the* Γ-*convergence* $F_p \xrightarrow{\Gamma} F$, *where*

$$F(u) = \begin{cases} 0 & \text{if } |u| \le 1 \text{ a.e.} \\ +\infty & \text{if not.} \end{cases}$$

Proof Let us first prove the Γ-liminf property, taking a sequence u_p with $F_p(u_p) \le C < \infty$ and assuming $u_p \rightharpoonup u$, we need to prove $|u| \le 1$. We fix an exponent p_0 and, for $p \ge p_0$, we write

$$\left(\fint_\Omega |u_p|^{p_0} dx\right)^{1/p_0} \le \left(\fint_\Omega |u_p|^p dx\right)^{1/p}$$
$$= \frac{(c(p)^{-1} F_p(u_p))^{1/p}}{|\Omega|^{1/p}} \le \left(\frac{C}{c(p)|\Omega|}\right)^{1/p}.$$

Using $c(p)^{1/p} \to 1$ we can take the limit as $p \to \infty$ of the right-hand side of the above inequality and obtain

$$\left(\fint_\Omega |u|^{p_0} dx\right)^{1/p_0} \le \liminf_p \left(\fint_\Omega |u_p|^{p_0} dx\right)^{1/p_0} \le 1,$$

where we used the lower semicontinuity, for fixed p_0, of the L^{p_0} norm w.r.t. the weak convergence in L^1. We can then take the limit $p_0 \to \infty$ and use the well-known fact that, for a fixed function on a set of finite measure, the L^p norm converges to the L^∞ norm as $p \to \infty$, to deduce $||u||_{L^\infty} \le 1$, which is equivalent to the Γ-liminf inequality.

For the Γ-limsup inequality we just need to prove that, given a function u with $|u| \leq 1$ a.e., we can find a recovery sequence. We choose $u_p = [pc(p)]^{-1/p} u$. This sequence converges to u since $p^{-1/p} \to 1$ and $c(p)^{-1/p} \to 1$; moreover we have $F_p(u) \leq \frac{|\Omega|}{p} \to 0$, which is the desired result. □

Remark 7.12 The most common choice for the coefficient $c(p)$ in the previous proposition is $c(p) = \frac{1}{p}$ but $c(p) = 1$ is also useful; both satisfy the assumption guaranteeing the Γ-convergence to F.

With essentially the very same proof we can also obtain the following.

Proposition 7.13 *Consider* $X = L^1(\Omega)$ *for a bounded convex domain* $\Omega \subset \mathbb{R}^d$ *and, for* $p \in [1, \infty)$, *define* $F_p : X \to \mathbb{R} \cup \{+\infty\}$ *via* $F_p(u) = c(p) \int_\Omega |\nabla u|^p$ *(a functional which takes the value* $+\infty$ *if* $u \in L^1 \setminus W^{1,p}$). *Assume that the coefficients* $c(p)$ *satisfy* $\lim_{p \to \infty} c(p)^{1/p} = 1$. *Then we have, as* $p \to \infty$, *the Γ-convergence* $F_p \xrightarrow{\Gamma} F$, *where*

$$F(u) = \begin{cases} 0 & \text{if } u \in \text{Lip}_1 . \\ +\infty & \text{if not.} \end{cases}$$

Proof Once we observe that, in a convex domain, the condition $u \in W^{1,\infty}$ and $|\nabla u| \leq 1$ a.e. is equivalent to $u \in \text{Lip}_1$, the proof is exactly the same as in Proposition 7.11. The recovery sequence can be taken to be, again, $u_p = [pc(p)]^{-1/p} u$, and for the Γ-liminf condition we just note that $u_p \to u$ in L^1 together with a bound on $\|\nabla u_p\|_{L^{p_0}}$ implies the weak convergence $u_p \rightharpoonup u$ in W^{1,p_0} and allows us to use the lower semicontinuity of the norm. We then obtain a uniform bound on $\|\nabla u\|_{L^{p_0}}$, independent of p_0, i.e. $\nabla u \in L^\infty$. □

Elliptic Functionals with Variable Coefficients We now consider a more surprising example. We consider a sequence of functions $a_n : [0, 1] \to \mathbb{R}$ such that $\Lambda_{min} \leq a_n \leq \Lambda_{max}$ for two strictly positive constants $\Lambda_{min}, \Lambda_{max}$. We then define a functional on $L^2([0, 1])$ via

$$F_n(u) := \begin{cases} \int_0^1 a_n(x) \frac{|u'(x)|^2}{2} \, dx & \text{if } u \in W^{1,2}, \ u(1) = 0, \\ +\infty & \text{if not.} \end{cases} \tag{7.1}$$

We now wonder what the Γ-limit of F_n could be. A natural guess is to assume, up to subsequences, that we have $a_n \xrightarrow{*} a$ in L^∞ and hope to prove that the functional where we replace a_n with a is the Γ-limit. If this was true, then we would also have Γ-convergence when adding $\int fu$ to the functional, for fixed $f \in L^2$ (since this is a continuous functional, and we can apply Proposition 7.6), and convergence of the minimizers. The minimizers u_n would be characterized by $(a_n u'_n)' = f$ with a transversality condition $u'_n(0) = 0$, so that this would imply $a_n(x) u'_n(x) = \int_0^x f(t) \, dt$ and hence $u'_n(x) = (a_n(x))^{-1} \int_0^x f(t) \, dt$. We then see that the weak convergence of a_n is not a good assumption, but we should rather require $a_n^{-1} \overset{*}{\rightharpoonup} a^{-1}$!!

We can now prove the following.

Proposition 7.14 *Assume that a_n is such that $\Lambda_{min} \leq a_n \leq \Lambda_{max}$ and $a_n^{-1} \overset{*}{\rightharpoonup} a^{-1}$ in L^∞. Let F_n be defined by (7.1) and F be defined analogously by replacing a_n with a. Then $F_n \overset{\Gamma}{\rightarrow} F$ (w.r.t. to the L^1 convergence).*

Proof To prove the Γ-liminf inequality, we write $F_n(u) = \int_0^1 L(u_n', a_n^{-1}) \, dx$, where $L : \mathbb{R} \times \mathbb{R}_+$ is given by $L(v, s) = \frac{|v|^2}{2s}$. We take a sequence $u_n \rightarrow u$ with $F_n(u_n) \leq C$. This bound implies that u_n is bounded in H^1 and then $u_n' \rightharpoonup u'$ in L^2. We can easily check that L is a convex function of two variables, for instance by computing its Hessian,[1] which is given by

$$D^2 L(v, s) = \begin{pmatrix} \frac{1}{s} & -\frac{v}{s^2} \\ -\frac{v}{s^2} & \frac{|v|^2}{s^3} \end{pmatrix} \geq 0.$$

Hence, by semicontinuity (see Proposition 3.9), we deduce from $a_n^{-1} \overset{*}{\rightharpoonup} a^{-1}$ and $u_n' \rightharpoonup u'$ that we have $\liminf F_n(u_n) \geq F(u)$.

To prove the Γ-limsup inequality, given u with $F(u) < +\infty$ (i.e. $u \in H^1$), we define u_n via $u_n' = \frac{a}{a_n} u'$ and $u_n(1) = 0$. We see that u_n' is bounded in L^2 and hence $u_n' \rightharpoonup v$. Integrating against a smooth test function and using $a_n^{-1} \overset{*}{\rightharpoonup} a^{-1}$ we see that we have $v = \frac{a}{a} u' = u'$, so that we obtain $u_n' \rightharpoonup u'$, the weak convergence being intended in the L^2 sense. Thanks to the final value $u_n(1) = u(1) = 0$, we deduce weak convergence of u_n to u in H^1, and strong in L^2. We then have

$$F_n(u_n) = \int \frac{1}{2} a_n u_n' \cdot u_n' = \int \frac{1}{2} a u' \cdot u_n' \rightarrow \int \frac{1}{2} a u' \cdot u' = F(u),$$

which proves the Γ-limsup inequality.   □

In the above proof, we strongly used the fact that we are in 1D, since we defined a recovery sequence by choosing u_n'. In higher dimensions, one could consider a functional of the form $\int \frac{a_n}{2} |\nabla u|^2$ but the functions $\frac{a}{a_n} \nabla u$ could not be gradients! Indeed, the higher-dimensional situation is much trickier, and it is (very partially) addressed in Exercise 7.11.

Remark 7.15 The above situation also includes one of the most studied cases, that of periodic homogenization. Assume $a_1 : \mathbb{R} \rightarrow [a_{min}, a_{max}]$ is a given 1-periodic function taking its values between a_{min} and a_{max}, with $a_{min} > 0$, and consider in particular the case where a_1 only takes values in $\{a_{min}, a_{max}\}$. Let us define the

---

[1] One can also refer to the computations in (4.18), as we are facing a perspective function.

quantity $r := |\{x \in [0, 1] : a_1(x) = a_{min}\}| \in (0, 1)$. Define $a_n(x) := a_1(nx)$. Then we have $F_n \xrightarrow{\Gamma} F$, where F is obtained using the constant function $a = (ra_{min}^{-1} + (1 - r)a_{max}^{-1})^{-1}$. This can be interpreted in terms of materials: a mixture of two different materials with conductivity a_{min} and a_{max} in proportions r and $1 - r$ produces a composite material whose conductivity is the (weighted) harmonic mean of the two conductivities.

7.3 Optimal Location and Optimal Quantization

This section is devoted to a very natural optimization problem: approximating a probability density with a finite number of points. This can be interpreted in different ways. The first comes from a classical optimization question in logistics or urban planning: suppose that we want to choose where to ideally put N post offices in a town Ω and that we know how the population (or the demand) is distributed in the domain Ω; where should we place the N offices? We suppose that we want to minimize the average distance for the population to reach the closest post office. In this case, if f is the population density, we solve

$$\min\left\{\int_\Omega d(x, S) f(x)\, dx : S \subset \Omega, \#S = N\right\},$$

where the set S stands for the position of the post offices. We can also decide to consider the average of a power of the distance, in which case we fix $p > 0$ and solve

$$\min\left\{\int_\Omega d(x, S)^p f(x)\, dx : S \subset \Omega, \#S = N\right\}. \tag{7.2}$$

Note that these optimization problems are exactly the ones presented in Sect. 6.3, but the constraint on S is different: we impose here the number of points, i.e. $\mathcal{H}^0(S)$, while in Sect. 6.3 we considered the measure $\mathcal{H}^1$ (together with a connectedness constraint, which is necessary for the existence).

Another interpretation of the same problem comes from signal processing. We can consider f as the probability distribution of a signal, supported on a whole domain Ω, and we want to approximate it with a probability measure concentrated on N points (compare with what we discussed in Sect. 6.3.1 about the 1D skeleton of an image); such a measure will be of the form $\sum_{j=1}^N a_j \delta_{x_j}$ and S will stand for $\{x_j : j = 1, \ldots, N\}$. Given S, the value $\int d(x, S)^p f(x)\, dx$ is indeed equal to $\min\{W_p(f, \nu)^p : \mathrm{spt}(\nu) \subset S\}$, where W_p is the Wasserstein distance of order p among probability measures (this only works for $p \geq 1$, as the Wasserstein distance is not defined for $p < 1$).

Box 7.1 Good to Know!—*Wasserstein Distances*

Given two probability measures $\mu, \nu \in \mathcal{P}(\Omega)$ and an exponent $p \geq 1$ we define

$$W_p(\mu, \nu) := \inf \left\{ \int_{\Omega \times \Omega} |x - y|^p \, d\gamma \; : \; \gamma \in \Pi(\mu, \nu) \right\}^{1/p},$$

where $\Pi(\mu, \nu)$ is the set of transport plans from μ to ν (i.e. measures on $\Omega \times \Omega$ with marginals μ and ν) see Sect. 4.6.

It is possible to prove that the above quantity, called the *Wasserstein distance* of order p between μ and ν, is indeed a distance on $\mathcal{P}(\Omega)$ if Ω is bounded (otherwise we need to restrict our attention to the set of measures μ with finite moment of order p, i.e. such that $\int |x|^p \, d\mu < +\infty$), and that it metrizes the weak-* convergence of probability measures (in duality with $C(\Omega)$) when Ω is compact. When μ has no atoms it is also possible to define W_p as follows

$$W_p(\mu, \nu) := \inf \left\{ \int_{\Omega} |x - T(x)|^p \, d\mu \; : \; T : \Omega \to \Omega, \; T_\# \mu = \nu \right\}^{1/p}$$

and an optimal map T, called the *optimal transport*, exists whenever $\Omega \subset \mathbb{R}^d$ and $\mu \ll \mathcal{L}^d$.

Given a closed set $S \subset \mathbb{R}^d$ and an absolutely continuous measure μ the problem

$$\min\{W_p(\mu, \nu) \; : \; \mathrm{spt}(\nu) \subset S\}$$

has a unique solution, given by $\nu = (\Pi_S)_\# \mu$, where Π_S is the projection onto S, which is well-defined Lebesgue a.e. (more precisely, the projection exists and is unique on any differentiability point of $x \mapsto d(x, S)$, which is a Lipschitz function, differentiable a.e. because of the Rademacher theorem), and the map $T = \Pi_S$ is the optimal transport from μ to ν.

For more details we refer to [177, Chapter 5].

Note that the weights a_j of each atom x_j are not prescribed, and will finally be equal to the measure (w.r.t. f) of the region which is projected on each x_j, i.e. $a_j = \int_{V_j} f(x) \, dx$ where

$$V_j = (\Pi_S)^{-1}(\{x_j\}) = \{x \; : \; |x - x_j| \leq |x - x_i| \text{ for all } i\}$$

is the so-called *Voronoi cell* of the point x_j. We then see that, once the points of S are known, the optimization among possible weights is easy,[2] while the optimization of the position of the points x_j is a non-convex optimization problem. The main unknown is hence the support S and we are interested in understanding how S covers Ω (or $\{f > 0\}$) as $N \to \infty$. It is indeed clear that, as $N \to \infty$, the optimal sets S_N become denser and denser and that the optimal value $\int d(x, S_N)^p f(x)\, dx$ tends to 0. We need to find the speed of convergence (i.e. find an exponent $\alpha > 0$ and a constant c such that $\int d(x, S_N)^p f(x)\, dx = O(N^{-\alpha})$ and $\lim_N N^\alpha \int d(x, S_N)^p f(x)\, dx = c$) and the limit distribution of S_N, i.e. the average number of points per unit volume when N is very large. Intuitively, one can guess that this density will be higher where f is larger, but we can also try to find an explicit expression in terms of f and, if possible, an exponent $\beta > 0$ such that it is of the order of f^β.

All these questions can be attacked via Γ-convergence tools.

We start by associating with every set S with $\#S = N$ the uniform probability measure on S, i.e. $\mu_S = \frac{1}{N} \sum_{y \in S} \delta_y \in \mathcal{P}(\Omega)$. Our question is to identify the limit as $N \to \infty$ of the measures μ_{S_N} where S_N is optimal. Identifying S with μ_S is natural and useful for two reasons; first, if we need to identify the limit density of the optimal set S_N what we need to find is exactly the weak-* limit of μ_{S_N}; second, this identification allows us to set all the optimization problems for different values of N on the same set $X = \mathcal{P}(\Omega)$.

We will indeed define a sequence of functionals F_N on the space of probability measures such that their minimization is equivalent to finding the optimal set S_N. We will then look for their Γ-limit F so that we can deduce that the optimal measures μ_{S_N} converge to the minimizer μ of F. If this minimizer is unique and easy to find, we have the answer to our asymptotical question.

We define now the functionals $F_N : \mathcal{P}(\Omega) \to \mathbb{R} \cup \{+\infty\}$ whose Γ-convergence will be the object of this section. We set

$$F_N(\mu) := \begin{cases} N^{p/d} \int_\Omega d(x, S)^p f(x)\, dx & \text{if } \mu = \mu_S \text{ with } \#S = N, \\ +\infty & \text{otherwise.} \end{cases}$$

This choice requires at least a clarification, i.e. why is the integral multiplied by $N^{p/d}$? This normalizing factor corresponds to the guess $\alpha = p/d$ for the speed of convergence and, without it, we could easily expect the Γ-limit to be the constant 0 functional. Now, when the Γ-limit is constant, every μ is a minimizer and we do not obtain any information on the limits of the minimizers μ_{S_N}. On the other hand, multiplying F_N by a multiplicative constant does not affect the minimizers, and the goal is to choose a good scaling so that the limit is non-trivial. Choosing

---

[2] We underline another possible quantization problem, where the unknown is still the set $S = \{x_1, \ldots, x_N\}$, but the weights are prescribed and equal to $1/N$, so that the problem becomes the minimization of $W_p(f, \nu)$ among empirical measures; in this case the Voronoi cells have to be replaced by a more involved notion, that of *Laguerre cells* (or *power cells*), which we will not develop here, see for instance Section 6.4.2 in [177].

$N^{p/d}$ is a sort of bet: if we are able to find a Γ-limit which is non-constant and whose minimal value is a finite strictly positive number, we also prove that we have $\int d(x, S_N)^p f(x)\, dx = O(N^{-p/d})$.

We consider the unit cube $[0,1]^d$ and define the constant

$$\theta_{p,d} := \inf \left\{ \liminf_N \ N^{1/d} \int_{[0,1]^d} d(x, S_N)^p\, dx, \ \#S_N = N \right\}$$

as well as, for technical reasons, the similar constant

$$\tilde{\theta}_{p,d} := \inf \left\{ \liminf_N \ N^{p/d} \int_{[0,1]^d} d(x, S_N \cup \partial[0,1]^d)\, dx, \ \#S_N = N \right\}.$$

Proposition 7.16 *We have $\theta_{p,d} = \tilde{\theta}_{p,d}$ and $0 < \theta_{p,d} < \infty$.*

Proof We have of course $\theta_{p,d} \geq \tilde{\theta}_{p,d}$. To prove the opposite inequality, fix $\varepsilon > 0$ and select a sequence of uniform grids on $\partial[0,1]^d$: decompose the boundary into $2d M^{d-1}$ small cubes, each of size $1/M$, choosing M such that $M^{-1} < \varepsilon N^{-1/d}$. We call such a grid G_N. Take a sequence S_N which almost realizes the infimum in the definition of $\tilde{\theta}_{p,d}$, i.e. $\#S_N = N$ and $\liminf_N N^{p/d} \int_{[0,1]^d} d(x, S_N \cup \partial[0,1]^d)^p\, dx \leq \tilde{\theta}_{p,d} + \varepsilon$. We then use

$$d(x, S_N \cup G_N) \leq d(x, S_N \cup \partial[0,1]^d) + \frac{\sqrt{d-1}}{M}.$$

Lemma 7.17 allows us to deduce from this, for $\varepsilon_1 > 0$,

$$d(x, S_N \cup G_N)^p \leq (1+\varepsilon_1) d(x, S_N \cup \partial[0,1]^d)^p + C(\varepsilon_1, p) \left(\frac{\sqrt{d-1}}{M} \right)^p.$$

We then obtain

$$\liminf_N \ N^{p/d} \int_{[0,1]^d} d(x, S_N \cup G_N)^p\, dx$$

$$\leq (1+\varepsilon_1)\tilde{\theta}_{p,d} + \varepsilon(1+\varepsilon_1) + \limsup_N C(d, p, \varepsilon_1) N^{p/d} \frac{1}{M^p}$$

$$\leq (1+\varepsilon_1)\tilde{\theta}_{p,d} + \varepsilon(1+\varepsilon_1) + C(d, p, \varepsilon_1)\varepsilon^p.$$

If we use $\#(S_N \cup G_N) \le N + 2dM^{d-1} = N + O(N^{(d-1)/d}) = N + o(N)$ we obtain a sequence of sets $\tilde{S}_N := S_N \cup G_N$ such that

$$\liminf_N \ (\#\tilde{S}_N)^{p/d} \int_{[0,1]^d} d(x, \tilde{S}_N)^p \, dx \le (1 + \varepsilon)(1 + \varepsilon_1)\tilde{\theta}_{p,d} + C(d, p, \varepsilon_1)\varepsilon^p.$$

Taking first the limit $\varepsilon \to 0$ and then $\varepsilon_1 \to 0$ we obtain $\theta_{p,d} \le \tilde{\theta}_{p,d}$.

In order to prove $\theta_{p,d} < +\infty$, just use a sequence of sets on a uniform grid in $[0,1]^d$: we can decompose the whole cube into M^d small cubes, each of size $1/M$, choosing M such that $M \approx N^{1/d}$.

In order to prove $\theta > 0$ we also use a uniform grid, but choosing M such that $M^d > 2N$. Then we take an arbitrary S_N with N points: in this case at least half of the cubes of the grid do not contain points of S_N. An empty cube of size δ contributes at least $C\delta^{d+p}$ in the integral, i.e. $cM^{-(d+p)}$. Since at least N cubes are empty we obtain $\theta_{p,d} \ge cN^{p/d} \cdot N \cdot M^{-(d+p)} = O(1)$. $\qquad\square$

Lemma 7.17 *Given $p > 0$ and $\varepsilon > 0$, there exists a finite constant $C(\varepsilon, p)$ such that we have*

$$(x + y)^p \le (1 + \varepsilon)x^p + C(\varepsilon, p)y^p \quad \text{for all } x, y \ge 0.$$

Proof Since the inequality is trivial for $y = 0$, it is enough to prove that we have

$$\sup_{x,y \ge 0, y > 0} \frac{(x + y)^p - (1 + \varepsilon)x^p}{y^p} < +\infty.$$

By homogeneity, we can restrict to $y = 1$ and thus we must consider

$$\sup_{x \ge 0} (x + 1)^p - (1 + \varepsilon)x^p.$$

Since we have $(x + 1)^p - (1 + \varepsilon)x^p \le 0$ for x large enough (depending on ε), the above sup is clearly a maximum and is thus finite, which proves the claim. $\qquad\square$

We then define the functional $F : \mathcal{P}(\Omega) \to \mathbb{R} \cup \{+\infty\}$ through

$$F(\mu) := \theta_{p,d} \int_\Omega \frac{f}{(\mu^{ac})^{p/d}} \, dx,$$

where μ^{ac} is the density of the absolutely continuous part of μ.

The role and the value of the constants $\theta_{p,d}$ deserve a small discussion. First of all, we note that in order to study the limit behavior of the minimization problem (7.2) we introduced some constants whose definition require us to solve a very similar problem. This may seem disappointing but, luckily, we do not need to compute the precise value of $\theta_{p,d}$ (we just need to be sure that it is a non-zero finite number) if we want to investigate the limit behavior of the minimizers. Indeed, they do not depend on any multiplicative constants in front of the functionals they minimize. This is also related to another interesting fact: the limit distribution of the N points of the sets S_N essentially depend on scaling (see [111] where a general framework is presented in order to obtain the asymptotical behavior of point configurations optimizing a large class of functionals, including the setting we present in this section) while the precise value of $\theta_{p,d}$ is instead related to how points arrange locally once their density is known. This is a different, and more geometrical question, which is not easy to answer: typically points are arranged on a regular lattice and we need to understand the best possible lattice configuration. In dimension $d = 2$ the answer is the same for every p, the lattice where each point is placed at the center of a regular hexagon is the optimal one, and this allows us to compute the value of $\theta_{p,2}$. The optimality of the hexagon can be seen as related to the fact that we are minimizing $\int_A |x|^p \, dx$ among sets A with fixed volume and which tile the space: the best A without the tiling constraint would be the ball, and the hexagon is in some sense the closest to the ball among the possible tilings. Many optimization problems among possible tilings have the hexagon as a solution in dimension 2, a fact which is known as the *honeycomb conjecture*, see also Sect. 7.5.

The Γ-convergence result that we will prove is the following.

Proposition 7.18 *Assume that Ω is a closed cube and that f is strictly positive and continuous. Then we have $F_N \xrightarrow{\Gamma} F$ in $\mathcal{P}(\Omega)$ (endowed with the weak-* convergence) as $N \to \infty$.*

Proof Let us start from the Γ-liminf inequality. Consider $\mu_N \xrightarrow{*} \mu$ and assume $F_N(\mu_N) \leq C$. In particular, we have $\mu_N = \mu_{S_N}$ for a sequence of sets S_N with $\#S_N = N$. Let us define the functions $\lambda_N(x) := N^{p/d} f(x) d(x, S_N)^p$. This sequence of functions is bounded in L^1, so we can assume that, as a sequence of positive measures, it converges weakly-* to a measure λ up to a subsequence. Choosing a subsequence which realizes the liminf, we will have $\liminf_N F_N(\mu_N) = \lambda(\Omega)$.

In order to estimate λ from below,[3] we fix a closed cube $Q \subset \Omega$. Let us call δ the size of this cube (its side, so that $|Q| = \delta^d$). We write

$$\lambda_N(Q) = N^{p/d} \int_Q f(x) d(x, S_N)^p \, dx$$

$$\geq (\min_Q f) \left(\frac{1}{\mu_N(Q)} \right)^{p/d} (\#S_N \cap Q)^{p/d} \int_Q d(x, (S_N \cap Q) \cup \partial Q)^p \, dx.$$

We note that the last part of the right-hand side recalls the definition of $\tilde{\theta}_{p,d}$. We also note that if we want to bound $\lambda_N(Q)$ from below we can assume $\lim_N \#S_N \cap Q = \infty$, otherwise, if the number of points in Q stays bounded, we necessarily have

---

[3] Note the analogy of the proof with that of the Gołąb theorem. Many lower semicontinuity or Γ-convergence theorems are proven in this way.

$\lambda_N(Q) \to \infty$. So, the sequence of sets $S_N \cap Q$ is admissible in the definition of $\tilde\theta_{p,d}$ after being scaled so as to fit the unit cube: indeed, if the unit cube in the definition of $\tilde\theta_{p,d}$ is replaced by a cube of size δ, the values of the integrals are multiplied by δ^{d+p}. We then have

$$\liminf_N (\#S_N \cap Q)^{p/d} \int_Q d(x, (S_N \cap Q) \cup \partial Q)^p \, dx \geq \delta^{d+p}\tilde\theta_{p,d} = \delta^{d+p}\theta_{p,d}$$

and hence

$$\liminf_N \lambda_N(Q) \geq \left(\min_Q f\right) \liminf_N \left(\frac{1}{\mu_N(Q)}\right)^{p/d} \delta^{d+p} \, \theta_{p,d}.$$

We now use the fact that, for closed sets, when a sequence of measures weakly converges, the mass given by the limit measure is larger than the limsup of the masses:

$$\lambda(Q) \geq \liminf_N \lambda_N(Q) \geq \left(\min_Q f\right) \left(\frac{1}{\mu(Q)}\right)^{p/d} \delta^{d+p}\theta_{p,d}.$$

This can be re-written as

$$\frac{\lambda(Q)}{|Q|} \geq \left(\min_Q f\right) \left(\frac{|Q|}{\mu(Q)}\right)^{p/d} \theta_{p,d}.$$

We now choose a sequence of cubes shrinking around a point $x \in \Omega$ and we use the fact that, for a.e. x, the ratio between the mass a measure gives to the cube and the volume of the cube tends to the density of the absolutely continuous part (see Lemma 7.19), thus obtaining (also using the continuity of f)

$$\lambda^{ac}(x) \geq \theta_{p,d}f(x) \left(\frac{1}{\mu^{ac}(x)}\right)^{p/d}.$$

This implies

$$\liminf_N F_N(\mu_N) = \lambda(\Omega) \geq \int_\Omega \lambda^{ac}(x) \, dx \geq F(\mu).$$

We now switch to the Γ-limsup inequality. Let us start from the case $\mu = \sum_i a_i \mathbb{1}_{Q_i}$, i.e. μ is absolutely continuous with piecewise constant density $a_i > 0$ on the cubes of a regular grid. In order to have a probability measure, we assume $\sum_i a_i |Q_i| = 1$. Fix $\varepsilon > 0$. Using the definition of $\theta_{p,d}$ we can find a finite set $S_0 \subset [0, 1]^d$ with $\#S_0 = N_0$ such that $N_0^{p/d} \int_{[0,1]^d} d(x, S_0)^p \, dx < \theta_{p,d}(1 + \varepsilon)$. We then divide each cube Q_i into M_i^d subcubes $Q_{i,j}$ of size δ_i on a regular grid, and on each subcube we put a scaled copy of S_0. We have $M_i^d \delta_i^d = |Q_i|$. We choose M_i

such that $N_0 M_i^d \approx a_i |Q_i| N$ (take M_i to be the integer part of $(a_i |Q_i| N / N_0)^{1/d}$)
so that, for $N \to \infty$, we have indeed $\mu_N \rightharpoonup \mu$ (where μ_N is the uniform measure
on the set S_N obtained by the union of all these scaled copies). Note that with this
construction, in general we do not have exactly N points in S_N, but we can complete
this set with arbitrary points, and the proportion of the number of points that we
added over N tends to 0. We now estimate

$$F_N(\mu_N) \leq N^{p/d} \sum_{i,j} \delta_i^{d+p} \theta_{p,d} (1+\varepsilon) N_0^{-p/d} \max_{Q_{i,j}} f.$$

For N large enough, the cubes $Q_{i,j}$ are small and we have

$$\max_{Q_{i,j}} f \leq (1+\varepsilon) \fint_{Q_{i,j}} f \, dx = (1+\varepsilon) \delta_i^{-d} \int_{Q_{i,j}} f \, dx.$$

Thus, we get

$$F_N(\mu_N) \leq N^{p/d} \theta_{p,d} (1+\varepsilon) N_0^{-p/d} \sum_{i,j} \delta_i^{p} \int_{Q_{i,j}} f \, dx.$$

Note that we have $\delta_i = |Q_i|^{1/d} / M_i \approx N_0^{1/d} N^{-1/d} a_i^{-1/d}$, whence

$$\limsup_{N} F_N(\mu_N) \leq \theta_{p,d} (1+\varepsilon) \sum_{i,j} \int_{Q_{i,j}} \frac{f}{a_i^{p/d}} \, dx = (1+\varepsilon) F(\mu).$$

This shows, ε being arbitrary, the Γ-limsup inequality in the case $\mu = \sum_i a_i \mathbb{1}_{Q_i}$.

We now need to extend our Γ-limsup inequality to other measures μ which are
not of the form $\mu = \sum_i a_i \mathbb{1}_{Q_i}$. We hence need to show that this class of measures
is dense in energy.

Take now an arbitrary probability μ with $F(\mu) < \infty$. Since f is assumed
to be strictly positive, this implies $\mu^{ac} > 0$ a.e. Take a regular grid of size
$\delta_k \to 0$, composed of k^d disjoint cubes Q_i and define $\mu_k := \sum_i a_i \mathbb{1}_{Q_i}$ with
$a_i = \mu(Q_i)/|Q_i|$ (one has to define the subcubes in a disjoint way, for instance
as products of semi-open intervals, of the form $[0, \delta_k)^d$). It is clear that we have
$\mu_k \rightharpoonup \mu$ since the mass is preserved in every cube, whose diameter tends to 0.

We then compute $F(\mu_k)$. We have

$$F(\mu_k) \leq \sum_i (\max_{Q_i} f) |Q_i| \left(\frac{\mu(Q_i)}{|Q_i|} \right)^{-p/d}.$$

We use the function $U(s) = s^{-p/d}$, which is decreasing and convex, together with Jensen's inequality, in order to obtain

$$\left(\frac{\mu(Q_i)}{|Q_i|}\right)^{-p/d} = U\left(\frac{\mu(Q_i)}{|Q_i|}\right) \le U\left(\fint_{Q_i} \mu^{ac}\,dx\right) \le \fint_{Q_i} U(\mu^{ac})\,dx.$$

This allows us to write

$$F(\mu_k) \le \sum_i \left(\max_{Q_i} f\right)|Q_i| \fint_{Q_i} U(\mu^{ac})\,dx = \sum_i (\max_{Q_i} f) \int_{Q_i} U(\mu^{ac})\,dx.$$

We finish by noting that, for $k \to \infty$, we have $\max_{Q_i} f \le (1 + \varepsilon_k)f$ on Q_i for a certain sequence of positive numbers $\varepsilon_k \to 0$ (depending on the modulus of continuity of f), so that $(\max_{Q_i} f) \int_{Q_i} U(\mu^{ac}) \le (1 + \varepsilon_k) \int_{Q_i} fU(\mu^{ac})$. We then get

$$F(\mu_k) \le (1 + \varepsilon_k)F(\mu),$$

which concludes the proof. $\square$

Lemma 7.19 *Let us denote by $Q(x, r)$ the cube $x + [-r, r]^d$, centered at x and with side $2r$, so that its volume is $(2r)^d$. Then, for every locally finite measure μ on $\mathbb{R}^d$ whose Radon–Nikodym decomposition is $\mu = \mu^{ac}\,dx + \mu^{sing}$, we have*

$$\lim_r \frac{\mu(Q(x, r))}{(2r)^d} = \mu^{ac}(x) \quad \text{for Lebesgue-a.e. } x.$$

Proof Because of the characterization of Lebesgue points (see Box 5.1), if x is a Lebesgue point of the L^1_{loc} function μ^{ac} we know that we have $\fint_{Q(x,r)} \mu^{ac} \to \mu^{ax}(x)$. Hence, it is enough to prove $\lim_r \mu^{sing}(Q(x, r))/(2r)^d = 0$ for a.e. x. Using $Q(x, r) \subset B(x, \sqrt{d}r)$ and $|B(x, \sqrt{d}r)| \le Cr^d$, it is enough to prove $\lim_r \mu^{sing}(B(x, r))/\omega_d r^d = 0$ for Lebesgue-a.e. x. We consider then $E_\varepsilon = \{x : \limsup_r \mu^{sing}(B(x, r))/\omega_d r^d \ge \varepsilon\}$ and we need to prove that E_ε is Lebesgue-negligible for every $\varepsilon > 0$. Let A be a Lebesgue-negligible set on which μ^{sing} is concentrated and set $B = E_\varepsilon \setminus A$. We have $\mu^{sing}(B) = 0$ since $B \cap A = \emptyset$; moreover, the theorem about densities and differentiation of measures presented in Box 6.7 provides $\varepsilon\mathcal{H}^d(B) \le \mu^{sing}(B) = 0$ so that $E_\varepsilon \setminus A$ is Lebesgue-negligible. Since A is also Lebesgue-negligible, this proves the claim. $\square$

Note that the assumption that Ω is a cube is just made for simplicity in the Γ-limsup, and that it is possible to get rid of it by suitably considering the "remainders" after filling Ω with cubes.

The above proof is a simplified version of that in [38] where f was only assumed to be l.s.c. (note that here we only used its continuity in the Γ-limsup part, since for the Γ-liminf the lower semicontinuity would have been enough). Actually, in [153] (which deals with the average distance problem presented in Sect. 6.3.1, but the answer to question 6 raised in [54] gives rise to a very similar problem to this one) even this assumption is removed, and f is only assumed to be L^1.

The consequences of the previous Γ-convergence proof include the two following results:

Proposition 7.20 *Assume that S_N is a sequence of optimizers for (7.2) with $N \to \infty$. Then, the sequence $\mu_N := \mu_{S_N}$ weakly-* converges to the measure μ which is absolutely continuous with density ρ equal to $cf^{d/(d+p)}$, where c is a normalization constant such that $\int \rho = 1$.*

Proof We just need to prove that this measure μ is the unique optimizer of F. First note that F can only be minimized by an absolutely continuous measure, as singular parts will not affect the value of the functional, so it is better to remove a possible singular part and use the same mass to increase the absolutely continuous part.

We use again the notation $U(s) = s^{-p/d}$, and write $\rho = cf^{d/(d+p)}$ as in the statement. Then we have, for $\mu \ll \mathcal{L}^d$,

$$F(\mu) = \theta_{p,d} \int_\Omega fU(\mu^{ac})\,dx = \theta_{p,d}c_0 \fint_\Omega U\left(\frac{\mu^{ac}}{\rho}\right)d\rho$$

$$\geq \theta_{p,d}c_0 U\left(\fint_\Omega \left(\frac{\mu^{ac}}{\rho}\right)d\rho\right)$$

$$= \theta_{p,d}c_0 U\left(\int_\Omega \mu^{ac}\,dx\right) = \theta_{p,d}c_0 U(1) = \theta_{p,d}c_0$$

where c_0 is such that $c_0\rho^{(d+p)/d} = f$, i.e.

$$c_0 = c^{-d/(d+p)} = \left(\int_\Omega f^{\frac{d}{d+p}}\,dx\right)^{\frac{d+p}{d}}.$$

The inequality follows from Jensen's inequality, and is an equality if and only if μ^{ac}/ρ is constant, which proves the claim since the mass has been fixed as 1. □

The following result is probably one of the most well-known in the field of quantization of measures, and is known as Zador's theorem (see [105]).

Proposition 7.21 *If we call m_N the minimal value of (7.2) then we have $m_N = O(N^{-p/d})$ and more precisely we have*

$$\lim_N N^{p/d}m_N = \theta_{p,d}\left(\int_\Omega f^{\frac{d}{d+p}}\,dx\right)^{\frac{d+p}{d}}.$$

Proof We apply again the properties of Γ-convergence. Since $m_N = N^{-p/d}\min F_N$ we just need to compute the limit of $\min F_N$, which equals $\min F$.

This computation is contained in the proof of Proposition 7.20, and we saw that the minimal value of F is $\theta_{p,d}c_0$, which proves the desired claim. □

These two results are standard in the subject and do not necessarily need to be expressed in terms of Γ-convergence. We refer for instance to the classical book [105], which discusses the topic of optimal quantization of measures from a different point of view. Yet, the language of Γ-convergence provided a unified framework to attack this and similar questions about the optimal location of resources, whether they are points, lines, small balls... the reference list includes, but is not limited to [38, 55, 56, 179]. We also underline [39], where the constraint on the cardinality of the set S is replaced by a constraint on the measure v. More precisely, instead of considering

$$\min\{W_p(f, v)^p : \#\operatorname{spt}(v) \le N\},$$

the authors of [39] considered problems of the form

$$\min\{W_p(f, v)^p : \mathcal{G}(v) \le N\},$$

where $\mathcal{G}$ is another functional defined on atomic measures, for instance

$$\mathcal{G}(v) := \begin{cases} \sum_k a_k^\alpha & \text{if } v = \sum_k a_k \delta_{x_k}, \\ +\infty & \text{if } v \text{ is not purely atomic}, \end{cases}$$

for an exponent $\alpha < 1$, which is one of the functionals studied in Sect. 3.3. Note that we can recover the standard case presented in this section when $\alpha = 0$. The approach via Γ-convergence of [39] is more involved than what we presented here, in particular because they consider an extended variable which also carries information on the local distribution of the masses of the atoms.

7.4 The Modica–Mortola Functional and the Perimeter

In this section we consider another example of Γ-convergence, and actually one of its most classical examples, as the title of the paper by Modica and Mortola [146, 147], where this result is proven, suggests. This result is concerned with the approximation of the perimeter functional via some functionals which are more typical for the calculus of variations, i.e. integral functionals involving the squared norm of the gradient. This Γ-convergence result is particularly interesting for numerical reasons; indeed, from a computational point of view, it is very difficult to handle perimeters, both because we are more accustomed to handling functions instead of sets, and because the perimeter is an energy which only involves a lower-dimensional subset, thus making the number of relevant pixels in the domain proportionally very small, and the computation unstable. Replacing the perimeter with a more standard functional involving derivatives of functions allows us to use more robust and more classical numerical methods. In some sense, the interest of the result presented in this section is exactly the opposite of that of the previous section: if in Sect. 7.3 we had a clear idea of the sequence of minimization problems that we wanted to study, and the goal was to investigate the asymptotic properties of the minimizers, in this section the important minimization problem is the limit one, and we want to deliberately build an approximating sequence which is easier to handle. On the other hand, both sections share this important aspect that the Γ-limit is quite different in nature than the approximating functionals.

We consider the following sequence[4] of functionals defined on $L^1(\Omega)$,

$$F_\varepsilon(u) := \begin{cases} \frac{\varepsilon}{2} \int_\Omega |\nabla u|^2 \, dx + \frac{1}{2\varepsilon} \int_\Omega W(u) \, dx & \text{if } u \in H^1_0(\Omega), \\ +\infty & \text{if not,} \end{cases}$$

where $W : \mathbb{R} \to \mathbb{R}$ is a continuous and bounded function satisfying $W(0) = W(1) = 0$ and $W > 0$ on $\mathbb{R} \setminus \{0, 1\}$. We denote by c_0 the constant given by $c_0 = \int_0^1 \sqrt{W}$.

We also define

$$F(u) := \begin{cases} c_0 \operatorname{Per}(A) & \text{if } u = \mathbb{1}_A \in \mathrm{BV}(\mathbb{R}^d), \\ +\infty & \text{if not.} \end{cases}$$

In this case we stress that the perimeter of A is computed inside the whole space, i.e. also considering $\partial A \cap \partial \Omega$.

We will prove the following result, which is due, as we already said, to Modica and Mortola [146].

Proposition 7.22 *Assume that Ω is a bounded convex set in $\mathbb{R}^d$. Then $F_\varepsilon \xrightarrow{\Gamma} F$ in $L^1(\Omega)$, as $\varepsilon \to 0$.*

Proof Let us start from the Γ-liminf inequality. Consider $u_\varepsilon \to u$ in L^1.

Note that the Young inequality $\frac{\varepsilon}{2}|a|^2 + \frac{1}{2\varepsilon}|b|^2 \geq ab$ provides the lower bound

$$F_\varepsilon(u_\varepsilon) \geq \int_\Omega \sqrt{W(u_\varepsilon)} |\nabla u_\varepsilon| \, dx = \int_\Omega |\nabla(\Phi(u_\varepsilon))| \, dx,$$

where $\Phi : \mathbb{R} \to \mathbb{R}$ is the function defined by $\Phi(0) = 0$ and $\Phi' = \sqrt{W}$. Note that, W being bounded, we have $|\Phi(u_\varepsilon)| \leq C|u_\varepsilon|$. This means that $\Phi(u_\varepsilon)$ is bounded in $\mathrm{BV}(\Omega)$ and, up to a subsequence, it converges strongly in L^1 to a function v. Up to another subsequence we also have pointwise convergence a.e., but we already had

---

[4] Throughout this section the approximation parameter is ε, which is a continuous parameter, so that when we speak about sequences or even about Γ-convergence we mean that we consider an arbitrary sequence $\varepsilon_n \to 0$.

$u_\varepsilon \to u$ in L^1 (hence a.e. up to subsequences), so that we get $v = \Phi(u)$. We then have $\nabla \Phi(u_\varepsilon) \to \nabla \Phi(u)$ in the sense of distributions and weakly as measures, and the lower semicontinuity of the norm implies $\Phi(u) \in BV(\Omega)$ and

$$||\nabla \Phi(u)||_{\mathcal{M}} \le \liminf_\varepsilon \int_\Omega |\nabla \Phi(u_\varepsilon)| \, dx \le \liminf_\varepsilon F_\varepsilon(u_\varepsilon).$$

On the other hand, since we can of course assume $F_\varepsilon(u_\varepsilon) \le C$, we also have $\int_\Omega W(u_\varepsilon) \le C\varepsilon$ and, by Fatou, $\int_\Omega W(u) = 0$, i.e. $u \in \{0, 1\}$ a.e. This means that we do have $u = \mathbb{1}_A$ for a measurable set $A \subset \Omega$. Note that in this case we have $\Phi(u) = \Phi(1)u = \Phi(1)\mathbb{1}_A$. Since we have $\Phi(1) = \int_0^1 \sqrt{W} > 0$, this implies $\mathbb{1}_A \in BV(\Omega)$, and we finally have

$$F(u) = \Phi(1) \operatorname{Per}(A) \le \liminf_\varepsilon F_\varepsilon(u_\varepsilon).$$

We now switch to the Γ-limsup inequality. We first consider the case $u = \mathbb{1}_A$ with A smooth and $d(A, \partial\Omega) > 0$. We need to build a recovery sequence u_ε. Let us define sd_A, the signed distance function to A given by

$$\operatorname{sd}_A(x) := \begin{cases} d(x, A) & \text{if } x \notin A, \\ -d(x, A^c) & \text{if } x \in A. \end{cases}$$

Take a function $\phi : \mathbb{R} \to [0, 1]$ such that there exist $L_\pm > 0$ with $\phi = 1$ on $(-\infty, L_-]$, $\phi = 0$ on $[L_+, +\infty)$, and $\phi \in C^1([-L_-, L_+])$. Define

$$u_\varepsilon = \phi\left(\frac{\operatorname{sd}_A}{\varepsilon}\right).$$

Note that $|\nabla \operatorname{sd}_A| = 1$ a.e. and hence we have $\varepsilon |\nabla u_\varepsilon|^2 = \frac{1}{\varepsilon}|\phi'|^2\left(\frac{\operatorname{sd}_A}{\varepsilon}\right)$ and

$$F_\varepsilon(u_\varepsilon) = \frac{1}{2\varepsilon} \int_\Omega \left(|\phi'|^2 + W\right)\left(\frac{\operatorname{sd}_A}{\varepsilon}\right) dx.$$

We now use the co-area formula (see Box 6.2) which provides the following equality, valid at least for Lipschitz functions $f, g : \Omega \to \mathbb{R}$:

$$\int g|\nabla f| \, dx = \int_{\mathbb{R}} dt \int_{\{f=t\}} g \, d\mathcal{H}^{d-1}.$$

We apply it to the case $f = sd_A$, for which the norm of the gradient is always 1. We then have

$$F_\varepsilon(u_\varepsilon) = \frac{1}{2\varepsilon} \int_{\mathbb{R}} \left(|\phi'|^2 + W(\phi) \right) \left(\frac{t}{\varepsilon} \right) \cdot \mathcal{H}^{d-1}(\{sd_A = t\}) \, dt$$

$$= \frac{1}{2} \int_{\mathbb{R}} \left(|\phi'|^2 + W(\phi) \right) (r) \cdot \mathcal{H}^{d-1}(\{sd_A = \varepsilon r\}) \, dr,$$

where the second equality comes from the change of variable $t = \varepsilon r$.

Since A is smooth we have, for every r, the convergence

$$\mathcal{H}^{d-1}(\{sd_A = \varepsilon r\}) \to \mathrm{Per}(A).$$

We can restrict the integral to $r \in [-L_-, L_+]$ which allows us to apply dominated convergence and obtain

$$\lim_\varepsilon F_\varepsilon(u_\varepsilon) = \mathrm{Per}(A) \int_{\mathbb{R}} \left(|\phi'|^2 + W(\phi) \right) (r) \, dr.$$

Moreover, it is clear that we have $u_\varepsilon \to \mathbb{1}_A$ in L^1 because of dominated convergence.

We now have to choose ϕ. Choose a function $\tilde{\phi}$ defined as a maximal solution of the Cauchy problem

$$\begin{cases} \tilde{\phi}' = -\sqrt{W(\tilde{\phi})}, \\ \tilde{\phi}(0) = \frac{1}{2}. \end{cases}$$

We necessarily have $\lim_{r \to -\infty} \tilde{\phi}(r) = 1$ and $\lim_{r \to +\infty} \tilde{\phi}(r) = 0$. The function $\tilde{\phi}$ is C^1 and monotone non-increasing.[5] Fix $\delta > 0$ and let $r_\pm$ be defined via $\tilde{\phi}(r_-) = 1-\delta$ and $\tilde{\phi}(r) = \delta$. We then take $\phi = \phi_\delta$ to be a function such that $\phi_\delta = \tilde{\phi}$ on $[r_-, r_+]$, $\phi_\delta \in C^1([r_- - 2\delta, r_+ + 2\delta])$, and $|\phi_\delta'| \le 1$ on $[r_- - 2\delta, r_-] \cup [r_+, r_+ + 2\delta]$, $\phi_\delta = 1$ on $(-\infty, r_- 2\delta)$ and $\phi_\delta = 0$ on $(R_+ + 2\delta, +\infty)$. We have

$$\int_{\mathbb{R}} \left(|\phi_\delta'|^2 + W(\phi_\delta) \right) (r) \, dr \le (1 + \sup W) 2\delta + \int_{r_-}^{r_+} \left(|\tilde{\phi}'|^2 + W(\tilde{\phi}) \right) (r) \, dr$$

$$\le C\delta + \int_{\mathbb{R}} \left(|\tilde{\phi}'|^2 + W(\tilde{\phi}) \right) (r) \, dr.$$

---

[5] Note that this function also coincides with the heteroclinic connection between the two wells 0 and 1 for the potential W, as described in Sect. 1.4.5.

Note that we have

$$\int_{\mathbb{R}} \left(|\tilde{\phi}'|^2 + W(\tilde{\phi}) \right)(r)\, dr = 2 \int_{\mathbb{R}} \sqrt{W(\tilde{\phi})} |\tilde{\phi}'|\, dr$$

$$= -2 \int_{\mathbb{R}} \frac{d}{dr} \left(\Phi \circ \tilde{\phi} \right) dr$$

$$= -2 \left(\Phi(\tilde{\phi}(+\infty)) - \Phi(\tilde{\phi}(-\infty)) \right)$$

$$= 2\Phi(1) = 2c_0,$$

which means

$$\lim_{\varepsilon} F_\varepsilon(u_\varepsilon) \leq (c_0 + C\delta) \operatorname{Per} A.$$

Since this holds for arbitrary $\delta > 0$, we obtain

$$(\Gamma\text{-}\limsup F_\varepsilon)(u) \leq c_0 \operatorname{Per} A = F(u).$$

We now need to extend our Γ-limsup inequality to other functions u which are not of the form $u = \mathbb{1}_A$ with A smooth and far from the boundary, but only $u = \mathbb{1}_A$ with A of finite perimeter. We hence need to show that the class $\mathcal{S}$ of indicators of smooth sets far from the boundary is dense in energy.

We start from a set A of finite perimeter with $d(A, \partial\Omega) > 0$ and set $u = \mathbb{1}_A$. The construction of Proposition 6.2 shows precisely that, if we take a smooth and compactly supported convolution kernel $\eta_n \rightharpoonup \delta_0$ defined by rescaling a fixed kernel η_1, and define $v_n := \eta_n * u$, we can chose $r_n \in (\delta, 1 - \delta)$ so that the set $A_n := \{v_n \geq r_n\}$ is a smooth set with $\operatorname{Per}(A_n) \leq \frac{\operatorname{Per} A}{1-2\delta}$ and $\mathbb{1}_{A_n} \to \mathbb{1}_A$ in L^1. This is not yet the desired sequence, since we have $F(u_n) \leq (1 - 2\delta)^{-1} F(u)$ instead of $\limsup_n F(u_n) \leq F(u)$, but this can be fixed easily. Indeed, for every δ this allows us to find $\tilde{u} \in \mathcal{S}$ with $\|\tilde{u} - u\|_{L^1}$ arbitrarily small and $F(\tilde{u}) \leq (1 - 2\delta)^{-1} F(u)$, which can be turned, using $\delta_n \to 0$, into a sequence which shows that $\mathcal{S}$ is dense in energy.

We now have to get rid of the assumption $d(A, \partial\Omega) > 0$. Using the fact that Ω is convex, and assuming without loss of generality that the origin 0 belongs to the interior of Ω, we can take $u = \mathbb{1}_A$ and define $u_n = \mathbb{1}_{t_n A}$ for a sequence $t_n \to 1$. The sets $t_n A$ are indeed far from the boundary. Moreover, we have $F(u_n) = c_0 \operatorname{Per}(t_n A) = t_n^{d-1} F(u) \to F(u)$. We are just left to prove $u_n \to u$ strongly in L^1. Because of the BV bound which provides compactness in L^1 we just need to prove any kind of weak convergence. Take a test function φ and compute $\int \varphi u_n = \int_{t_n A} \varphi = (t_n)^d \int_A \varphi(t_n y)\, dy \to \int_A \varphi$, as soon as φ is continuous. This shows the weak convergence in the sense of measures $u_n \rightharpoonup u$ and, thanks to the BV bound, strong L^1 convergence, and concludes the proof. $\qquad \square$

We do not detail here some interesting variants of the above Γ-convergence result, where one could impose a volume constraint on A in the perimeter minimization (as for the isoperimetric problem of Sect. 6.1), which corresponds to a constraint on $\int u$ in the approximation, or prescribe boundary values. Note that, in general, imposing constraints (which are stable under the convergence used in the Γ-convergence) do not affect the Γ-liminf part, but make it harder to prove the Γ-limsup part, since we need to produce a recovery sequence which satisfies them. Moreover, it is not possible to evoke Proposition 7.6, since the function G imposing the constraints is not continuous (as it takes the values 0 and $+\infty$). Yet, in many simple cases it is not hard to slightly modify the unconstrained recovery sequence in order to preserve the constraints.

7.5 Discussion: Phase-Field Approximation of Free Discontinuity and Optimal Network Problems

This section discusses some recent and less recent works on the approximation of variational problems involving energy terms concentrated on lower-dimensional objects, in the same spirit as the Modica–Mortola result of the previous section. This kind of approximation is usually referred to as *phase-field* approximation, since the sharp behavior $\{0, 1\}$ (belonging or not to a set) is replaced by a continuous variable $u \in [0, 1]$, which recalls what is usually done in phase transition models, where instead of just considering water and ice, intermediate states, corresponding to different proportions of melted ice, are considered. This terminology was introduced by Van der Waals, who was the first to propose an energy of the form of the Modica–Mortola energy in this setting.

Optimal Partitions, and a Particular Case of the Steiner Problem We already mentioned in a different setting the so-called *honeycomb conjecture*, a name which encompasses several optimal partition questions where a large domain (and ideally the whole space) has to be partitioned into pieces of unit volume so as to minimize a global energy. The most well-known problem among them consists in minimizing the average perimeter: how do we tile $\mathbb{R}^d$ by a partition into cells A_i of unit volume so that the asymptotic average perimeter of the partition, defined via

$$\lim_{R\to\infty} \frac{\sum_i \mathcal{H}^{d-1}(\partial A_i \cap B(0, R))}{|B(0, R)|}$$

is minimal? (Note that $|B(0, R)|$ also approximately equals the number of cells in the same ball, since cells have volume 1.) In dimension two the question dates back to antiquity and the conjecture that the hexagonal tiling (exactly the one used by bees) should be optimal is usually attributed to Pappus of Alexandria (fourth century). An affirmative answer was first given by Fejes Tóth [86], in the restricted case where each cell is required to be a convex polygon, while the general case was proven many years later by Thomas C. Hales [108]. On the other hand, in higher dimensions the picture is rather unclear. The problem when $d = 3$ was first raised by Lord Kelvin (the same Lord Kelvin who gave his name to the temperature unit) in 1894, and he conjectured that a certain tiling made of suitably adjusted truncated octahedra (whose faces are 8 slightly curved hexagons and 6 squares) may be optimal. This tiling satisfied suitable first-order optimality conditions and was therefore a satisfactory competitor. Yet, in [197], Denis Weaire and Robert Phelan found a counter-example, proving that a tiling made of a block of 8 polyhedra of two different types, one with 14 faces and the other with 12, gave a better average perimeter, even if the improvement was only of about 0.3%.

It is natural to attack the question numerically. Of course it is impossible to really attack the asymptotic question, but one can fix a large domain Ω with volume $N \in \mathbb{N}$ and look for the solution of the problem

$$\min \left\{ \sum_{i=1}^{N} \mathrm{Per}(A_i) \ : \ |A_i| = 1, \ (A_i)_i \text{ is a partition of } \Omega \right\}.$$

Note that it is irrelevant whether the perimeters of A_i are defined as $\mathrm{Per}(A_i; \Omega)$ or $\mathrm{Per}(A_i; \mathbb{R}^d)$, since the boundary $\partial\Omega$ is fixed and would be fully covered by $\bigcup_i \partial A_i$.

This can be clearly approximated via a suitable modification of the Modica–Mortola functional by studying

$$\min \left\{ \begin{array}{l} \sum_{i=1}^{N} \dfrac{\varepsilon}{2} \displaystyle\int_{\Omega} |\nabla u_i|^2 \mathrm{d}x + \dfrac{1}{2\varepsilon} \int_{\Omega} W(u_i)\mathrm{d}x \ : \ u_i \in H_0^1(\Omega), \\[2mm] \displaystyle\int u_i = 1, u_1 + \cdots + u_N = 1 \end{array} \right\},$$

which is what is done in [163] for numerical purposes. It is shown in this paper that both Kelvin's solution and the Weaire–Phelan solution can be found, depending on the initialization of the optimization algorithm (the optimization problem being non-convex, the output can depend on the initialization). This is remarkable, due to the very small gap between the energies of the two configurations.[6]

Yet another remarkable fact is that the same approach can be used to attack a very particular instance of the Euclidean Steiner tree problem[7] (see Sect. 6.3.1). Indeed, assume that we want to connect a set D of N points which lie on the boundary of a

---

[6] According to some texts on the internet, after the publication of [163] a new configuration better than that of Weaire–Phelan was discovered in [161], but after a careful look at the latter paper it seems that this is not exactly what is claimed in it.

[7] This very natural approach is due to É. Oudet, but is not contained in any particular paper, despite being the starting point of many oral presentations on the subject.

convex domain $\Omega \subset \mathbb{R}^2$. Since the competitors Σ connecting them will be trees, i.e. with no loops, we can prove that they determine a partition of Ω into N sets, and the total perimeter of the sets in the partition will be equal to twice the length of Σ plus the length of $\partial\Omega$. Hence, we can find a solution of the Steiner problem by solving

$$\min\left\{\sum_{i=1}^{N}\frac{\varepsilon}{2}\int_{\Omega}|\nabla u_i|^2\,dx + \frac{1}{2\varepsilon}\int_{\Omega}W(u_i)\,dx \ : \ \begin{array}{l} u_i \in H^1(\Omega), \\ u_1 + \cdots + u_N = 1, \\ \mathrm{Tr}[u_i] = 1 \text{ on } \Gamma_i \end{array}\right\},$$

where Γ_i is the part of $\partial\Omega$ bounded by two consecutive points of D.

The Mumford–Shah Functional and the Ambrosio–Tortorelli Approximation
The calculus of variations has had a very strong impact in some parts of applied mathematics related, for instance, to image analysis and reconstruction, and it would have been a pity to end a book on it without mentioning any of these ideas, one of the most well-known being the so-called Mumford–Shah functional (see [155]). The starting point to arrive to this functional is a general procedure in signal processing: when observing some data g we can expect that these data have been noised, and we would like to denoise them; to do so, we look for another signal u which minimizes a sum $d(u, g) + R(u)$, where $d(u, g)$ is a term called the *data fidelity*, which ensures that u is as close as possible to the observed data, while R is a regularizing term which ensures that u fits as much as possible the a priori properties of the signal we are looking for (for instance being continuous, being sparse...). For image analysis, we identify images with functions defined on a (typically rectangular) domain and valued in $[0, 1]$ (for gray-scale images, otherwise we can consider functions valued in $[0, 1]^3$ for color pictures), and the a priori idea that we have is that the images should be smooth, except at contours which are the boundaries of different objects. A reasonable choice of a regularization term is then

$$R(u) := \inf\left\{\int_{\Omega\setminus S}|\nabla u|^2\,dx + \lambda\mathcal{H}^{d-1}(S) \ : \ u \in H^1(\Omega\setminus S)\right\}, \qquad (7.3)$$

where the infimum is among all possible closed sets S such that $u \in H^1(\Omega \setminus S)$, so that if there is no such set with finite $\mathcal{H}^{d-1}$ measure the value of R is $+\infty$. In practice, the functions u with $R(u) < +\infty$ are smooth outside a small jump set of dimension $d - 1$, which corresponds to the desired contours. The parameter λ tunes the relative importance of the two parts of the functional, so that a very small value of λ allows for larger and larger jump sets.

It is possible to re-express a relaxed version of the functional R with the language of BV functions.

> ## Box 7.2 Important Notion—*Decomposition of the BV Derivative, and SBV Functions*
>
> **Definition** Given a function $u \in BV(\Omega)$ we say that a point x_0 belongs to the jump set J_u if there exists a unit vector v and two constants $u^{\pm}$ with $u^+ \neq u^-$ such that
>
> $$\lim_{r \to 0} \fint_{B(x_0,r) \cap \{x:(x-x_0)\cdot \pm v>0\}} |u - u^{\pm}| \, dx = 0.$$
>
> We can prove that $|\nabla u| \geq |u^+ - u^-| \mathcal{H}^{d-1}_{|J_u}$, so that J_u is a countable union of sets with finite $\mathcal{H}^{d-1}$ measure. Note that at every point of J_u the vector v will be unique and will also play the role of the normal to J_u.
>
> Given $u \in BV(\Omega)$ we decompose its distributional gradient $\nabla u \in \mathcal{M}(\Omega)^d$ as
>
> $$\nabla u = (\nabla u)^{ac} + (\nabla u)^{cant} + (\nabla u)^{jump},$$
>
> where $(\nabla u)^{ac}$ is the absolutely continuous part of the measure ∇u but the singular part is also decomposed into two sub-parts, $(\nabla u)^{jump}$ which is the restriction of ∇u to the set J_u, and $(\nabla u)^{cant}$, which is the remaining ("Cantor") part. In dimension $d = 1$ this corresponds to the decomposition of a measure into its absolutely continuous part, atomic part, and the remaining singular part, as done in Sect. 3.3. In the general case atoms have been replaced with $(d - 1)$-dimensional objects. The name "jump" comes from the fact that atoms in u' in 1D exactly correspond to jumps, and "Cantor" refers to the devil's staircase function u introduced by Cantor, whose derivative u' has no absolutely continuous part (indeed, $u' = 0$ a.e.) nor atoms (indeed, u is continuous) but u is non-constant.
>
> **Definition** A function $u \in BV(\Omega)$ is said to belong to the space $SBV(\Omega)$ (the set of special functions of bounded variations) if $(\nabla u)^{cant} = 0$.
>
> We refer to [14] for the theory of BV functions and for more details.

With this language, we can then choose

$$R(u) = \int_{\Omega} |(\nabla u)^{ac}|^2 \, dx + \lambda \mathcal{H}^{d-1}(J_u).$$

Of course this expression does not coincide with the one in (7.3) since the set J_u has no reason to be closed, but one has to interpret this as a suitable relaxed version of (7.3).

The Mumford–Shah problem then consists in solving

$$\min \left\{ \alpha \int_\Omega |u - g|^2 \, dx + R(u) \ : \ u \in SBV(\Omega) \right\},$$

a problem which can be re-written in two variables as

$$\min_{u, S : u \in H^1(\Omega \setminus S)} \alpha \int_\Omega |u - g|^2 \, dx + \int_{\Omega \setminus S} |\nabla u|^2 \, dx + \lambda \mathcal{H}^{d-1}(S).$$

It is important to stress that the existence of an optimal u is not straightforward, as the functional that is minimized does not fit into classical lower semicontinuity theorems, except for $d = 1$, where $\mathcal{H}^{d-1}(S) = \#S$ and we can see $R(u)$ as a particular case of the functionals studied in Theorem 3.28 applied to u'. For the higher-dimensional case, a different semicontinuity theorem is necessary, and this is part of a whole theory of SBV functions developed by Ambrosia. We refer to [9, 10] for existence results in this setting and to [14] for a detailed exposition of the theory. As an example, we mention the following non-trivial statement: given a sequence of functions $u_n \in SBV(\Omega)$ with $||(\nabla u_n)^{ac}||_{L^p}$, $||u_n||_{L^\infty}$ and $\mathcal{H}^{d-1}(J_{u_n})$ all uniformly bounded in n (for some $p > 1$), there exists a subsequence strongly converging in L^1 to a function $u \in SBV(\Omega)$ and each of these three quantities is lower semicontinuous along this sequence.

The numerical treatment of the Mumford–Shah minimization problem is made difficult by the role played by the jump set, which concentrates an important part of the energy, despite being Lebesgue-negligible. On a space discretization of the problem (actually, image problems are discrete in space, since the number of pixels in the image, though large, is finite), this means that an important part of the energy is contained in a very small proportion of pixels. A phase-field approximation for this problem was necessary, and has been developed by Ambrosia and Tortorelli in [16, 17]. The main idea is the Γ-convergence of the following functional of two variables:

$$R_\varepsilon(u, \varphi) := \int_\Omega \varphi |\nabla u|^2 \, dx + \lambda \left(\frac{\varepsilon}{2} \int_\Omega |\nabla \varphi|^2 \, dx + \frac{1}{2\varepsilon} \int_\Omega (1 - \varphi)^2 \, dx \right).$$

When $\varepsilon \to 0$, the variable φ tends to be as close as possible to the constant 1, but the first term favors the fact that it could take the value 0. At the points where $\varphi = 0$ or φ is very small, the condition $u \in H^1$ coming from the first term is no longer enforced. As a result, when optimizing this functional and sending $\varepsilon \to 0$, the variable φ takes values very close to 0 on a set S on which u is finally allowed to jump, and the second term, which closely resembles the Modica–Mortola functional, penalizes the $\mathcal{H}^{d-1}$ measure of this set S. The precise statement which can be proven is the Γ-convergence of R_ε to R in $L^2(\Omega)$, which then allows us to add the data fidelity term and obtain a useful result in order to approximate the Mumford–Shah functional. Note that here we do not use a double-well potential W as in Sect. 7.4. The reason

for this is that we expect $\varphi = 1$ at the limit (hence, we want one only well), but the
other term (the one involving $\varphi|\nabla u|^2$) acts as a second well favoring $\varphi = 0$.

Because of the possible jumps on $\{\varphi = 0\}$, these problems are also called *free
discontinuity problems* (the discontinuity region being part of the optimization). See
[41] for the general theory on this approximation.

Approximation of the Steiner Problem and of Connected Networks   The
Ambrosio–Tortorelli result reveals a very clever idea behind this approximation: if
a condition has to be enforced everywhere except at a small set S, it is possible
to penalize it with a weight φ (so that we morally have $S = \{\varphi = 0\}$) and
add a Modica–Mortola-like functional $\frac{\varepsilon}{2} \int_\Omega |\nabla \varphi|^2 + \frac{1}{2\varepsilon} \int_\Omega (1 - \varphi)^2$ on φ, which
imposes that at the limit we have $\varphi = 1$ a.e. and we pay for the measure $\mathcal{H}^{d-1}(S)$.
This idea could be applied to other optimization problems, one of the simplest
being the average distance problem presented in Sect. 6.3.1, at least in its variant
where the length is penalized instead of being constrained. Indeed, the quantity
$\int_\Omega d(x, \Sigma) f(x) \, dx$ could be written as (see Sect. 7.3)

$$\min \left\{ W_1(f, \nu) : \text{spt}\, \nu \subset \Sigma \right\}.$$

This quantity, using the bidual problem of the optimal transport for the linear cost
presented in Sect. 4.6, could in turn be written as

$$\min \left\{ \|\mathbf{v}\|_{L^1} : \nabla \cdot \mathbf{v} = f - \nu, \text{spt}\, \nu \subset \Sigma \right\} = \min \left\{ \|\mathbf{v}\|_{L^1} : \nabla \cdot \mathbf{v} = f \text{ outside } \Sigma \right\},$$

where the last equality comes from the elimination of the variable ν: if the
divergence has to be equal to $f - \nu$ but ν is an arbitrary measure concentrated
on Σ, then this is equivalent to requiring that the divergence equals f outside Σ.

A tempting idea would then be to solve a minimization problem in the variable
$\mathbf{v}$ and φ, including a term like $\|\mathbf{v}\|_{L^1}$, a term of the form $\frac{1}{\varepsilon} \int \varphi \, d|\nabla \cdot \mathbf{v} - f|$, and a
Modica–Mortola term which mimics the length of the set $S = \{\varphi = 0\}$. The only
problem here[8] consists in the fact that we are forgetting the connectedness constraint
on S. This can be fixed by using the distance function d_φ associated with the weight
φ (see Sects. 1.4.4 and 4.6)

$$d_\varphi(x, y) := \inf \left\{ \int_0^1 \varphi(\gamma(t))|\gamma'(t)| \, dt : \gamma(0) = x, \gamma(1) = y \right\}.$$

Indeed, imposing $d_\varphi(x, x_0)$ to be small somewhere does not only mean that φ is
small at x, but also that a whole path can be found which connects x to x_0 where φ
is small, and this can be used to prove connectedness. More precisely, it is possible
to prove the following result (see [32]): if μ_ε is a family of positive measures on

---

[8] Besides the fact that we expect a similar construction could only work in the case $d = 2$, since
Modica–Mortola functionals provide $\mathcal{H}^{d-1}$ terms, and we want to obtain $\mathcal{H}^1$.

$\Omega \subset \mathbb{R}^2$ and $\varphi_\varepsilon \in H^1(\Omega) \cap C(\overline{\Omega})$ satisfy $\varphi_\varepsilon = 1$ on $\partial\Omega$ and $0 \le \varphi_\varepsilon \le 1$ in Ω, if $c_\varepsilon \to 0$ and x_ε is sequence of points of Ω such that

1. $x_\varepsilon \to x_0$ for some $x_0 \in \overline{\Omega}$,
2. $\mu_\varepsilon \overset{*}{\rightharpoonup} \mu$ for some measure μ on Ω,
3. $d_{\varphi_\varepsilon}(\cdot, x_\varepsilon)$ converges uniformly to some 1-Lipschitz function d on Ω,
4. we have $\sup_{\varepsilon > 0} S_\varepsilon(\varphi_\varepsilon, \mu_\varepsilon, x_\varepsilon) < +\infty$, where

$$S_\varepsilon(\varphi, \mu, x_0) := \frac{\varepsilon}{2} \int_\Omega |\nabla \varphi|^2 \, dx + \frac{1}{2\varepsilon} \int_\Omega (1 - \varphi)^2 \, dx + \frac{1}{c_\varepsilon} \int_\Omega d_\varphi(x, x_0) \, d\mu(x),$$

then

1. the compact set $K := \{x \in \overline{\Omega} \; ; \; d(x) = 0\}$ is connected,
2. $x_0 \in K$,
3. $\mathrm{spt}(\mu) \subset K$,
4. we have $\mathcal{H}^1(K) \le \liminf_{\varepsilon \to 0} S_\varepsilon(\varphi_\varepsilon, \mu_\varepsilon, x_\varepsilon)$.

We then obtain an approximation for the average distance problem

$$\min \left\{ \int_\Omega d(x, \Sigma) f(x) \, dx + \mathcal{H}^1(\Sigma) \quad : \quad \Sigma \subset \Omega \text{ compact and connected} \right\}$$

which is given by

$$\min \int_\Omega d|\mathbf{v}| + \frac{\varepsilon}{2} \int_\Omega |\nabla \varphi|^2 \, dx + \frac{1}{2\varepsilon} \int_\Omega (1 - \varphi)^2 \, dx + \frac{1}{c_\varepsilon} \int_\Omega d_\varphi(x, x_0) \, d|\nabla \cdot \mathbf{v} - f|,$$

where the unknowns are the vector field $\mathbf{v}$, the weight function φ, and the base point x_0 (indeed, if we fixed x_0 we would obtain the problem where we impose that Σ is connected and contains x_0). For fixed $\varepsilon > 0$ the existence of a minimizer for this problem is not straightforward, because of the term in d_φ. Indeed, as it is defined as an infimum, it is easy to see that $\varphi \mapsto d_\varphi(x, y)$ is u.s.c. for most reasonable notions of convergence on φ, but lower semicontinuity is more delicate. A simple result that one can prove (see [32]) is the following: if $\varphi_n \to \varphi$ uniformly and there exists a lower bound $c > 0$ such that $\varphi_n \ge c$ for every n, then $d_{\varphi_n}(x, y) \to d_\varphi(x, y)$ for every pair (x, y). Counterexamples exist without the lower bound, and it is not clear whether the weak H^1 convergence (that we can easily obtain in this setting because of the penalization $\varepsilon \int |\nabla \varphi|^2$) can help. In order to obtain the existence of the minimizers for fixed $\varepsilon > 0$ it is then possible to restrict to functions φ satisfying a lower bound $\varphi \ge c'_\varepsilon$, and to add a term of the form $\varepsilon^m \int |\nabla \varphi|^p$ for $p > d$ and m large enough, which is enough to guarantee compactness in C^0.

It is then easy to find a variant for the optimal compliance problem (see again Sect. 6.3). We consider the Neumann case, i.e. the problem we want to approximate is

$$\min \left\{ C_p(\Sigma) + \lambda \mathcal{H}^1(\Sigma) \ : \ \Sigma \subset \overline{\Omega} \text{ compact and connected} \right\},$$

where

$$C_p(\Sigma) := -\inf \left\{ \frac{1}{p} \int_{\Omega \setminus \Sigma} |\nabla u|^p \, dx + \int_\Omega f u \, dx \ : \ u \in C_c^\infty(\overline{\Omega} \setminus \Sigma) \right\}.$$

It is annoying that C_p is defined as a sup (as a negative inf), which prevents us from obtaining a min-min formulation; yet, the convex minimization problem which defines it can be treated by convex duality and transformed into a maximization, since we can prove that we have

$$\inf \left\{ \frac{1}{p} \int_{\Omega \setminus \Sigma} |\nabla u|^p \, dx + \int_\Omega f u \, dx, \ : \ u \in C_c^\infty(\overline{\Omega} \setminus \Sigma) \right\}$$

$$= \sup \left\{ -\frac{1}{p'} \int_{\Omega \setminus \Sigma} |v|^{p'} \, dx \ : \ \nabla \cdot v = f - v, \, \mathrm{spt}\, v \subset \Sigma) \right\}.$$

This suggests that we should simply consider

$$\min \int_\Omega \frac{1}{p'} |v|^{p'} dx + \frac{\varepsilon}{2} \int_\Omega |\nabla \varphi|^2 dx + \frac{1}{2\varepsilon} \int_\Omega (1-\varphi)^2 dx + \frac{1}{c_\varepsilon} \int_\Omega d_\varphi(x, x_0) \, d|\nabla \cdot v - f|.$$

Finally, these ideas can be used to attack the simplest and most natural optimization problem among compact connected sets, which is the Euclidean Steiner tree problem (see [138] besides [32]). The precise result, which unfortunately cannot be expressed explicitly in terms of Γ-convergence (because the variable v has disappeared), is the following.

Theorem 7.23 *Let Ω' be the convex hull of the set D and $\Omega \supset \Omega'$ a larger convex open set. For all $\varepsilon > 0$ let φ_ε be a minimizer of $\varphi \mapsto S_\varepsilon(\varphi, \mu, x_0)$, where $x_0 \in D$ and μ is an arbitrary measure such that $\mathrm{spt}\, \mu \supset D \setminus \{x_0\}$, among the H^1 continuous functions bounded from below by ε, from above by 1, and equal to 1 on $\partial\Omega$. Consider the sequence of functions $d_{\varphi_\varepsilon}(\cdot, x_0)$, which are 1-Lipschitz and converge, up to subsequences, to a certain function d. Then the set $K := \{d = 0\}$ is compact, connected and is a solution to the Steiner Problem (6.1).*

All these results can then be used for numerical computations (see [32], but also [33]), but of course they require us to handle the computation of d_φ and of its gradient with respect to φ. This can be done, for instance, via the so-called *Fast Marching* method (see Rouy and Tourin [174] for a consistent discretization, [182] for an efficient algorithm, and [25] for the computation of the gradient).

Branched Transportation Networks In this very last part of the section we consider a minimal flow problem, where, contrary to what happens in traffic congestion (Sect. 4.6), joint transportation on a common path is encouraged. This is reasonable when we consider "economy of scale" principles: when a single planner needs to organize a transport (building a transportation network, for instance), the total cost happens to be *sub-additive* (the cost of moving two objects cannot exceed the sum of the two costs), and in many cases it has *decreasing marginal costs* (i.e. the cost of adding a unit to a given background quantity is a decreasing function of the background, which means that the cost is actually concave).

Note that we are considering here the transport cost as a function of the mass, and not of the length. The silent assumption is that the cost will be linear in the length (is the case in all minimal flow problems). Many branching structures transporting different fluids, such as road systems, communication networks, river basins, blood vessels, leaves, trees and so on... may be easily thought of as coming from a variational principle for a cost of this kind. In all these cases, the shape of the optimal solution for this kind of problem follows the same principle: particles are collected at some points, move together as much as possible, and then branch towards their different destinations. This is why these problems are nowadays known as *branched transportation problems*.

The formalization of the discrete framework of these problems is very classical in optimization and operational research.

Translating into our language the work of Gilbert [99] (see also [100], these two works presenting this question as an extension of Steiner's minimal length problem), we consider two atomic probability measures $\mu = \sum_{i=1}^{m} a_i \delta_{x_i}$ and $\nu = \sum_{j=1}^{n} b_j \delta_{y_j}$ supported in some compact subset $\Omega \subset \mathbb{R}^d$, and we want to connect them through a weighted oriented graph $G = (s_k, \theta_k)_k$, where the s_k are oriented segments $s_k = (p_k, q_k)$ in $\mathbb{R}^d$ and $\theta_k \in \mathbb{R}_+$ is a weight associated with the segment s_k. We denote by $\hat{s}_k$ the orientation $\frac{q_k - p_k}{|q_k - p_k|}$ of s_k and by $|s_k|$ its length $|q_k - p_k|$.

We count as *incoming* the mass conveyed by the segments $s_k = (p_k, q_k)$ reaching x, i.e. such that $x = q_k$, increased by the mass given by μ, i.e. increased by a_i if $x = x_i$. Similarly, the *outgoing mass* at x is the total weight of the segments $s_k = (p_k, q_k)$ leaving x, i.e. such that $p_k = x$, increased by the mass taken out by ν, i.e. by b_j if $x = y_j$. The set of graphs irrigating ν from μ is denoted by $G(\mu, \nu)$.

For $\alpha \in [0, 1]$, we define the cost of α-irrigation of a graph G, also called the *Gilbert energy*, as

$$\mathcal{E}_\alpha(G) = \sum_{k=1}^{n} \theta_k^\alpha |s_k|,$$

which means that the cost of moving a mass m along a segment of length l is $m^\alpha l$.

Given μ, ν atomic probabilities, we want to minimize this energy among all graphs sending μ to ν, which reads

$$\min \ \{\mathcal{E}_\alpha(G) : G \in G(\mu, \nu)\}.$$

Note that for $\alpha \in [0, 1]$ the function $t \mapsto t^\alpha$ is concave and subadditive; it is strictly concave and strictly subadditive for $\alpha \in (0, 1)$. In this way larger tubes bringing the mass from μ to ν are preferred to several smaller ones transporting the same total mass. As for the case of the Steiner problem, it can be seen that optimal graphs have no cycles. It is possible to prove optimality conditions on the free junctions in the graph, which generalize the $120°$ condition of the Steiner problem.

More recently Xia, in [200], has proposed a new formalization leading to generalizations of this problem to arbitrary probability measures μ and ν, and this gave rise to a wide literature on the subject (see [27] or [177, Section 4.4]). The main idea was to translate the Kirchhoff law constraint into a divergence constraint.

Given a finite graph $G = (s_k, \theta_k)_k$, we associate with G the vector measure

$$\mathbf{v}_G := \sum_k \theta_k \hat{s}_k \mathcal{H}^1_{|s_k} \in \mathcal{M}^d(\Omega).$$

We can express both the cost and the constraint in terms of $\mathbf{v}_G$ only. First note that $G \in G(\mu, \nu)$ (i.e. it satisfies mass balance conditions for irrigating ν from μ) if and only if $\nabla \cdot \mathbf{v}_G = \mu - \nu$. Then, we also write

$$\mathcal{E}_\alpha(G) = \sum_{k=1}^n \theta_k^\alpha \, |s_k| = \int_{\text{spt}\,\mathbf{v}_G} |\theta(x)|^\alpha \ d\mathcal{H}^1(x),$$

where θ is the density of $\mathbf{v}_G$ w.r.t. the $\mathcal{H}^1$ measure.

This suggests a continuous Eulerian formulation of Gilbert's problem by generalizing the Gilbert energy, the minimization problem becoming

$$\min \ \{M_\alpha(\mathbf{v}) : \mathbf{v} \in \mathcal{M}^d_{\text{div}}(\Omega), \ \nabla \cdot \mathbf{v} = \mu - \nu\}. \tag{7.4}$$

Obviously, one needs to define the energy M_α (which is also called the "α-mass" of $\mathbf{v}$). Xia's original idea is to define it by relaxation, i.e.

$$M_\alpha(\mathbf{v}) := \inf \left\{ \liminf_n \mathcal{E}_\alpha(G_n) : G_n \in G(\mu_n, \nu_n), \mathbf{v}_{G_n} \rightharpoonup \mathbf{v} \text{ in } \mathcal{M}^d_{\text{div}}(\Omega) \right\},$$

where the convergence in $\mathcal{M}^d_{\text{div}}(\Omega)$ means $\mathbf{v}_{G_n} \rightharpoonup \mathbf{v}$ and $\nabla \cdot \mathbf{v}_{G_n} = \mu_n - \nu_n \rightharpoonup \nabla \cdot \mathbf{v}$ as measures.

As a consequence of a famous rectifiability theorem due to B. White [198], one can prove that $M_\alpha(\mathbf{v}) < +\infty$ implies that $\mathbf{v}$ is a measure concentrated on a set E of dimension 1 and satisfying

$$M_\alpha(\mathbf{v}) = \int_E \left(\frac{d\mathbf{v}}{d\mathcal{H}^1}\right)^\alpha d\mathcal{H}^1.$$

It can also be proven that, when μ and ν are both atomic measures, we retrieve Gilbert's original problem.

From the point of view of the practical search for solutions, branched transport problems lie in between combinatorial optimization (they are similar to the Steiner minimal connection problem, which is, on a graph, known to be NP-hard) and continuous minimal flow problems. The numerical methods for finding a solution have to face at least two difficulties: concavity (and hence the high number of local minimizers) and singularity (i.e. working with singular structures such as networks instead of using L^p or smoother functions, as in the congested traffic case).

Inspired by the previous works on phase-field approximation of lower-dimensional objects, in [164, 176] an approximation for branched transport has been proposed in the two-dimensional case $\Omega \subset \mathbb{R}^2$. Let $\mathcal{M}^d_{\mathrm{div}}(\Omega)$ be the space of vector measures on Ω with divergence which is also a measure (endowed with the weak convergence of both the measures and their divergences). Consider on $\mathcal{M}^d_{\mathrm{div}}(\Omega)$ the functionals

$$M^\alpha_\varepsilon(\mathbf{v}) = \varepsilon^{\alpha-1} \int_\Omega |\mathbf{v}(x)|^\beta \, dx + \varepsilon^{\alpha+1} \int_\Omega |\nabla\mathbf{v}(x)|^2 \, dx, \qquad (7.5)$$

defined on $\mathbf{v} \in H^1(\Omega; \mathbb{R}^2)$ and set to $+\infty$ outside $H^1 \cap \mathcal{M}^d_{\mathrm{div}}(\Omega)$. The exponent $\beta = \frac{4\alpha-2}{\alpha+1}$ is fixed via suitable scaling computations.

Compared to the Modica–Mortola case, here the double-well potential is replaced with a concave power. Note that concave powers, in their minimization, if the average value for u is fixed in a region (which is in some sense the meaning of weak convergence), prefer either $u = 0$ or $|u|$ to be as large as possible, i.e. there is sort of a double well at zero and infinity.

Theorem 7.24 *Assume $d = 2$ and $\alpha \in (\frac{1}{2}, 1)$: then we have Γ-convergence of the functionals M^α_ε to cM^α, with respect to the convergence in $\mathcal{M}^d_{\mathrm{div}}(\Omega)$, as $\varepsilon \to 0$, where c is given by*

$$c = \alpha^{-1}\left(4c_0\alpha/(1-\alpha)\right)^{1-\alpha}, \text{ where } c_0 = \int_0^1 \sqrt{t^\beta - t} \, dt.$$

It is interesting to observe that the above Γ-convergence result is a sort of higher-dimensional counterpart of a result of Bouchitté et al. [37], who found (in a more general framework, with applications to the stability of charged droplets) the Γ-limit of the following energy

$$F_\varepsilon(u) := \varepsilon^{\alpha-1} \int_{\mathbb{R}} W(u) \, dx + \varepsilon^{\alpha+1} \int_{\mathbb{R}} |u'|^2 \, dx$$

whenever $\alpha \in (0, 1)$ is a given exponent and $W : \mathbb{R}_+ \to \mathbb{R}_+$ is a cost function such that $W(0) = 0$, $W'(0) > 0$ and $W(s) \approx s^\beta$ for $s \to \infty$, where $\beta = \frac{4\alpha-2}{\alpha+1}$. The Γ-limit is a functional F on measures of the same form as those studied in Sect. 3.3, and more precisely

$$F(u) := \begin{cases} \sum_i g(a_i) & \text{if } u = \sum_i a_i \delta_{x_i} \\ +\infty & \text{if } u \text{ is not purely atomic,} \end{cases}$$

where $g(s)$ is proportional to s^α. Theorem 7.24 differs from the result of [37] in that it concerns one-dimensional energies in a 2D setting, while [37] concerns 0-dimensional energies (the limit measures are forced to be purely atomic) in a 1D setting. The 2D statement can be guessed (but not easily proven) from the 1D result via a slicing procedure.

As a consequence of the approximation result contained in Theorem 7.24, É. Oudet performed efficient numerical simulations by minimizing the functionals M_ε^α (more precisely: first a large value of ε is fixed, so that the problem is more or less convex, and a minimizer is found by gradient descent, then, following a continuation method, the value of ε is reduced and at every step the gradient descent starts from the previously found minimizer). In the end we find a "well-chosen" local minimizer, quite satisfactory in practice. This has been used in [168] to solve a shape optimization problem consisting in finding the set with unit volume which is the closest in the sense of branched transport to the Dirac mass δ_0, a set which is conjectured to have fractal boundary.

7.6 Exercises

Exercise 7.1 Let $(F_n)_n$ be a sequence of functionals defined on a common metric space X, such that $F_n \leq F_{n+1}$ for every n. Assume that each F_n is l.s.c. and set $F := \sup_n F_n$. Prove $F_n \xrightarrow{\Gamma} F$.

Exercise 7.2 Consider a continuous function $F : X \to \mathbb{R}$ defined on a metric space (X, d) and a function $G : X \to [0, +\infty]$. Define $F_n^+ := F + \frac{1}{n}G$ and $F_n^- := F - \frac{1}{n}G$.

1. Prove that we have $F_n^+ \xrightarrow{\Gamma} F$ if and only if the set $\{G < +\infty\}$ is dense.
2. Prove that we have $F_n^- \xrightarrow{\Gamma} F$ if and only if G is finite everywhere and locally bounded around each point of X.

Exercise 7.3 Let $(F_n)_n$ be a sequence of functionals defined on a common compact metric space X. Prove that there exists a subsequence F_{n_k} and a function F such that $F_{n_k} \xrightarrow{\Gamma} F$.

Exercise 7.4 Prove the result of Proposition 7.11 when $|\Omega|$ is not of finite measure, but has σ-finite measure.

Exercise 7.5 Let us define the following functionals on $X = L^2([-1, 1])$

$$F_n(u) := \begin{cases} \frac{1}{2n} \int_{-1}^{1} |u'(t)|^2 \, dt + \frac{1}{2} \int_{-1}^{1} |u(t) - t|^2 \, dt & \text{if } u \in H_0^1([-1, 1]), \\ +\infty & \text{otherwise}; \end{cases}$$

together with

$$H(u) = \begin{cases} \frac{1}{2} \int_{-1}^{1} |u(t) - t|^2 \, dt & \text{if } u \in H_0^1([-1, 1]), \\ +\infty & \text{otherwise}; \end{cases}$$

and $F(u) := \frac{1}{2} \int_{-1}^{1} |u(t) - t|^2 \, dt$ for every $u \in X$.

1. Prove that, for each n, the functional F_n is l.s.c. for the L^2 (strong) convergence.
2. Prove that F is l.s.c. for the same convergence, but not H.
3. Find the minimizer u_n of F_n over X.
4. Find the limit as $n \to \infty$ of u_n. Is it a strong L^2 limit? Is it a uniform limit? A pointwise a.e. limit?
5. Find the Γ-limit of F_n.
6. Does the functional H admit a minimizer in X?

Exercise 7.6 Let $\Omega \subset \mathbb{R}^d$ be an open domain. On the space $H_0^1(\Omega)$ consider the sequence of functionals

$$F_\varepsilon(u) = \int_\Omega \left[\frac{|\nabla u(x)|^2}{2} + \frac{\sin(\varepsilon u(x))}{\varepsilon} \right] dx.$$

1. Prove that, for each $\varepsilon > 0$, the functional F_ε admits at least a minimizer u_ε.
2. Prove that the minimizers u_ε satisfy $\|\nabla u_\varepsilon\|_{L^2}^2 \le 2\|u_\varepsilon\|_{L^1}$ and that the norm $\|u_\varepsilon\|_{H_0^1}$ is bounded by a constant independent of ε.
3. Find the Γ-limit F_0, in the weak H_0^1 topology, of the functionals F_ε as $\varepsilon \to 0$.
4. Characterize via a PDE the unique minimizer u_0 of the limit functional F_0.
5. Prove $u_\varepsilon \rightharpoonup u_0$ in the weak H_0^1 topology.
6. Prove that the convergence $u_\varepsilon \to u_0$ is actually strong in H_0^1.

Exercise 7.7 Prove the convergence in the last question of Exercise 5.9 using Γ-convergence.

Exercise 7.8 Given a sequence of functions $a_n : \Omega \to \mathbb{R}$ which are bounded from below and above by two strictly positive constants a_-, a_+ and a function a bounded by the same constants, prove that we have

$$\begin{cases} a_n \overset{*}{\rightharpoonup} a \\ a_n^{-1} \overset{*}{\rightharpoonup} a^{-1} \end{cases}$$

if and only if we have $a_n \to a$ in L^2 (and hence in all L^p spaces for $p < \infty$).

Exercise 7.9 Consider the following sequence of minimization problems, for $n \geq 0$,

$$\min\left\{ \int_0^1 \left(\frac{|u'(t)|^2}{\frac{3}{2} + \sin^2(nt)} + \left(\frac{3}{2} + \sin^2(nt) \right) u(t) \right) dt : u \in H^1([0,1]), u(0) = 0 \right\},$$

calling u_n their unique minimizers and m_n their minimal values.

Prove that we have $u_n(t) \to t^2 - 2t$ uniformly, and $m_n \to -2/3$.

Exercise 7.10 Consider the functions $a_n : [0,1] \to \mathbb{R}$ given by $a_n(x) = a(nx)$ where $a = 2\sum_{k \in \mathbb{Z}} \mathbb{1}_{[2k,2k+1]} + \sum_{k \in \mathbb{Z}} \mathbb{1}_{[2k-1,2k]}$. Given $f \in L^1([0,1])$ with $\int_0^1 f(t)\,dt = 0$ and $p \in (1, +\infty)$, compute

$$\lim_{n \to \infty} \min\left\{ \int_0^1 \left(\frac{1}{p} a_n |u'(t)|^p dt + f(t)u(t) \right) dt : u \in W^{1,p}([0,1]) \right\}.$$

Exercise 7.11 Let $\Omega \subset \mathbb{R}^d$ be the unit square $\Omega = (0,1)^2$ and $S = \{0\} \times (0,1) \subset \partial\Omega$. Consider the functions $a_n : \Omega \to \mathbb{R}$ defined via

$$a_n(x,y) = \begin{cases} A_0 & \text{if } x \in [\frac{2k}{2n}, \frac{2k+1}{2n}) \text{ for } k \in \mathbb{Z} \\ A_1 & \text{if } x \in [\frac{2k+1}{2n}, \frac{2k+2}{2n}) \text{ for } k \in \mathbb{Z}, \end{cases}$$

where $0 < A_0 < A_1$ are two given values. On the space $L^2(\Omega)$ (endowed with the strong L^2 convergence), consider the sequence of functionals

$$F_n(u) = \begin{cases} \int_\Omega a_n |\nabla u|^2 & \text{if } u \in X, \\ +\infty & \text{if not,} \end{cases}$$

where $X \subset L^2(\Omega)$ is the space of functions $u \in H^1(\Omega)$ satisfying $\mathrm{Tr}[u] = 0$ on S.

Set for convenience

$$A_* := \left(\frac{\frac{1}{A_0} + \frac{1}{A_1}}{2} \right)^{-1} \quad \text{and} \quad A_\diamond := \frac{A_0 + A_1}{2}.$$

Find the Γ-limit of F_n. Is it of the form $F(u) = \int a|\nabla u|^2$ for $u \in X$?

Exercise 7.12 Let u_n, φ_n be two sequences of functions belonging to $H^1([0, 1])$. Assume

$$u_n \rightharpoonup u_0, \quad \varphi_n \rightharpoonup \varphi_0, \quad E_n(u_n, \varphi_n, [0, 1]) \le C,$$

where E_n is the energy defined for every interval $J \subset [0, 1]$ through

$$E_n(u, \varphi, J) := \int_J \left(\varphi(t)|u'(t)|^2 + \frac{1}{2n}|\varphi'(t)|^2 + \frac{n}{2}|1 - \varphi(t)|^2 \right) dt,$$

$C \in \mathbb{R}$ is a given constant, the weak convergence of u_n and v_n occur in $L^2([0, 1])$, and u_0 is a function which is piecewise C^1 on $[0, 1]$ (i.e. there exists a partition $0 = t_0 < t_1 < \cdots < t_N = 1$ such that $u_0 \in C^1((t_i, t_{i+1}))$, and the limits of u_0 exist and are finite at $t = t_i^\pm$ but $u_0(t_i^-) \ne u_0(t_i^+)$ for $i = 1, \ldots, N-1$).

Denote by $\mathcal{J}$ the family of all the intervals J compactly contained in one of the open intervals $(t_i, t_{i+1}[)$. Also assume, for simplicity, that we have $u_n \rightharpoonup u_0$ in $H^1(J)$ for every interval $J \in \mathcal{J}$.

Prove that we necessarily have

1. $\varphi_0 = 1$ a.e. and $\varphi_n \to \varphi_0$ strongly in L^2.
2. $\liminf_{n \to \infty} E_n(u_n, \varphi_n, J) \ge \int_J |u_0'(t)|^2 dt$ for every interval $J \in \mathcal{J}$.
3. $\liminf_{n \to \infty} E_n(u_n, \varphi_n, J) \ge 1$ for every interval J containing one of the points t_i.
4. $C \ge \liminf_{n \to \infty} E_n(u_n, \varphi_n, [0, 1]) \ge \int_0^1 |u_0'(t)|^2 dt + (N-1)$.

Exercise 7.13 Let u_ε be solutions of the minimization problems P_ε given by

$$P_\varepsilon := \min \left\{ \int_0^\pi \left(\frac{\varepsilon}{2}|u'(t)|^2 + \frac{1}{2\varepsilon}\sin^2(u(t)) + 10^3|u(t) - t| \right) dt : u \in H^1([0, \pi]) \right\}.$$

Prove that u_ε converges strongly (the whole sequence, not up to subsequences!!) in L^1 to a function u_0 as $\varepsilon \to 0$, find this function, and prove that the convergence is actually strong in all the L^p spaces with $p < \infty$. Is it a strong L^∞ convergence?

Hints

Hint to Exercise 7.1 Use $F_n \geq F_{n_0}$ and the semicontinuity of each F_{n_0}.

Hint to Exercise 7.2 For F_n^+ the only difficult part is the Γ-limsup. Use Proposition 7.3. For F_n^- the only difficult part is the Γ-liminf.

Hint to Exercise 7.3 Consider the sets $\mathrm{Epi}(F_n) \cap (X \times [0, M])$, and prove that any local Hausdorff limit of epigraphs is an epigraph.

Hint to Exercise 7.4 Consider a sup of restrictions to finite-measure subsets.

Hint to Exercise 7.5 The key point to realize is that the Γ-limit of F_n is F and not H.

Hint to Exercise 7.6 Use a suitable Taylor expansion of the sine function to justify that the Γ-limit is the pointwise limit. The strong H^1 convergence comes from the convergence of the minimal value.

Hint to Exercise 7.7 Use the variational formulation of Exercise 5.9 and prove Γ-convergence noting that we indeed have strong convergence of the coefficients in the integrand.

Hint to Exercise 7.8 Use an inequality of the form $\frac{1}{t} \geq \frac{1}{s} - \frac{1}{s^2}(t - s) + c|t - s|^2$.

Hint to Exercise 7.9 Use Proposition 7.14.

Hint to Exercise 7.10 Again, use Proposition 7.14.

Hint to Exercise 7.11 For the Γ-liminf, separately deal with $\partial_x u_n$ (the number A_* will appear) and $\partial_y u_n$ (here the number $A_\diamond$ will appear). For the Γ-limsup, use the antiderivative in one direction.

Hint to Exercise 7.12 Adapt the result that we briefly sketched for the Ambrosio–Tortorelli approximation.

Hint to Exercise 7.13 Adapt the proof of the Modica–Mortola result to this case with infinitely many wells and find the minimizer of the limit functional, which is a combination of the number of jumps and the L^1 distance from the identity.

References

1. R.A. Adams, *Sobolev Spaces*. Pure and Applied Mathematics, vol. 65 (Academic Press, New York-London, 1975)
2. G. Alberti, On the structure of singular sets of convex functions. Calc. Var. Part. Differ. Equ. **2**, 17–27 (1994)
3. G. Alberti, L. Ambrosio, A geometrical approach to monotone functions in $\mathbb{R}^d$. Math. Z. **230**, 259–316 (1999)
4. G. Alberti, L. Ambrosio, P. Cannarsa, On the singularities of convex functions. Manuscripta Math **76**, 421–435 (1992)
5. G. Alberti, M. Ottolini, On the structure of continua with finite length and Gołąb's semicontinuity theorem. Nonlinear Anal. Theory Methods Appl. **153**, 35–55 (2017)
6. J.-J. Alibert, B. Dacorogna, An example of a quasiconvex function that is not polyconvex in two dimensions. Arch. Ration. Mech. Anal. **117**, 155–166 (1992)
7. N.D. Alikakos, G. Fusco, On the connection problem for potentials with several global minima. Indiana Univ. Math. J. **57**(4), 1871–1906 (2008)
8. F. Almgren, Optimal isoperimetric inequalities. Indiana Univ. Math. J. **35**(3), 451–547 (1986)
9. L. Ambrosio, A compactness theorem for a new class of functions of bounded variation. Boll. Un. Mat. Ital. **3-B**, 857–881 (1989)
10. L. Ambrosio, Variational problems in SBV and image segmentation. Acta Appl. Math. **17**, 1–40 (1989)
11. L. Ambrosio, A. Carlotto, A. Massaccesi, *Lectures on Elliptic Partial Differential Equations* (Scuola Normale Superiore, Pisa, 2018)
12. L. Ambrosio, A. Figalli, On the regularity of the pressure field of Brenier's weak solutions to incompressible Euler equations. Calc. Var. PDE **31**(4), 497–509 (2008)
13. L. Ambrosio, A. Figalli, Geodesics in the space of measure-preserving maps and plans. Arch. Ration. Mech. Anal. **194**, 421–462 (2009)
14. L. Ambrosio, N. Fusco, D. Pallara, *Functions of Bounded Variation and Free Discontinuity Problems*. Oxford Mathematical Monographs (The Clarendon Press, Oxford University Press, New York, 2000)
15. L. Ambrosio, P. Tilli, *Topics on Analysis in Metric Spaces*. Oxford Lecture Series in Mathematics and its Applications, vol. 25 (Oxford University Press, Oxford, 2004).
16. L. Ambrosio, V.M. Tortorelli, Approximation of functionals depending on jumps by elliptic functionals via Γ-convergence. Commun. Pure Appl. Math. **43**(8), 999–1036 (1990)
17. L. Ambrosio, V.M. Tortorelli, On the approximation of free discontinuity problems. Boll. Un. Mat. Ital. B (7) **6**(1), 105–123 (1992)

18. P. Antonopoulos, P. Smyrnelis, On minimizers of the Hamiltonian system $u'' = \nabla W(u)$ and on the existence of heteroclinic, homoclinic and periodic orbits. Indiana Univ. Math. J. **65**(5), 1503–1524 (2016)
19. S.N. Armstrong, C.K. Smart, An easy proof of Jensen's theorem on the uniqueness of infinity harmonic functions. Calc. Var. Part. Differ. Equ. **37**, 381–384 (2010)
20. H. Attouch, G. Buttazzo, G. Michaille, *Variational Analysis in Sobolev and BV Spaces: Applications to PDEs and Optimization* (SIAM, Philadelphia, 2006)
21. M. Bardi, I. Capuzzo-Dolcetta, *Optimal Control and Viscosity Solutions of Hamilton–Jacobi–Bellman Equations* (Birkhäuser, Basel, 2008)
22. M. Beckmann, A continuous model of transportation. Econometrica **20**, 643–660 (1952)
23. M. Beckmann, C. McGuire, C. Winsten, *Studies in the Economics of Transportation* (Yale University Press, New Haven, 1956)
24. J.-D. Benamou, Y. Brenier, A computational fluid mechanics solution to the Monge–Kantorovich mass transfer problem. Numer. Math. **84**, 375–393 (2000)
25. F. Benmansour, G. Carlier, G. Peyré, F. Santambrogio, Fast marching derivatives with respect to metrics and applications. Numer. Math. **116**(3), 357–381 (2010)
26. J. Bernoulli, Problema novum ad cujus solutionem Mathematici invitantur. Acta Eruditorum (1696)
27. M. Bernot, V. Caselles, J.-M. Morel, *Optimal Transportation Networks, Models and Theory.* Lecture Notes in Mathematics, vol. 1955 (Springer, Berlin, 2008)
28. S. Bernstein, Sur la nature analytique des solutions des équations aux dérivées partielles du second ordre. (French). Math. Ann. **59** (1–2), 20–76 (1904)
29. S. Bernstein, Über die isoperimetrische Eigenschaft des Kreises auf der Kugeloberfläche und in der Ebene. Math. Ann. **60**, 117–136 (1905)
30. B. Bojarski, T. Iwaniec, p-harmonic equation and quasiregular mappings, in *Partial Differential Equations (Warsaw 1984)*, vol. 19 (Banach Center Publications, Warszawa, 1987), pp. 25–38
31. T. Bonnesen, Über die isoperimetrische Defizite ebener Figuren. Math. Ann. **91**, 252–268 (1924)
32. M. Bonnivard, A. Lemenant, F. Santambrogio, Approximation of length minimization problems among compact connected sets. SIAM J. Math. Anal. **47**(2), 1489–1529 (2015)
33. M. Bonnivard, E. Bretin, A. Lemenant, Numerical approximation of the Steiner problem in dimension 2 and 3. Math. Comput. **89**(321), 1–43 (2020)
34. J. Borwein, V. Montesinos, J.D. Vanderwerff, Boundedness, differentiability and extensions of convex functions. J. Convex Anal. **13**(3/4), 587 (2006)
35. G. Bouchitté, G. Buttazzo, New lower semicontinuity results for nonconvex functionals defined on measures. Nonlinear Anal. **15**, 679–692 (1990)
36. G. Bouchitté, G. Buttazzo, Characterization of optimal shapes and masses through Monge–Kantorovich equation. J. Eur. Math. Soc. **3**(2), 139–168 (2001)
37. G. Bouchitté, C. Dubs, P. Seppecher, Transitions de phases avec un potentiel dégénéré à l'infini, application à l'équilibre de petites gouttes. C. R. Acad. Sci. Paris Sér. I Math. **323**(9), 1103–1108 (1996)
38. G. Bouchitté, C. Jimenez, M. Rajesh, Asymptotique d'un problème de positionnement optimal. C. R. Acad. Sci. Paris Ser. I Math. **335**, 1–6 (2002)
39. G. Bouchitté, C. Jimenez, M. Rajesh, Asymptotic analysis of a class of optimal location problems. J. Math. Pures Appl. **95**(4), 382–419 (2011)
40. A. Braides, *Γ-Convergence for Beginners* (Clarendon Press, Oxford, 2002)
41. A. Braides, *Approximation of Free-Discontinuity Problems.* Lecture Notes in Mathematics, vol. 1694 (Springer-Verlag, Berlin, 1998)
42. A. Brancolini, G. Buttazzo, Optimal networks for mass transportation problems. ESAIM Control Optim. Calc. Var. **11**(1), 88–101 (2005)
43. A. Brancolini, G. Buttazzo, F. Santambrogio, Path functionals over Wasserstein spaces. J. Eur. Math. Soc. **8**(3), 415–434 (2006)

44. L. Brasco, G. Carlier, F. Santambrogio, Congested traffic dynamics, weak flows and very degenerate elliptic equations. J. Math. Pures Appl. **93**(6), 652–671 (2010)
45. Y. Brenier, Décomposition polaire et réarrangement monotone des champs de vecteurs. (French). C. R. Acad. Sci. Paris Sér. I Math. **305**(19), 805–808 (1987)
46. Y. Brenier, Polar factorization and monotone rearrangement of vector-valued functions. Commun. Pure Appl. Math. **44**, 375–417 (1991)
47. Y. Brenier, Minimal geodesics on groups of volume-preserving maps and generalized solutions of the Euler equations. Commun. Pure Appl. Math. **52**(4), 411–452 (1999)
48. H. Brezis, *Analyse fonctionnelle* (French). Collection Mathématiques Appliquées pour la Maîtrise. Théorie et Applications (Masson, Paris, 1983)
49. H. Brezis, J.-M. Coron, E.H. Lieb, Harmonic maps with defects. Commun. Math. Phys. **107**, 649–705 (1986)
50. D. Bucur, G. Buttazzo, *Variational Methods in Some Shape Optimization Problems*. Appunti dei Corsi Tenuti da Docenti della Scuola (Scuola Normale Superiore, Pisa, 2002)
51. B. Bulanyi, A. Lemenant, Regularity for the planar optimal p-compliance problem. ESAIM Control Optim. Calc. Var. **27**, 35 (2021)
52. G. Buttazzo, *Semicontinuity, Relaxation, and Integral Representation in the Calculus of Variations*. Longman, London (1989)
53. G. Buttazzo, M. Giaquinta, S. Hildebrandt, *One-Dimensional Variational Problems* (Clarendon Press, Oxford, 1999)
54. G. Buttazzo, É. Oudet, E. Stepanov, Optimal transportation problems with free Dirichlet regions, in *Variational Methods for Discontinuous Structures*. PNLDE, vol 51. (Birkhäuser, Basel, 2002), pp. 41–65
55. G. Buttazzo, F. Santambrogio, Asymptotical compliance optimization for connected networks. Netw. Heterog. Media **2**(4), 761–777 (2007)
56. G. Buttazzo, F. Santambrogio, N. Varchon, Asymptotics of an optimal compliance-location problem. ESAIM Contr. Opt. Calc. Var. **12**(4), 752–769 (2006)
57. G. Buttazzo, E. Stepanov, Optimal transportation networks as free Dirichlet regions for the Monge–Kantorovich problem. Ann. Sc. Norm. Super. Pisa Cl. Sci. (5) **2**(4), 631–678 (2003)
58. L. Caffarelli, X. Cabré, *Fully Nonlinear Elliptic Equations* (AMS, Providence, 1995)
59. P. Cannarsa, C. Sinestrari, *Semiconcave Functions, Hamilton–Jacobi Equations, and Optimal Control* (Birkhäuser, Basel, 2004).
60. P. Cardaliaguet, Notes on Mean Field Games (from P.-L. Lions' lectures at Collège de France), available at https://www.ceremade.dauphine.fr/~cardaliaguet/
61. P. Cardaliaguet, A.R. Mészáros, F. Santambrogio, First order Mean Field Games with density constraints: pressure equals price. SIAM J. Contr. Opt. **54**(5), 2672–2709 (2016)
62. G. Carlier, *Classical and Modern Optimization*. Advanced Textbooks in Mathematics (World Scientific, Singapore, 2021)
63. G. Carlier, C. Jimenez, F. Santambrogio, Optimal transportation with traffic congestion and Wardrop equilibria. SIAM J. Control Optim. **47**(3), 1330–1350 (2008)
64. A. Cellina, On the bounded slope condition and the validity of the Euler Lagrange equation. SIAM J. Contr. Opt. **40**(4), 1270–1279 (2001)
65. A. Chambolle, J. Lamboley, A. Lemenant, E. Stepanov, Regularity for the optimal compliance problem with length penalization. SIAM J. Math. Anal. **49**(2), 1166–1224 (2017)
66. F. Clarke, *Functional Analysis, Calculus of Variations and Optimal Control* (Springer, Berlin, 2013)
67. B. Dacorogna, *Direct Methods in the Calculus of Variations*, 2nd edn. (Springer Science & Business Media, New York, 2007)
68. B. Dacorogna, P. Marcellini, A counterexample in the vectorial calculus of variations, in *Material Instabilities in Continuum Mechanics*, ed. by J.M. Ball (Oxford University Press, Oxford, 1988), pp. 77–83
69. G. Dal Maso, *An Introduction to Γ-Convergence* (Birkhäuser, Basel, 1992)
70. G. Dal Maso, J.M. Morel, S. Solimini, A variational method in image segmentation: Existence and approximation results. Acta Math. **168**, 89–151 (1992)

71. G. David, *Singular Sets of Minimizers for the Mumford–Shah Functional*. Progress in Mathematics, vol. 233 (Birkhäuser Verlag, Basel, 2005)

72. E. De Giorgi, Sull'analiticità delle estremali degli integrali multipli. Atti della Accademia Nazionale dei Lincei. Rendiconti. Classe di Scienze Fisiche, Matematiche e Naturali, Serie VIII (in Italian) **20**, 438–441 (1956)

73. E. De Giorgi, Sulla differenziabilità e l'analiticità delle estremali degli integrali multipli regolari. Memorie dell'Accademia delle scienze di Torino **3**, 25–43 (1957)

74. E. De Giorgi, Sulla proprietà isoperimetrica della ipersfera nella classe degli insiemi aventi frontiera orientata di misura finita. Atti Accad. Naz. Lincei Mem. Cl. Sc. Fis. Mat. Natur. Sez. I (8) **5**, 33–44 (1958)

75. E. De Giorgi, Un esempio di estremali discontinue per un problema variazionale di tipo ellittico. *Bollettino dell'Unione Matematica Italiana*, Serie IV (in Italian) **1**, 135–137 (1968)

76. E. De Giorgi, T. Franzoni, Su un tipo di convergenza variazionale. Atti Acc. Naz. Lincei Rend. **58**(8), 842–850 (1975)

77. E. Di Nezza, G. Palatucci, E. Valdinoci, Hitchhiker's guide to the fractional Sobolev spaces. Bull. Sci. Math. **136**(5), 521–573 (2012)

78. A. Dumas, and F. Santambrogio, Optimal trajectories in L^1 and under L^1 penalizations. Preprint https://cvgmt.sns.it/paper/6023/

79. S. Dweik, F. Santambrogio, L^p bounds for boundary-to-boundary transport densities, and $W^{1,p}$ bounds for the BV least gradient problem in 2D. Calc. Var. PDEs **58**, Article number 31 (2019)

80. J. Eells, Jr., J.H. Sampson, Harmonic mappings of Riemannian manifolds. Am. J. Math. **86**(1), 109–160 (1964)

81. I. Ekeland, R. Temam, *Convex Analysis and Variational Problems*. Studies in Mathematics and Its Applications (Elsevier, Amsterdam, 1976

82. L. Esposito, N. Fusco, C. Trombetti, A quantitative version of the isoperimetric inequality: the anisotropic case. Ann. Sci. Norm. Super. Pisa, Cl. Sci. (5) **4**, 619–651 (2005)

83. L.C. Evans, *Partial Differential Equations*. Graduate Studies in Mathematics, 2nd edn. (AMS, Providence, 2010)

84. L.C. Evans, R.F. Gariepy, *Measure Theory and Fine Properties of Functions*. Textbooks in Mathematics, Revised edn. (CRC Press, Boca Raton, 2015)

85. G. Faber, Beweis, dass unter allen homogenen Membranen von gleicher Fläche und gleicher Spannung die kreisfӧrmige den tiefsten Grundton gibt Sitzungsber. Bayer. Akad. Wiss. München, Math.-Phys. Kl. 169–172 (1923)

86. L. Fejes Tóth, Über das kürzeste Kurvennetz, das eine Kugeloberfläche in flächengleiche konvexe Teile zerlegt. Math. Naturwiss. Anz. Ungar. Akad. Wiss. **62**, 349–354 (1943)

87. A. Figalli, F. Maggi, A. Pratelli, A mass transportation approach to quantitative isoperimetric inequalities. Inventiones Mathematicae **182**(1), 167–211 (2010)

88. B. Fuglede, Stability in the isoperimetric problem. Bull. Lond. Math. Soc. **18**, 599–605 (1986)

89. B. Fuglede, Stability in the isoperimetric problem for convex or nearly spherical domains in $\mathbb{R}^n$. Trans. Am. Math. Soc. **314**, 619–638 (1989)

90. G. Fusco, G.F. Gronchi, M. Novaga, On the existence of heteroclinic connections. Sao Paulo J. Math. Sci. **12**(1), 68–81 (2018)

91. N. Fusco, The classical isoperimetric theorem. Rend. Accad. Sci. Fis. Mat. Napoli **71**, 63–107 (2004)

92. N. Fusco, F. Maggi, A. Pratelli, The sharp quantitative isoperimetric inequality. Ann. Math. **168**, 941–980 (2008)

93. G. Galilei, *Discorsi e dimostrazioni matematiche intorno a due nuove scienze attenenti alla meccanica e i movimenti locali* (Ludovico Elzeviro, Leiden, 1638)

94. M. Ghisi, M. Gobbino, The monopolist's problem: existence, relaxation and approximation. Calc. Var. Part. Differ. Equ. **24**(1), 111–129 (2005)

95. M. Giaquinta, *Multiple Integrals in the Calculus of Variations and Nonlinear Elliptic Systems*. Annals of Mathematics Studies, vol. 105 (Princeton University Press, Princeton, 1983)

96. M. Giaquinta, S. Hildebrandt, *Calculus of Variations I and II* (Springer, Berlin, 2004)

97. G. Giaquinta, L. Martinazzi, *An Introduction to the Regularity Theory for Elliptic Systems, Harmonic Maps and Minimal Graphs*. Lecture Notes SNS (Springer, Berlin, 2012)

98. D. Gilbarg, N.S. Trudinger, *Elliptic Partial Differential Equations of Second Order*. Grundlehren der Mathematischen Wissenschaften, vol. 224 (Springer-Verlag, Berlin, 1977)

99. E.N. Gilbert, Minimum cost communication networks. Bell Syst. Tech. J. **46**, 2209–2227 (1967)

100. E.N. Gilbert, H.O. Pollak, Steiner minimal trees. SIAM J. Appl. Math. **16**, 1–29 (1968)

101. E. Giusti, *Direct Methods in the Calculus of Variations* (World Scientific, Berlin, 2003)

102. E. Giusti, M. Miranda, Un esempio di soluzioni discontinue per un problema di minimo relativo ad un integrale regolare del calcolo delle variazioni. Bollettino dell'Unione Matematica Italiana, Serie IV (in Italian) **2**, 1–8 (1968)

103. S. Gołąb, Sur quelques points de la théorie de la longueur. Ann. Soc. Polon. Math. **7**, 227–241 (1929)

104. W. Górny, P. Rybka, A. Sabra, Special cases of the planar least gradient problem. Nonlinear Analysis **151**, 66–95 (2017)

105. S. Graf, H. Luschgy, *Foundations of Quantization for Probability Distributions* (Springer-Verlag, Berlin, 2000)

106. L. Grafakos, *Classical Fourier Analysis*. Graduate Texts in Mathematics (Springer, Berlin, 2014)

107. A. Granas, J. Dugundji, *Fixed Point Theory*. Springer Monographs in Mathematics (Springer-Verlag, New York, 2003)

108. T.C. Hales, The Honeycomb Conjecture. Discrete Comput. Geom. **25**(1): 1–22 (2021)

109. R.R. Hall, W.K. Hayman, A.W. Weitsman, On asymmetry and capacity. J. d'Analyse Math. **56**, 87–123 (1991)

110. Q. Han, F. Lin, *Elliptic Partial Differential Equations*. Courant Lecture Notes, vol. 1, 2nd edn. (American Mathematical Society, Providence, 2011)

111. D. Hardin, E.B. Saff, O. Vlasiuk, Asymptotic properties of short-range interaction functionals. Preprint, https://arxiv.org/abs/2010.11937 (2020)

112. F. Hélein, J.C. Wood, Harmonic maps, in *Handbook on Global Analysis*, ed. by D. Krupka, D. Saunders (Elsevier, Amsterdam, 2008), pp. 417–491

113. A. Henrot, M. Pierre, *Variation et optimisation de formes: Une analyse géométrique*. Mathématiques & Applications, vol. 48 (Springer, Berlin, 2005)

114. A. Henrot, M. Pierre, *Shape Variation and Optimization: A Geometrical Analysis*. Tracts in Mathematics, vol. 28 (EMS, Helsinki, 2018)

115. D. Hilbert, *Mathematische Probleme*. Nachrichten von der Königlichen Gesellschaft der Wissenschaften zu Göttingen, Mathematisch-Physikalische Klasse (in German) **3**, 253–297 (1900)

116. M.-C. Hong, On the minimality of the p-harmonic map $\frac{x}{|x|} : B^n \to S^{n-1}$. Calc. Var. **13**, 459–468 (2001)

117. M. Huang, R.P. Malhamé, P.E. Caines, Large population stochastic dynamic games: closed-loop McKean–Vlasov systems and the Nash certainty equivalence principle. Commun. Inf. Syst. **6**(3), 221–252, (2006)

118. A. Hurwitz, Sur quelques applications géométriques des séries de Fourier, in *Mathematische Werke. Erster Band Funktionentheorie* (Springer, Basel, 1932)

119. T. Iwaniec, J.J. Manfredi, Regularity of p-harmonic functions on the plane. Rev. Mat. Iberoamericana **5**(1–2), 1–19 (1989)

120. C. Jimenez, Dynamic formulation of optimal transport problems. J. Convex Anal. **15**(3), 593–622 (2008)

121. F. John, L. Nirenberg, On functions of bounded mean oscillation. Commun. Pure Appl. Math. **14**(3), 415–426 (1961)

122. J. Jost. Equilibrium maps between metric spaces. Calc. Var. Par. Differ. Equ. **2**, 173–204 (1994)

123. M. Junianus Justinus, *Epitoma Historiarum Philippicarum Pompei Trogi*, III or IV century CE

124. L. Kantorovich, On the transfer of masses. Dokl. Acad. Nauk. USSR **37**, 7–8 (1942)
125. R. Karp, Reducibility among combinatorial problems, in *Complexity of Computer Computations* (Plenum Press, New York, 1972), pp. 85–103
126. V.S. Klimov, On the symmetrization of anisotropic integral functionals. Izv. Vyssh. Uchebn. Zaved. Mat. **8**, 26–32 (1999)
127. N.J. Korevaar, R.M. Schoen, Sobolev spaces and harmonic maps for metric space targets. Commun. Anal. Geom. **1**(3–4), 561–659 (1993)
128. H. Knothe, Contributions to the theory of convex bodies. Mich. Math. J. **4**, 39–52 (1957)
129. E. Krahn, Über eine von Rayleigh formulierte Minimaleigenschaft des Kreises. Math. Ann. **94**, 97–100 (1925)
130. E. Krahn, Über Minimaleigenschaften der Kugel in drei und mehr Dimensionen. Acta Commun. Univ. Tartu (Dorpat) **A9**, 1–44 (1926). English transl., Ü. Lumiste and J. Peetre (eds.), Edgar Krahn, 1894–1961, A Centenary Volume, IOS Press, 1994, Chap. 6, pp. 139–174
131. O.A. Ladyzhenskaya, N.N. Ural'tseva, *Linear and Quasilinear Elliptic Equations* (Academic Press. New York/London, 1968)
132. J.-M. Lasry, P.-L. Lions, Jeux à champ moyen. I. Le cas stationnaire. C. R. Math. Acad. Sci. Paris **343**(9), 619–625 (2006)
133. J.-M. Lasry, P.-L. Lions, Jeux à champ moyen. II. Horizon fini et contrôle optimal. C. R. Math. Acad. Sci. Paris **343**(10), 679–684 (2006)
134. J.-M. Lasry, P.-L. Lions, Mean-field games. Jpn. J. Math. **2**, 229–260 (2007)
135. H. Lavenant, F. Santambrogio, Optimal density evolution with congestion: L^∞ bounds via flow interchange techniques and applications to variational mean field games. Commun. Partial Differ. Equ. **43**(12), 1761–1802 (2018)
136. A. Lemenant, About the regularity of average distance minimizers in $\mathbb{R}^2$. J. Convex Anal. **18**(4), 949–981 (2011)
137. A. Lemenant, A presentation of the average distance minimizing problem. Zap. Nauchn. Sem. S.-Peterburg. Otdel. Mat. Inst. Steklov. (POMI) **390**, 117–146 (2011). (Teoriya Predstavlenii, Dinamicheskie Sistemy, Kombinatornye Metody. XX) **308**
138. A. Lemenant, F. Santambrogio, A Modica–Mortola approximation for the Steiner problem. C. R. Math. Acad. Sci. Paris **352**(5), 451–454 (2014)
139. F.-H. Lin, Une remarque sur l'application $x/|x|$. C. R. Acad. Sc. Paris **305**, 529–531 (1987)
140. P. Lindqvist, Notes on the p-Laplace equation. Available on-line at the page https://folk.ntnu.no/lqvist/p-laplace.pdf
141. P.-L. Lions, Series of lectures on mean filed games, *Collège de France*, Paris, 2006-2012, video-recorderd and available at the web page http://www.college-de-france.fr/site/audio-video/
142. F. Maggi, *Sets of Finite Perimeter and Geometric Variational Problems*. Cambridge Studies in Advanced Mathematics, vol. 135 (Cambridge University Press, Cambridge, 2012)
143. J. Marcinkiewicz, Sur l'interpolation d'operations. C. R. Acad. Sci. Paris **208**, 1272–1273 (1939)
144. G. Mazanti, F. Santambrogio, Minimal-time mean field games. Math. Mod. Meth. Appl. Sci. **29**(8), 1413–1464 (2019)
145. V.D. Milman, G. Schechtman, *Asymptotic Theory of Finite-Dimensional Normed Spaces*, with an appendix by M. Gromov. Lecture Notes in Mathematics, vol. 1200 (Springer-Verlag, Berlin, 1986)
146. L. Modica, S. Mortola, Un esempio di Γ-convergenza. (Italian). Boll. Un. Mat. Ital. B **14**(1), 285–299 (1977)
147. L. Modica, S. Mortola, Il limite nella Γ-convergenza di una famiglia di funzionali ellittici. Boll. Un. Mat. Ital. A (5), **14**(3), 526–529 (1977)
148. D. Monderer, L.S. Shapley, Potential games. Games Econ. Behavior **14**, 124–143 (1996)
149. G. Monge, Mémoire sur la théorie des déblais et des remblais. *Histoire de l'Académie Royale des Sciences de Paris* (1781), pp. 666–704

150. A. Monteil, F. Santambrogio, Metric methods for heteroclinic connections. Math. Methods Appl. Sci. **41**(3), 1019–1024 (2018)
151. A. Monteil, F. Santambrogio, Metric methods for heteroclinic connections in infinite dimensional spaces. Indiana Univ. Math. J. **69**(4), 1445–1503 (2020)
152. C. Morrey, Quasi-convexity and the lower semicontinuity of multiple integrals. Pac. J. Math. **2**(1), 25–53 (1952)
153. S. Mosconi, P. Tilli, Γ-Convergence for the irrigation problem. J. Conv. Anal. **12**(1), 145–158 (2005)
154. J. Moser, A new proof of de Giorgi's theorem concerning the regularity problem for elliptic differential equations. Commun. Pure Appl. Math. **13**(3), 457–468 (1960)
155. D. Mumford, J. Shah, Optimal approximations by piecewise smooth functions and associated variational problems. Commun. Pure Appl. Math. **XLII**(5), 577–685 (1989)
156. D. Musielak, Dido's Problem. When a myth of ancient literature became a problem of variational calculus. Preprint, https://arxiv.org/abs/2301.02917
157. J. Nash, Equilibrium points in n-person games. Proc. Natl. Acad. Sci. USA **36**(1), 48–49 (1950)
158. J. Nash, Non-cooperative games. Ann. Math. **54**(2), 286–295 (1951)
159. J. Nash, Parabolic equations. Proc. Natl. Acad. Sci. USA **43**(8), 754–758 (1957)
160. J. Nash, Continuity of solutions of parabolic and elliptic equations. Am. J. Math. **80**(4), 931–954 (1958)
161. E. Opsomer, N. Vandewalle, Novel structures for optimal space partitions. New J. Phys. **18**(10), 103008 (2016)
162. M.J. Osborne, A. Rubinstein, *A Course in Game Theory* (The MIT Press, Cambridge, 1994).
163. É. Oudet, Approximation of partitions of least perimeter by Γ-convergence: around Kelvin's conjecture. Exp. Math. **20**(3), 260–270 (2011)
164. É. Oudet, F. Santambrogio, A Modica–Mortola approximation for branched transport and applications. Arch. Ration. Mech. Anal. **201**(1), 115–142 (2011)
165. M.R. Pakzad, Existence of infinitely many weakly harmonic maps from a domain in $\mathbb{R}^n$ into $\mathbb{S}^2$ for non-constant boundary data. Calc. Var. **13**, 97–121 (2001)
166. E. Paolini, E. Stepanov, Existence and regularity results for the Steiner problem. Calc. Var. Partial Differ. Equ. **46**(3), 837–860 (2012)
167. E. Paolini, E. Stepanov, Y. Teplitskaya, An example of an infinite Steiner tree connecting an uncountable set. Adv. Calc. Var. **8**(3), 267–290 (2015)
168. P. Pegon, F. Santambrogio, Q. Xia, A fractal shape optimization problem in branched transport. J. Math. Pures Appl. **123**, 244–269 (2019)
169. I.G. Petrowsky, Sur l'analyticité des solutions des systèmes d'équations différentielles. (French). Recueil Mathématique (Matematicheskii Sbornik) **5**(47), 3–70 (1939)
170. G. Pólya, G. Szegő, *Isoperimetric Inequalities in Mathematical Physics*. Annals of Mathematical Studies, vol. 27 (Princeton University Press, Princeton, 1951)
171. A. Prosinski, F. Santambrogio, Global-in-time regularity via duality for congestion-penalized mean field games. Stochastics **89**(6–7), 923–942 (2017)
172. F. Rindler, *Calculus of Variations* (Springer, Berlin, 2018)
173. R.T. Rockafellar, *Convex Analysis* (Princeton University Press, Princeton, 1970)
174. E. Rouy, A. Tourin, A viscosity solution approach to shape from shading. SIAM J. Numer. Anal. **29**, 867–884 (1992)
175. F. Santambrogio, Inégalités Isopérimétriques quantitatives via le transport optimal, (d'après A. Figalli, F. Maggi et A. Pratelli). Proceedings of the *Bourbaki Seminar*, 2011. Astérisque **348**, Exposé no. 1034 (2012)
176. F. Santambrogio, A Modica–Mortola approximation for branched transport. Comptes Rendus Acad. Sci. **348**(15–16), 941–945 (2010)
177. F. Santambrogio, *Optimal Transport for Applied Mathematicians*. Progress in Nonlinear Differential Equations and Their Applications, vol. 87 (Birkhäuser, Basel, 2015)
178. F. Santambrogio, Lecture notes on variational mean field games, in *Mean Field Games*, ed. by P. Cardaliaguet and A. Porretta. Lecture Notes in Mathematics Book Series (LNMCIME), vol. 2281 (Springer, 2020), pp. 159–201

179. F. Santambrogio, P. Tilli, Blow-up of optimal sets in the irrigation problem. J. Geom. Anal. **15**(2), 343–362 (2005)

180. F. Santambrogio, A. Xepapadeas, A.N. Yannacopoulos, Rational expectations equilibria in a Ramsey model of optimal growth with non-local spatial externalities. J. Math. Pures Appl. **140**, 259–279 (2020)

181. R. Schoen, K. Uhlenbeck, A regularity theory for harmonic maps. J. Differ. Geom. **17**(2), 307–335 (1982)

182. J.A. Sethian, *Level Set Methods and Fast Marching Methods* (Cambridge University Press, Cambridge, 1999)

183. D. Slepčev, Counterexample to regularity in average-distance problem. Ann. Inst. H. Poincaré Anal. Non Linéaire **31**(1), 169–184 (2014)

184. C. Sourdis, The heteroclinic connection problem for general double-well potentials. Mediterranean J. Math. **13**(6), 4693–4710 (2016)

185. S. Spagnolo, Sulla convergenza di soluzioni di equazioni paraboliche ed ellittiche. Ann. Scuola Norm. Sup. Pisa, Cl. Sci. III **22**(4), 571–597 (1968)

186. G. Stampacchia, On some regular multiple integral problems in the calculus of variations. Commun. Pure Appl. Math. **16**, 383–421 (1963)

187. G. Stampacchia, The spaces $L^{p,\lambda}$, $N^{p,\lambda}$ and interpolation. Ann. Scuola Norm. Sup. Pisa **19**, 443–462 (1965)

188. P. Sternberg, Vector-valued local minimizers of nonconvex variational problems. Rocky Mountain J. Math. **21**(2), 799–807 (1991)

189. A. Takayama, *Mathematical Economics*, 2nd edn. (Cambridge University Press, New York, 1985)

190. P. Tilli, Some explicit examples of minimizers for the irrigation problem. J. Convex Anal. **17**(2), 583–595 (2010)

191. P. Tilli, D. Zucco, Asymptotics of the first Laplace eigenvalue with Dirichlet regions of prescribed length. SIAM J. Math. Anal. **45**(6), 3266–3282 (2013)

192. K. Uhlenbeck, Regularity for a class of non-linear elliptic systems. Acta Math. **138**, 219–240 (1977)

193. J. Van Schaftingen, Anisotropic symmetrization. Ann. l'I.H.P. Anal. Non linéaire **23**(4), 539–565 (2006)

194. P. Vergilius Maro, *Aeneid*, between 29 and 19 BC

195. C. Villani, *Topics in Optimal Transportation*. Graduate Studies in Mathematics (AMS, Providence, 2003)

196. J.G. Wardrop, Some theoretical aspects of road traffic research. Proc. Inst. Civ. Eng. **2**, 325–378 (1952)

197. D. Weaire, R. Phelan, A counter-example to Kelvin's conjecture on minimal surfaces. Philos. Mag. Lett. **69**, 107–110 (1994)

198. B. White, Rectifiability of flat chains. Ann. Math. (2) **150**(1), 165–184 (1999)

199. G. Wulff, Zur Frage der Geschwindigkeit des Wachsturms und der Auflösung der Kristallfläschen. Z. Kristallogr. **34**, 449–530 (1901)

200. Q. Xia, Optimal Paths related to transport problems. Commun. Cont. Math. **5**(2), 251–279 (2003)

201. A. Zuniga, P. Sternberg, On the heteroclinic connection problem for multi-well gradient systems. J. Differ. Equ. **261**(7), 3987–4007 (2016)

202. A. Zygmund, On a theorem of Marcinkiewicz concerning interpolation of operations. J. Math. Pures Appl. Neuvième Série **35**, 223–248 (1956)

Index

© The Author(s), under exclusive license to Springer Nature Switzerland AG 2023
F. Santambrogio, *A Course in the Calculus of Variations*, Universitext,
https://doi.org/10.1007/978-3-031-45036-5

Printed in the United States
by Baker & Taylor Publisher Services